CAMBRIDGE LIBRARY COLLECTION

Books of enduring scholarly value

Mathematical Sciences

From its pre-historic roots in simple counting to the algorithms powering modern desktop computers, from the genius of Archimedes to the genius of Einstein, advances in mathematical understanding and numerical techniques have been directly responsible for creating the modern world as we know it. This series will provide a library of the most influential publications and writers on mathematics in its broadest sense. As such, it will show not only the deep roots from which modern science and technology have grown, but also the astonishing breadth of application of mathematical techniques in the humanities and social sciences, and in everyday life.

Memoir of Augustus De Morgan

The fame of Augustus De Morgan (1806–71), a brilliant mathematician and logician, has been eclipsed by that of his son, the celebrated ceramicist William De Morgan. However, as readers of this *Memoir* will discover, De Morgan senior enjoyed an equally distinguished, if turbulent, career. Collated by his wife, and published in 1882, nine years after his death, the *Memoir of Augustus de Morgan* chronicles the varied life of an under-appreciated genius. Biographical narrative is interleaved with his own correspondence, revealing a humorous and warm personality as well as an exceptional intellect. As the *Pall Mall Gazette* told its readers, 'quaint and original to the last, every word of De Morgan's correspondence is well worth reading'. Although rich in detail about his work and publications, Sophia Elizabeth's affectionate account of her husband is also sympathetic and witty, making it an ideal introduction to one of Britain's greatest minds.

Cambridge University Press has long been a pioneer in the reissuing of out-of-print titles from its own backlist, producing digital reprints of books that are still sought after by scholars and students but could not be reprinted economically using traditional technology. The Cambridge Library Collection extends this activity to a wider range of books which are still of importance to researchers and professionals, either for the source material they contain, or as landmarks in the history of their academic discipline.

Drawing from the world-renowned collections in the Cambridge University Library, and guided by the advice of experts in each subject area, Cambridge University Press is using state-of-the-art scanning machines in its own Printing House to capture the content of each book selected for inclusion. The files are processed to give a consistently clear, crisp image, and the books finished to the high quality standard for which the Press is recognised around the world. The latest print-on-demand technology ensures that the books will remain available indefinitely, and that orders for single or multiple copies can quickly be supplied.

The Cambridge Library Collection will bring back to life books of enduring scholarly value (including out-of-copyright works originally issued by other publishers) across a wide range of disciplines in the humanities and social sciences and in science and technology.

Memoir of Augustus De Morgan

With Selections from His Letters

Sophia Elizabeth De Morgan

CAMBRIDGE UNIVERSITY PRESS

Cambridge, New York, Melbourne, Madrid, Cape Town, Singapore,
São Paolo, Delhi, Dubai, Tokyo

Published in the United States of America by Cambridge University Press, New York

www.cambridge.org
Information on this title: www.cambridge.org/9781108014472

This edition first published 1882
This digitally printed version 2010

ISBN 978-1-108-01447-2 Paperback

AUGUSTUS DE MORGAN

A De Morgan

FROM A PHOTOGRAPH BY MAYALL LONDON.

MEMOIR

OF

AUGUSTUS DE MORGAN

BY HIS WIFE

SOPHIA ELIZABETH DE MORGAN

WITH SELECTIONS FROM HIS LETTERS

LONDON
LONGMANS, GREEN, AND CO.
1882

PREFACE.

I NEED hardly say that in the following pages I have not attempted a scientific memoir. My object has been to supply that part of my husband's life the material for which would not be within the reach of another biographer.

The selection from his letters might have been much larger, if I could in all cases have inserted those of his correspondents. Without these many would have been incomprehensible. As it is, I may have over-estimated the attention which readers will be disposed to give to them. My rule in choosing the letters has been to take those which are most characteristic of the writer, and in this way to give to readers already acquainted with him through his writings a more familiar knowledge of him as a man.

His connection with University College, and the events which led to his leaving it, are necessarily made prominent. So long a time has elapsed since their occurrence, and I have known so little during that time of the Institution, that I cannot even surmise how the present Council would in like circumstances share the convictions or confirm the action of its predecessors. After the lapse of sixteen years I trust that the narrative will provoke no revival of the somewhat acrimonious controversy which ensued. It might perhaps have been in some ways better that Mr. De Morgan should have published a fuller statement of his views at the time, and have thus left less to be done by his biographer. But he had several reasons for not doing this. He refrained partly from reluctance to add to the censures which were being pronounced on the College, perhaps too emphatically, even by well-wishers, and re-echoed by its enemies with unconcealed satisfaction ; partly by the feeling that he had made no sacrifice of a pecuniary nature in resigning his Professorship ; but, as I think, chiefly from weariness and disappointment, and from a desire to have done with the Institution as soon as possible. Nothing, not even a distinct recantation of the measure which made him leave, would have induced him to resume his chair, for he would have held such a recantation to be but another concession to expediency in deference to the storm unexpectedly raised.

Should any portion of what I have written appear uncalled for, it must be remembered that I could not touch my husband's side of the question without placing the whole before my readers. The insertion of the lengthy justification of the Council by members of the Senate will, I trust, exempt me from the charge of having suppressed arguments on the other side.

SOPHIA ELIZABETH DE MORGAN.

Cheyne Row,
Chelsea, 1882.

CONTENTS.

SECTION I.

FROM 1806 TO 1827.

SECTION II.

FROM 1827 TO 1831.

SECTION III.

FROM 1831 TO 1836.

SECTION IV.

SECTION V.

FROM 1836 TO 1846.

SECTION VI.

SECTION VII.

FROM 1846 TO 1855.

SECTION XI.

FROM 1866 TO 1871.

SECTION XII.

MEMOIR

OF

AUGUSTUS DE MORGAN.

SECTION I.

AUGUSTUS DE MORGAN was born in the year 1806 at Madura, in the Madras Presidency. His father Lieutenant-Colonel De Morgan had held appointments, some of them staff situations, at several stations in India; and at the time of his fifth child's birth had chosen Madura in preference to Vellore on account of its comparative quietness. This choice was fortunate, for the battalion of Colonel De Morgan's regiment commanded by Colonel Fanshawe was at Vellore during the time of the mutiny of the native troops; and thus escaped the terrible outbreak in which several English officers lost their lives, and Colonel Fanshawe was murdered. Even at the quieter stations there was cause for alarm from the general disaffection of the native troops, and my husband's mother told how she, being then near her confinement, saw Colonel De Morgan, when the sentries were changed, creep out of bed to listen to the Sepoys, that he might learn if any plot were in agitation, about which information might be given with the password. 1806.

Voyage to England.

When Augustus was seven months old his father and mother came to England with three children, two daughters and the infant. They sailed in the *Duchess*

1807. *of Gordon,* one of a convoy of nearly forty ships. The Commodore, Captain (afterwards Admiral) Sir John Beaufort, was on friendly terms with my husband in after life. But Mr. De Morgan had no suspicion of having sailed under his convoy until after Mrs. De Morgan's death, when the notice of it was found in her journal. When Admiral Beaufort heard of this he wrote, in a letter dated Oct. 10, 1857, five weeks before his own death,—

Admiral Beaufort. 'Our co-residence for three or four months, not in the same street or village or county, but in the same track along the ocean, is an amusing link in our two life-threads; but not the less flattering to me as being claimed by you, and as finding myself one of the *dramatis personæ* in your mother's journal of the *Jane, Duchess of Gordon.* You most correctly picture us as being 'at the two ends of the chain,' for while it was my post to lead that gigantic fleet of upwards of thirty large vessels, I well remember that she was in all cases the sternmost, in spite of the number of hoarse hints that were given her through our guns. Passengers, even ladies, are never very tender in their criticisms on the poor commodore, and it would be charming to see how your mother retaliated for his above coarse language by her sharp and witty castigation.'

1808. Colonel De Morgan settled at Worcester with his wife and children, but returned to India in 1808 alone. Some disturbances in the Madras Army, causing the suspension from command of several officers, including himself, gave him much trouble and anxiety for some time; but the affair, which was settled by an inquiry at the India House, resulted in his complete and honourable acquittal. On his return to England in 1810, the family lived in the north of Devonshire; first at Appledore, then at Bideford, then at Barnstaple. In 1812, one daughter having died, and two sons been born, they settled at Taunton in Somersetshire. The father again left England for Madras, and took the command of a battalion at

Quilon. Being ordered home ill with liver complaint, he left Madras in 1816, and died near St. Helena, on his way to England. 1816.

In a list given of his schools and instructors by Mr. De Morgan, his father's name occurs as his first teacher. He was then four years old, and learnt 'reading and numeration.' The heading of one column in this list, 'Age of the Victim,' shows in a half-serious, half-humorous way the idea 'the Victim' retained of his early schooling. He did not mean that it was worse in his case than in that of other boys, and he always spoke gratefully of his father; but he was no exception to the rule that most children, especially those of great intellectual promise, are more or less victims to our unenlightened methods of education. Of these exceptional children I have heard him say that those have the best chance who have the least teaching.

At Barnstaple he learned, from a Miss Williams, reading, writing, and spelling; at Taunton, being between seven and eight, from Mrs. Poole, reading, writing, arithmetic, and (very) general knowledge. He always retained a painful remembrance of this school. The Rev. J. Fenner, a Unitarian minister, was for a short time his teacher. The pupil was at that time about nine years old, and added Greek and Latin to his other studies. Mr. Fenner was the uncle of Henry Crabb Robinson, who died in 1867, aged ninety-one, and who had been at one time a pupil in the school. The next two teachers were, at Blandford, the Rev. T. Keynes, Independent minister; and at Taunton, the Rev. H. Barker, Church of England clergyman, at whose school he was taught Latin, Greek, Euclid, Algebra, and a little Hebrew. His last schoolmaster, a clever man, and one of whom, though he was not a high mathematician, his pupil always spoke with respect, was the Rev. J. Parsons, M.A., formerly Fellow of Oriel. At Mr. Parsons' school, at Redland, near Bristol, Latin, Greek, and mathematics were taught. Mr. De Morgan

1813. Teachers.

1820. was at this school from the age of fourteen to sixteen and a half, and at this period of his life his mathematical powers were first developed. It was strange that among so many teachers the germ of mathematical ability should have been so long unnoticed. It could not be quite latent or quite unformed in the brain of a boy of fourteen; it can only be supposed that the routine of school teaching smothered and hid it from observation. Education means drawing out; it often is keeping in, and it is well when it is no worse. In this case it was good for the pupil and for mathematics that the early germ should be left to its own resources of natural growth, uncrippled and undistorted by mistaken systems of teaching. It was accidentally developed, and indeed made known to its possessor by the observation of a dear old friend, Mr. Hugh Standert, of Taunton. Seeing the boy very busy making a neat figure with ruler and compasses, and finding that the essence of the proposition was supposed to lie in its accurate geometrical drawing, he asked what was to be done. Augustus said he was *drawing mathematics.* 'That's not mathematics,' said his friend; 'come, and I will show you what is.' So the lines and angles were rubbed out, and the future mathematician, greatly surprised by finding that he had missed the aim of Euclid, was soon intent on the first demonstration he ever knew the meaning of. I do not think, from what I have heard him say, that Mr. Standert was instrumental in further bringing out the latent power. But its owner had become in some degree aware of the mine of wealth that only required working, and as some mathematics was taught at Mr. Parsons' school, the little help that was needed was soon turned to profit. He soon left his teacher behind, and from that time his great delight was to work out questions which were often as much his own as their solution.

Mathematical talent.

I can only find one little mention of his first going to his school in his own handwriting. In a letter to Dean

Peacock in 1852, he says *àpropos* of Robert Young, of whom he had been writing,— 1820.

'When I was sent to school near Bristol in 1820, I was consigned to R. Young, who especially warned me not to walk in my sleep, as there were no leads outside the windows; they had been removed. The consequence was, that though I never walked in my sleep before or since that I remember, I was awakened by the wind blowing on me, and found myself before the open window, with my knee on the lower ledge. I crept back to bed, leaving the window open, and the family being alarmed by the noise, came into my room, and found me asleep and the window open, so that as their fenestral logic did not reason both ways, they forgot that the leads were not there, and searched the whole house for thieves.'

Mr. Parsons' school.

Mr. Robert Reece, his old schoolfellow and constant friend of forty years, writes concerning these early school days:—

'I entered Mr. Parsons' school at Redland, near Bristol, on August 12, 1819. I think dear De Morgan came among us at the latter end of the following year, or in January 1821.

'He was certainly a fine stout fellow for his age, and at once took a high place in the school. He had a grievous infirmity, the loss of one of his eyes,[1] which provoked all kinds of gibes and practical jokes among the boys.'

Mr. Reece has told me how these cruel practical jokes were put an end to. One lad was in the habit of playing a trick upon his schoolfellow which deserves a worse name than thoughtlessness. He would come up stealthily to De Morgan's blind side, and holding a sharp-pointed penknife to his cheek, speak to him suddenly by name. De Morgan on turning round received the point of the knife in his face. His friend Reece agreed with him that until the aggressor should receive a sound thrashing he

[1] From birth. Both eyes were affected with the 'sore eye' of India, and the left was saved.

1820–21. School days.

would never desist from his cruelty. 'But how,' said the tormented lad, 'can I catch him to thrash him? I cannot see him. He comes up, and is gone before I can lay hands on him.' 'But you *shall,*' said Reece; and the arrangement was made. Reece, knowing that when his friend was quietly reading, as he often did, at a desk so placed that his blind side was near the door, the enemy would be likely to approach, hid himself in such a way that when the boy entered he could shut the door and prevent escape. All happened as he expected. De Morgan sat down at his desk with a book before him. Very soon the cowardly aggressor came quietly in, pointed his knife at his cheek, and said suddenly, 'De Morgan!' His intended victim did not turn round as he had done before, and in a moment the lad, a stout boy of fourteen, was seized behind by Reece, who gave him over to receive the 'sound thrashing' which De Morgan administered, and which proved effectual in making him keep the peace from that time.

Mr. Reece tells how he and his friend, with another boy of similar tastes, contrived a late reading party, unsanctioned by the master. One of the three asked Mr. Parsons to lend them Scott's poems, at that time just published. Having got 'The Lady of the Lake,' they waited till all the other boys were in bed, the lights out, and all things quiet; then De Morgan produced a match pistol and a tinder, snapped a spark and lit the candle, and then read to his two companions till all three were too sleepy to take an interest in Ellen and Roderick Dhu. I do not mention this as an example to be followed, but I hope my readers will forgive them.

Mr. Reece says, 'I was impressed with his wonderful ability from the first, and I courted him, and gave him my admiration and my love. In return, he became attached to me, and invited or permitted me to sit by him in play-hours. He never joined in the sports of the boys, owing to his infirmity. He had a remarkable talent for drawing caricatures, of the kind that Gilray was so famous for. I took great interest in these drawings, and I had the privilege of suggesting a subject now and then. Two of them I remember—the one, "Charon's boat," with figures; the other, "The Devil," with the three black graces, Law, Physic, and Divinity. It seems an odd thing to record, but I well remember that I was advanced in "Bland's Quadratic Equations" when De Morgan took up that well-known elementary book, "Bridge's Algebra," for the first time. But it was so. He read Bridge's book like a novel. In less than a month he had gone through that treatise and dashed into Bland, and so got out of sight, as far as I was concerned. It is scarcely necessary to say that all his school work was admirably performed. Mr. Parsons had the highest opinion of him.'[1]

1822.

Caricatures.

Mr. Parsons being a good classical scholar, his aim was rather to make his boys good classics than mathematicians. If the mathematical power had not made itself apparent, and taken the place of all other interests in the pupil's mind, his studies would probably have taken the direction desired by his master. As it was, he was a good Greek and Latin scholar, and his classical reading was wide and varied. The teaching at Redland School was good, but abuses creep in everywhere. Here is an account of a way of saying the lessons, given to me by my husband in explanation of some remarks I had written on education. Like the midnight entertainment before mentioned, the story will afford hints to teachers rather than an example for pupils:—

'An ingenious application of the logical fallacy of a part for the whole was invented by schoolboys by the help of Providence, to moderate a mischief which would otherwise have been severely felt. It was thought necessary that boys should learn by heart Latin and Greek

[1] The writer of the above died two years after his friend, his affection for whom was one of the strongest feelings in his mind while consciousness remained.

1822. Schoolboy tricks.

verses, *to strengthen* the memory. The poor ignorant Virgil and Homer scanners, and their subordinate Euclid and algebra drillers, had not the smallest idea that *a memory* is an adjunct of each faculty, that the training of one is of little or no help to another, and that the memory of words, which they over-cultivated, differs widely among young people. The allowance was forty lines a day, Latin and Greek alternately, for five days in the week, the whole two hundred to be repeated in one lot on Saturday. There was as much difference between the boys in the rapidity of committing to memory, as between the two pilgrims who went peashod to Loretto, the one with hard peas, the other with boiled, in his shoes. But the boys had the sense to *learn their parts*, as the actors do, and again, like the actors, *they learnt the cues.* This was carried on, at a school at which I was, year after year, without a single detection. Even the *contretemps* which arose when a boy was ill on a Saturday, or when one who had been ill on a week-day came in on the Saturday, were adroitly got over. I am perfectly satisfied that the master, an old Fellow of Oriel, was a party to the whole proceeding, as a means of reconciling the appearances demanded by opinion with the amount of word-catching which he thought sufficient. And judging by what I have heard of other schools, I suspect that such connivance was not infrequent.'

The boys of Mr. Parsons' school attended St. Michael's Church, Bristol. Having heard something from Mr. De Morgan of his juvenile delinquencies, arising from thinking more of mathematics than of the scarcely audible sermon, I searched out the school pew during a visit to Bristol, and there found, neatly marked on the oak wainscot partition, the first and second propositions of Euclid and one or two simple equations, with the initials A. De M. They were made in rows of small holes, pierced with the sharp point of a shoe-buckle, and are by this time probably repaired and cleaned away.

The testimony of Mr. Reece to the affection felt for their schoolfellow by most of his companions has been confirmed to me by one or two of the few who remain in this world, and I find in letters from friends many little confirmations. 1822.

'I have known more of you than you of me,' his friend Mr. Leslie Ellis wrote to him during the long and suffering illness which preceded his death. 'Even while you were yet at Mr. Parsons' and I was a child I had heard of you, and of course in later years I have heard of you very often; but though everybody spoke well of you, I was left to find out for myself how kind you could be to a sick man—how kind, I think I must infer, to all about you.'

And another time,—

'I had, since your recollection of Parsons, two brothers there, and I remember my father speaking of having seen you, and saying that the usher complained that you were "such a glutton," meaning in the matter of reading; but I cannot recollect whether he spoke of mathematical or classical reading, or of both.'

But the boy was probably, at school, very like what he was at home, when his mother, who loved him fondly, described him as a quiet, thoughtful boy, occasionally but not often irritable, and never so well pleased as when he could get her to listen to his reading and explanations, and 'always speculating on things that nobody else thought of, and asking her questions far beyond her power to answer.'

One element of his early teaching strongly tinged his character in after life. Col. De Morgan, who was a strictly religious man, of a rather evangelical, as it is falsely called, turn of feeling, was premature, seeing the sensitiveness and grasp of the mind he had to deal with, in inculcating rigid doctrines, and insisting on formal observances. The religious training of his son thus begun, was continued, after his father left England, by his excellent mother, who, with the best intentions in the

1822. Religious education. world, was unable to adapt the spiritual food to the needs of the recipient. He was made to learn by heart and repeat long Scripture lessons, so chosen that their meaning and connection with each other and with himself were quite imperceptible; indeed, I have heard him say that, from frequent repetition, the words and phrases became meaningless to him. He was taken to church twice in the week, three times on Sunday, and required to give an abstract of every sermon he heard. Being thus administered, religion could not fail to become a source of misery. Sunday was the one wretched day of the week, to be got over somehow, and church was a place of penance. A worse result of the system even than this was the confusing together in an honest young mind all ideas of right and wrong, truth and falsehood, in connection with religion. The awful description of the Devil and his doings, and the eternal burning to be undergone by all who did not believe what he could not then, and never was able to believe in the prescribed form, were set in the boy's mind against Jesus Christ's declaration that His Father was a God of love, and that repentance was the only condition of forgiveness. For a boy like Augustus De Morgan, whose clear perceptions, love of truth, and readiness to venerate, rendered him sensitive to every word spoken by those whom he loved, or who were in authority over him, such mental antagonisms must have been the cause of great anguish, and he could only escape, after he once began to think, by dismissing the whole from his mind. The problem of how to reconcile the Divine idea of God's love with the human notion of God's justice was a harder one than he ever met with in after life, and he gave it up as insoluble. Not being yet able to detect the logical fallacies and critical errors which formed part of the arguments used to convince him, he could only receive those arguments in silence, but without assent. Happily the evil corrected itself, and no harm was done in the end. His innate sense of relation-

ship to his heavenly Father was too strong to allow him 1822.
to become atheistical, and his reasoning power too sound to allow him to be sceptical as to the Christian revelation. But the process of pulling down and building up took time, and it was years before the impressions of his childhood could pass away, and the natural, healthy working of the religious spirit could begin. Such an experiment is a dangerous one for parents to try, and the greater the early indications of religious feeling in a child, the more cautious and forbearing should they be in their direction of it.

One lasting injury done to him by the compulsory attendance so often at public worship, was his inability in after life to listen for any time to speaking or preaching. He said that the old troubles of the three services on Sunday, and the 'dreary sermons' came back to him, and to get rid of these memories he thought of something different from what was being said.

In February 1823 he entered Trinity College, Cam- 1823.
bridge. His old schoolmaster, Mr. Parsons, with other friends, had counselled his pupil's reading for honours in Classics; and Mrs. De Morgan's wish was that her son should enter the Church as an Evangelical clergyman. She had had all the responsibility of her children's education, and, looking to the success of her eldest at Cambridge as a most important element of his future welfare, naturally trusted to the advice of her friends, and believed that all attention given to Mathematics beyond what was needed for his examination would be so much labour lost. He was but sixteen years and a half old when he went to Cambridge, entering at a by-term. Though always studious and persevering, yet at his first examination, when his attention had been divided between the classical reading he had forced himself to attend to, and the Mathematics which he loved, he stood at the top of the second class only. But his failure, as she considered it, caused his mother great anxiety, and her letters to him

Age on entering.

1823. at the time are filled with earnest entreaties 'not to disregard good advice,' 'not to be so wilful,' &c.

It is certain that her exhortations were not needed. He *had* exerted himself, as his tutor's letter to him shows: 'I am sorry that I cannot congratulate you on being in the first class, though your merits and exertions richly deserved it.' We may remember that he was at this time under seventeen, that he had been at college only three months, and that he beat by many places his competitor of his own year, two years his senior in age. Moreover, he had been urged by his tutor, against his own conviction, to go in for examination at this time.

It appears from some of his mother's letters that he had in reply assured her that he would comply with her wish with respect to his reading. But it cannot be wondered at that the University lectures opened the field into which he had long desired to enter. It was like new life to him when he listened to Dr. Peacock's explanations, and followed up the study he loved under the guidance of one who knew how to show the way. From the conflict between his own inclinations and the wishes of his friends it is certain that his path could not be quite smooth, but happily the University courses made it better during the second than in the first year. A greater amount of Mathematics was then required in the college examination, and he was found at the head of the first class. Mr. Higman, his tutor, wrote to his mother: 'Notwithstanding my disappointment last year, I had formed such a very favourable opinion of Mr. De Morgan's talent, and was so much pleased with his industry and the implicit attention he paid to every direction that I gave him, that I felt perfectly assured that he would, on the next trial, when less depended on Classics, distinguish himself in a very extraordinary manner. Nor have my prognostics with regard to his success proved deceitful; he is not only in our first class, but far, very far, the first in it.'

1824. Tutor's opinion of his capabilities.

For the first two years of his Cambridge life, owing to

the difficulty of getting rooms in Trinity College, he was in lodgings. After this his rooms were over the gateway. At this time his mother wrote to him: 'I hope I am mistaken in supposing from your letter that you go entirely to "the chapel," and not with Mr. and Mrs. ——, to hear the Gospel on Sundays. . . . You are very young, my love, and will be likely to go wrong from being left to yourself so soon if you do not take advantage of the experience of those who have gone before you. I have less fear for you than I should have for many youths of your age, because you are studious and steady, because you love your mother tenderly, but above all, because you are the child of many prayers; but I shall be most anxious if you do not hear dear Mr. Simeon.'

1824.

His mother's anxiety.

In another letter she says, speaking of the same friends who had assisted him in small money arrangements at Cambridge: 'Mrs. —— tells me you are like a man of fifty in settling your accounts with her for things she has bought. Dear *own* son of your father and mother, go on through life with the same scrupulous punctuality; it will be a means of keeping you from spending extravagantly; it will make you respected and beloved, and preserve you from that sort of carelessness which brings many young men to ruin.'

I would not put on record expressions showing the intense anxiety of a most energetic and loving mother for a beloved child, except to afford an instance of how the very best intentions may be acted on in such a way as to frustrate their own fulfilment. Mrs. De Morgan had put some books, of what would now be called a 'Low Church' tendency, into a box with other things for her son, accompanying them with a letter, from which I extract the following:—

Religious reading.

'I am so anxious that you should read occasionally the books I send (unknown to you), and was so fearful you might endeavour to persuade yourself and me that you had no time for such studies, that I thought the best way

1824. Doctrines of salvation.

would be to say nothing about it *viva voce*, but to send them uninvited; and, having determined on that, an explanatory note became necessary. I beseech you, for my sake, to read with attention these books, to utter a prayer over them whenever you open them, "that they may be blessed to you as they have been to thousands," many of whom are now rejoicing in heaven, where you *wish* to go, but where you never can go while you remain wilfully ignorant of your state by nature, and of your need of a Saviour. Your believing an atonement necessary in a *general* sense [1] will not avail you. You must go *by* yourself and *for* yourself to Christ for pardon and grace, and until you do this you may rest assured you are in a most awful state—liable to be hurled into everlasting torment by every little accident, every disease, nay, even by a crumb of bread going the wrong way. Can you wonder that a mother, doting as I do on you, feels miserable when she contemplates a beloved child wantonly sporting on the edge of so tremendous a precipice? . . . Can you picture to yourself any agonies like those which would take possession of your mind were you assured that before to-morrow morning you would be standing at the tremendous bar of an *angry God*?'

The young man thus appealed to was dutiful and affectionate, and these exhortations troubled him much. His reason and instinctive love of God told him that they must arise from misinterpretations of Scripture, and from human notions of Divine things. In many less logical and fearless minds they would have produced disgust with religion altogether; but the intellect of the future logician was too clear to confound the thing itself with its abuses, or with the misrepresentations of ill-judging advocates.

[1] From this expression, and from what I have heard, I conclude that Mr. De Morgan had assured his mother of his belief in the atonement in the Scripture sense, namely, the reconciliation or at-one-ment of sinning and repentant man to a loving God, not the reconciling of an angry God to mankind, in consequence of intellectual belief. This was his creed in after life.

These had not even the effect of making him dismiss the matter from his mind, for during the whole of his college life his mind was actively employed on questions connected with theology and philosophy. He never saw the gospel in any other light than as a professing declaration of God's love and mercy, but it was some time before he was convinced of its historical truth and supramundane origin. 1825.

In 1825 Mr. De Morgan was again high in his class. He had had an illness, perhaps from reading too much and too late at night; and his mother, whose gratification was damped by her great anxiety about his health, writes to him: 'You are much higher than I expected from your humble account of yourself, and I rely on your letting me know if you should suffer materially.' In April a Trinity scholarship was awarded him. After this time some friends must have made his mother anxious by accounts of his general and discursive reading, for she writes: 'I have heard of you lately as a man who reads much, but who is not likely *to do* much, because he will not conform to the instructions of those who could assist him.' The indocility to which she refers consisted in extensive Mathematical reading beyond the bounds marked out by his tutors, and in the study of Metaphysics, Mental Philosophy, and even Theology. Berkeley's writings attracted him strongly; the immateriality of Berkeley's doctrine being suited to a mind instinctively resting upon a spiritual Father, and believing that we depend on His sustaining power as well for absolute existence as for support and guidance through life. It is far from improbable that Berkeley's speculations, falling in in a great degree with his own, gave a strong bias to his subsequent thoughts on metaphysical questions.

Scholarship.

He never forgot what he owed to his teachers in the University. These were, as entered in his own book, his college tutor J. P. Higman, Archdeacon Thorp, G. B. Airy, A. Coddington, H. Parr Hamilton (Dean of Salis-

1825. bury), G. Peacock (Dean of Ely), and W. Whewell (afterwards Master of Trinity). With all of these gentlemen he kept up a friendship and correspondence during their joint lives.

Early friendships. His college friends of nearly his own age were William Heald, William Mason, Arthur Neate, and Thomas Falconer, all of Trinity College. Among those whose friendship he valued, none were more esteemed by him than his teachers—Dean Peacock, Dr. Whewell, Mr. Coddington, and Dr. Thorp, afterwards Archdeacon Thorp. Mr. Heald, afterwards Rector of Birstal, in Yorkshire, died in 1875. Mr. Mason, Rector of Pickhill, near Thirsk, died in 1873, and his companion and chum of Trinity College days, Arthur Neate, died at his rectory, Alvescot, near Oxford, in 1870.

Some peculiarities in his college life were well known to Cambridge men of his year. The habit of reading through great part of the night, and, in consequence, getting up very late the next day, was notorious; and fellow-collegians, coming home from a wine party at four in the morning, might find him just going to bed. One of these, better known in the University for *rows* than for reading, has told me how often he himself, being late next day from a different cause, has gone into De Morgan's rooms, just below his own, and begged for an air on the

Love of music. flute to 'soothe a headache.' His flute, which he played exquisitely, was a great source of pleasure to himself and his friends. He was a member of the 'Camus,' a musical club so called from the initials of its designation—Cambridge Amateur Musical Union Society; and their meetings, and those at the houses of a few musical families, were his chief recreation. He was a born musician. His mother said that when listening to the piano, even when a very little child, a discordant note would make him cry out and shiver. I must not omit to record his insatiable appetite for novel-reading, always a great relaxation in his leisure time, and doubtless a useful rest to an over-active brain in the case of one who did not care for riding or boating. Let it be good or bad in a literary point of view, almost any work of fiction was welcome, provided it had plenty of incident and dialogue, and was not over-sentimental. He told me that he soon exhausted the stores of the circulating library at Cambridge. Like his schoolfellows, his college friends loved him for his genial kindness, unwillingness to find fault, and quiet love of fun, always excepting practical jokes, with which he had no patience at all. 1826–27.

Intention to study medicine.

During the last year and a half of his stay at Cambridge Mr. De Morgan had some thoughts of becoming a physician. With his views on religion, his ordination was out of the question; but he liked the study of medicine, and some friends advised him to read it with a purpose. This intention did not last long. His old friend, Mr. Hugh Standert, of Taunton, knew by experience what was generally needed for success in medical practice, and an acquaintance from infancy made him believe that Augustus was not pliant enough, and could not, or would not, be sufficiently ready to adapt himself to the fancies and peculiarities he would meet with to make him a popular doctor. Whether or not he had any special genius for medicine is uncertain. His mother agreed with Mr. Standert, and urged upon her son that his success in medicine might depend on an amount of toleration for ignorance and folly which, with his 'hatred of everything low,' he would find a great trial. She begged him to 'throw physic to the dogs,' and to turn his thoughts to law. He complied, but did not like his destination. Events proved that he was right, that he had not found his proper place in the world's workshop.

1827. Degree.

In 1827 he took the degree of fourth wrangler, the order being Gordon, Turner, Cleasby, De Morgan. This place, as one of his scientific biographers truly says, 'did not declare his real power, or the exceptional aptitude of his mind for mathematical study.' He had

1826. been expected to be senior or second wrangler, and his lower degree was attributed by his contemporaries and competitors to the wide mathematical reading by which he was often led away from the course prescribed for examination. This failure, in a possibly fallacious test, was his own early, but unintentional, protest against competitive examinations; for which he felt excessive disapprobation even before his experience as a teacher showed him not only their mischievous effect upon mind and health, but their insufficiency to determine the real worth of a candidate for honours. In saying this I do not detract from the merit of the gentlemen who stood above Mr. De Morgan in the Tripos of 1826. All three distinguished themselves in after life, but as he was undoubtedly the first in mathematical ability, it is likely that their precedence of him might be due to the fact that his love of the study led him to read more widely and discursively than his friends on the very subject on which excellence was to be tested.

At the time of his taking his B.A. degree he came to live with his two brothers, mother, and sister in London. He had determined to go to the Bar, and was beginning his legal studies, but he very much preferred teaching mathematics to reading law. Something like the objection urged by his friends to medicine was uppermost in his mind, and he feared or imagined that in practising at the Bar he might find it difficult to satisfy both his clients and his conscience. But these scruples were overcome, and he entered at Lincoln's Inn.

SECTION II.

1827 TO 1831.

It was at this time that he became acquainted with my father, William Frend. They first met at the office of the Nautical Almanac, of which their common friend, Lieutenant Stratford, R.N., had been recently appointed Comptroller. Mr. Frend and Mr. Stratford were both members of the old Mathematical, and subsequently of the Astronomical Society. Though my father was, even at that time, far behind Mr. De Morgan as a mathematician, the two had a good deal of mathematics in common. My father had been second wrangler in a year in which the two highest were close together, and was, as his son-in-law afterwards described him, an exceedingly clear thinker and writer. It is possible, as Mr. De Morgan said, that this mental clearness and directness may have caused his mathematical heresy, the rejection of the use of negative quantities in algebraical operations; and it is probable that he thus deprived himself of an instrument of work, the use of which might have led him to greater eminence in the higher branches. This same heresy gave occasion to many amusing arguments and discussions. But between these two sympathy in matters of morals and principle formed a stronger bond than similarity of pursuit. My father had sacrificed good prospects as a clergyman to his conscientious scruples about the doctrines of the Established Church, as expressed in the Creeds and Articles, and had been through life an earnest advocate of religious liberty. These cir-

1827.

William Frend.

1827. Family of the writer.

cumstances won for him at once the respect and esteem of Mr. De Morgan, who, like himself, had thrown off the restraints of a creed which he could not hold, and which he refused to profess without holding. I think my father, who was a good Hebrew scholar, afterwards helped him to clear away some of the doubts and difficulties resulting from mistranslations of Scripture, and fostered by the early teaching of a sect not critically learned.

We were living at Stoke Newington, in one of those old houses with wooded grounds, of which so few remain near London. It had formerly belonged to Daniel Defoe, and Isaac Watts had inhabited it. In my father's time it was the scene of many a pleasant gathering of men and women of all degrees of intellectual ability, and of almost every shade of political and religious opinion. The spot where the old house stood has become the centre of a district of streets and shops, built where the tall trees grew, and nothing now remains to commemorate its existence but the name of Defoe Street.

Mr. De Morgan first came to our house with Mr. Stratford. He then looked so much older than he was that we were surprised by hearing his real age—just twenty-one. I was nineteen. We soon found out that this 'rising man,' of whom great things were expected in science, and who had evidently read so much, could rival us in love of fun, fairy tales, and ghost stories, and even showed me a new figure in cat's cradle. He was in person very like what he continued through life, but paler, probably from the effects of his recent Cambridge reading. His hair and whiskers were very thick and curly; he was not bald till thirty years after. I remember his having a slight pleasure in saying things which startled formal religionists, but which we, who were not formal, soon understood to mean what they expressed, and no more. These sayings were humorous, and like the half-mischievous jests of a very young man. It was easy to see that a deep religious feeling underlay the contempt for observance which his early training had caused, and that his consciousness of the care and fatherhood of the Almighty was a sacred thing belonging to himself alone, not to be profaned by contact with human forms or inventions. My father, who, like people who have made their own belief, was a little impatient in argument, at first thought him an unbeliever; and so, in a certain sense, he was; but it was only in such things as he could not find a reason for believing. I mention religious questions because they entered much into our thoughts and conversation at that time. As to the Gospels, he waited for a better and more critical understanding of them than could be gained from his first instructors, and this a rather extensive reading of theology enabled him to acquire before he left this world. When I first knew him, I was puzzled by such books as Volney's 'Ruins of Empires,' Sir. W. Drummond's writings, and other works of antiquarian research, to which a great interest in our friend Godfrey Higgins's investigations had led me. Mr. De Morgan showed me the scientific errors of some of these writers, and the insufficiency of their theories to account for all that they have tried to explain. He was well informed in Eastern astronomy and mythology, and saw that much of modern doctrine has gained something of its form, at least, from ancient symbolism.[1] Lieut.-Col. Briggs, his uncle by marriage, had begun his 'Ferishta,' and his nephew's interest in the work had brought him much into the society of Oriental scholars. The ancient grandeur and simplicity of the East at once excited and satisfied his imagination. He sometimes said that India with its skies and mountains 'might be really worth looking at,' whereas he never saw any

1827.

Antiquarian science.

[1] All scholars must see that the time is approaching when a better knowledge of ancient religions will show that they have been misunderstood, and that they are not entirely fictitious or entirely astronomical. If this were the place it would not be difficult to show the connection of all.

1827. scenery in England than which he could not picture to himself something infinitely grander. He was proud of his birth in the sacred city of Madura, and at one time longed to visit his native country, and fancied that every one had the same instinctive desire. Luckily, his doing so when young was prevented by the defect of sight which justified his mother in refusing a cadetship for him.

Music.

During the ten years which preceded our marriage, his delightful flute, accompanied by my sister on the piano, was a great pleasure to us. I lost the gratification of accompanying him then, and it was afterwards a sorrow to me that I was not a musician. Our acquaintance began just before he became a candidate for the Professorship of Mathematics, but he, like my father, took an interest in the foundation of the new University, of which, indeed, my father had been one of the first projectors.

London University.

It has been observed that when the time is ripe for bringing forward any measure, ideas come at the same time to more than one mind fitted to receive them, and it is often difficult to find the author of the first suggestion. This is especially true in the case of the foundation of large institutions. In what follows I do not mean to assert that my father was the first suggester of a college or university in London, but, being one of the few persons now living who can remember the beginning of University College and the expressed designs and hopes of its founders, I venture to give, more in detail than the scope of a biography would justify, a short account of its origin; and in thus contributing my share of its history I must speak of that part which I best remember.

About or before the year 1820, some liberal-minded men, after long pondering on the disabilities of Jews and Dissenters in gaining a good education, came to the conclusion that as the doors of the two Universities were closed against them, the difficulty could best be met by establishing a University in which the highest academical teaching should be given without reference to religious differences. As this could not be done in an institution in which the pupils resided without excluding religion altogether from education, a necessary condition of the establishment was the daily attendance of students on college lectures, so that while living under their parents' roof they might be brought up in the religion of the family.

1827.

First suggestions.

My father's ideas of the proposed institution had been embodied in some letters signed 'Civis,' and published in a monthly periodical[1] edited by Mr. John Thelwall, somewhere about 1819, which did not survive a third number. The writer was well qualified by his own academical status, and by the subsequent abandonment of Church preferment which led him into connection with intelligent Dissenters, to estimate the value of University training, and the great loss and deprivation sustained by young men every way qualified to profit by it who were unable from religious belief to receive it. He looked forward to the day when all forms of religion should be held equal within the walls of the noble institution which he contemplated, in which good conduct and compliance with rules should be the only conditions of admission.

A short time after the publication of the letters referred to, Mr. Thomas Campbell, the poet, first visited their writer, and informed him that Lord Brougham (then Mr. Brougham) and Dr. Birkbeck, with himself and one or two others, believed that the time for making the attempt was come. I was about twelve years old when Mr. Brougham dined with my father to consult upon it. Some meetings took place, other liberal men joined them, and after some delay the first active committee was formed.

Mr. Frend was prevented by long and severe illness

[1] I have tried in vain to find the title of this periodical, which is not in the British Museum. It must have appeared between 1818 and 1822.

1827. from taking an active part in the first movements, but he joined the general committee on his recovery, became a shareholder, and on the election of a council and officers was appointed one of four auditors.

The establishment of University College, called at first the London University, promised to fulfil the hopes of all friends of education, and was hailed as a forerunner of religious freedom. My father naturally took the liveliest interest in its progress. Mr. De Morgan welcomed the opening of the College, as not only meeting a great want of the time, but as offering to himself a prospect of leaving the study of Law, which he did not like, for the teaching and pursuit of Science. When the time came for the appointment of Professors he sent his name in as a candidate for the Mathematical chair. He was one of thirty-two candidates. The committee for examining testimonials found among his the highest certificates from Dr. Thorp, Dr. Peacock, Professor Airy, Professor Coddington, and others, his Cambridge teachers. He was much younger than any of his competitors, but his election to the chair of Mathematics was made unanimously, and afterwards confirmed by the Council on February 23, 1828, being formally communicated to him without delay.

Election to Professorship.

It was a little characteristic incident connected with the appointment of the future Mathematical Professor, that while the election was going on in one part of the college, and he with some others of the candidates were in the common room, he took up a volume lying on the table, which proved to be Miss Porter's 'Field of the Forty Footsteps.' The scene of this novel is laid in the fields which formed the site of the building and its surroundings. It was said that, some years before, the marks of the weird 'forty footsteps' might still be seen in the ground, but builders and stonemasons had effectually removed them, and fanciful comparisons were drawn between the effacement of these marks of the brothers' rivalry and the barbarity of their lady love as the new foundations arose, and the disappearance of crime and ignorance under the work which the College had to do. The love of fiction was strong enough in the candidate's mind to make him forget his interest in what was going on, and he had run through the volume before a whisper reached his ears as to the result of the election. 1827.

In looking at the past history of an institution it is useful to trace not only the successes, but the mistakes which have caused failure and disturbance; for even in cases where present prosperity may lead to imitation, a statement of errors committed and corrected will be as a chart of the rocks to be avoided hereafter. I shall try to give a truthful sketch of the early history of the College; not entirely omitting those elements in its formation which created discord in the first years, and which had some share long after in the disastrous termination of my husband's connection with it. Had he lived long enough he would have himself done this, far better than any one else. His pen was held for a time by consideration for contemporaries, most of whom are now gone. Circumstances connected with his memory have arisen since he was taken from us which make it imperative on me to do the work which he left undone.

State of education at this time.

To learn this history fairly we must look back to the state of education, and to the needs and disabilities which led to the foundation of the London University. These disabilities and needs were felt, not so much by highly educated academical men wishing for a cheap school for their sons, as by the great body of enlightened Jews and Dissenters, held back by religious tests from sharing in University advantages, but intelligent enough to perceive the value of what they lost, and rich enough to supply the want for themselves. The wealth of this party was of course represented by commercial men. To these must be added some parents living in London and the neighbourhood who could not afford to send their sons to college, and to whom the attendance on daily lectures

1828. while living at home seemed more desirable. A few of the most liberal thinkers of the time gave their best help to the completion of the design, but the large body of men who had been trained under University discipline held aloof from an institution from which religious tests were excluded, and which might at some time compete with the two Universities, bound up as they were by old usage with the interests of the Established Church, whose foundations were laid in the time of a church older still.

Founders of London University.

Thus, with the exception of the few enlightened scholars who generally held out a hand to their less fortunate brethren, the founders of the London University were either liberal politicians, not always familiar with the details of academical discipline, or mercantile men, who, with the best possible intentions, had no experience of the best way of securing concord and due balance in the relations of governing body, teacher, and pupil.

Constitution.

The Deed of Settlement of the London University bears date 1826. The Institution was a proprietary one, the funds being raised partly by shares, partly by subscriptions. The management was vested in a council of twenty-four gentlemen chosen from among the proprietors, and a general meeting of proprietors formed the highest court of appeal. The Professors were elected by the Council; and a Warden, who was to be the medium of communication between the Council and Professors, and superintendent of the household department, was appointed. The duties of the Professors were confined to their class-rooms, in which, as it afterwards appeared, they were not absolute.

It would have been well for the infant institution if a piece of advice given by Mr. De Morgan long after, and in a different connection,[1] could have been acted on at this time. 'Never begin,' he said, 'by drawing up con-

[1] On the establishment of the Ladies' College, Bedford Square.

stitutions. They are sure to prove clogs on the wheel. 1828.
Let the work begin in good earnest, and with no needless machinery. If it is done well you will soon see what is wanted, and the constitution will be formed by meeting the needs as they arise.' The founders of University College, as of other public institutions, had not grasped the idea of this natural growth, and the effect of their arrangements was to put a clog upon the wheels, which shook the whole vehicle, and well-nigh overturned it at first going off.

As I cannot enter into the history of this institution farther than is necessary to explain my husband's connection with it, no names except those which belong to that part of the history will be brought forward.

The design of the London University, as set forth in pamphlets, speeches, and the general understanding of the time, and repeated many years later in an official document,[1] was to provide a liberal education in Classics, Mathematics, Physical Science, and Medicine, without regard to religious distinction either in teacher or pupil.

The teaching was to be given in lectures attended daily by students, and the only condition of entry, beside the fee, was good conduct and compliance with the rules laid down for the maintenance of order in the college.

In conformity with this avowed principle of religious
neutrality we find, among the Professors first chosen, Professors.
three Clergymen of the Church of England, one Independent minister, a Jewish gentleman, who in his place of Hebrew professor taught the reading of the Old Tes-

[1] No reference whatever is made to religion in the *Deed of Settlement, Regulations,* or *By-Laws.* In these it is stated that the object of the University is to afford an education in Mathematical and Physical Science, Classics and Medicine. The absolute determination to leave the subject of religion entirely untouched appears negatively from these documents, but positively from all the addresses given in the institution, in newspaper articles, and in the general understanding of all the parties connected with it, a great number of whom, being rich Dissenters, watched the proceedings with a jealous eye.

1828. tament, and other gentlemen nominally churchmen, but whose religious views were known to vary from strict orthodoxy to the widest latitudinarianism.

The appointment to the Mathematical Professorship pleased some of Mr. De Morgan's Cambridge friends, who spoke and wrote of it as a boon to the London University and to himself. It would be untrue to say that all his friends rejoiced in it, for his own family and near relations, who had anticipated a brilliant success for him at the Bar, felt that to take a position as yet doubtful, with a greater doubt of fitting remuneration, was really a sacrifice on his part. My father shared in this feeling, and, in reply to the expression of it, Mr. De Morgan wrote:—

Letter to Mr. Frend.

'You seem to fancy that I was going to the Bar from choice. The fact is, that of all the professions which are called learned, the Bar was the most open to me; but my choice will be to keep to the sciences as long as they will feed me. I am very glad that I can sleep without the chance of dreaming that I see an "Indenture of Five Parts," or some such matter, held up between me and the *Mécanique Céleste*, knowing all the time that the dream must come true.'

One false step due to the tendency in young associations to frame constitutions before their needs are known, was the appointment of a Warden for the new University. The next error arose from the same cause, and showed the inability of the governing body to perceive what was due to men of worth and education, if they meant such men to give them the weight of their character and influence. As a good friend to the new College wrote to Mr. De Morgan, speaking of two influential members of Council: 'A. believes that the University depends on the Professors, B. that the Professors depend on the University.' Unfortunately the A.s were in the minority.

Appointment

Mr. De Morgan received the official notice of his

appointment on the day of the election, and was informed at the same time that 'a formal certificate of appointment will be prepared, in which the duties of the Professors will be specified, and they will be required to sign an acceptance of the authority of the Council and of the rules of the University on receiving them.' 1828.

Undue restraints on Professors.

These conditions and obligations were not such as could be accepted by men accustomed to academical discipline, and who knew the value of their work. They were the work of a governing body new to its own duties, and to the claims and rights of those for whom they were composed. But after a strong remonstrance the Professors were enabled to hold their diplomas on a simple declaration of adherence to the constitution as set forth in the Deed of Settlement. The classes opened on the following November, and on the 5th the Professor of Mathematics gave his introductory lecture, when, as he says, he 'began to teach himself to better purpose than he had been taught, as does every man who is not a fool, let his former teachers be what they may.'[1]

Introductory lecture.

This lecture 'On the Study of Mathematics' takes a much wider view of that study, and its effects upon the mind, than its title alone would imply. It is an essay upon the progress of knowledge, the need of knowledge, the right of everyone to as much knowledge as can be given to him, and the place in mental development which the culture of the reasoning power ought to hold. It is not only a discourse upon mental education, but upon mind itself. It was the work of a young man of twenty-two years and four months old, and the earnestness and sanguineness of youth may be seen in the strong deter-

[1] In this year he published a translation of the first three chapters of Bourdon's *Algebra*. This was afterwards superseded in his class-room by his own *Arithmetic and Algebra*. In his own copy of Bourdon is inscribed, after his name, 'Aged 22 years and 2 months, being the first work he ever published.—A. De M., Aug. 26, 1846.'

1828. mination with which his work was begun, and the high hopes which he felt of the work the University had to do. How well *his* part was done, after years, and the consenting voice of many pupils whose own work bore the fruit of his teaching, have given proof.

What part he had in obtaining for the College its subsequent reputation it is not for me to say. How he was repaid, the judgment of the future must determine.

The Mathematical class during the first session consisted of nearly one hundred pupils. In the next year there was an increase of numbers. The Professor gave two lectures every day, the first from nine till ten A.M., the second from three to four in the afternoon. After each lecture he remained for a time at his desk, in order that pupils who had found any part obscure might come to him to have their difficulties cleared up. In this way the two lectures occupied about three hours in the day, and the pupils' exercises which were to be examined rather less than an hour more.

Day school. Various proposals had been made by the most active among the Professors for improving the condition of the institution; among those which were carried into effect were the foundation of a day school in connection with the University, and the annual distribution of prizes and honours. But, as might be expected from the elements of which the new institution was founded, it could not go on long smoothly. Troubles began soon after the opening, due to arrangements which resulted from the formation of a constitution and laws before the working necessities of the institution could be known; and all the misapprehensions which soon arose among the component members were traceable to this cause. These were set forth chiefly in the following pamphlets, printed for private circulation:—

1. 'A Letter to the Shareholders and Councillors of the University.'

2. 'Statements respecting the University of London, prepared at the desire of the Council by Nine of the Professors.' 1828.

3. 'Letter to the Council of the University of London, by Leonard Horner, Warden of the University.'

4. 'Observations on a Letter, &c., by L. Horner, Esq., &c., &c., by Nine Professors.'

1829. Colonel Briggs.

During the vacation of 1829, Mr. De Morgan spent a few weeks in Paris, chiefly at the house of Colonel, afterwards General John Briggs, of the Madras Army. Col. Briggs and his father, Dr. Briggs, also of the Indian army, had married two Miss Dodsons, sisters of Mr. De Morgan's mother. They were, therefore, his uncles by marriage. Col. Briggs, who, as a young man, had served under Sir John Malcolm during the time of the dissolution of the Mahratta Confederacy, afterwards held successively a diplomatic post in Persia, and that of Resident at Nagpoor, and finally, for a short time, the place of Senior Commissioner of Mysore. He was an able officer and an indefatigable student of Eastern language, history, and Science. His work on the Land Tax of India was one of the earliest protests against some points of British misrule in the East. In the bringing out of this work his nephew Augustus gave him a good deal of assistance. Besides this work, General Briggs was the author of 'Letters on India,' an excellent guide for young men entering the army, even now when the army is under different rule; and besides the 'Ferishta,' already mentioned, he translated the work by Ghulam Hussein on the 'Decay of the Mogul Empire.' His knowledge of Eastern languages and Science had brought him and our friend Godfrey Higgins into intimate acquaintance. They visited my father together at Stoke Newington, and their animated discussions were always amusing and often instructive, though the two had a tendency to differ about Hindoo temples and topes and remains, which Mr. Higgins declared had been built and decorated

1829. according to his theory of ancient Astronomy, and Colonel Briggs as firmly maintained were not so when he last saw them.

Acquaintances in Paris.

The visit to Paris was paid just before Colonel Briggs left for India. It was a time of great enjoyment, and besides the society of his uncle's family and the pleasures of Paris life, then quite new to him, whose time had been altogether given to study, he made acquaintance with many of the scientific men and scholars of the time. Among these were MM. Hachette (with whom he corresponded till M. Hachette's death in 1832), Biot, Chladni, the Duc de Broglie, and others. With M. Quêtelet he became acquainted two years later. M. Bourdon, whose work on Algebra he had translated, was in Paris, but the two never met.

India.

My husband's interest in his birthplace had always been kept alive by intercourse with his many relations there, some of whom were in the Madras army, some in the Civil Service. It is well known how frequent were the disputes and jealousies among the servants of the East India Company. Col. De Morgan had suffered much from accusations made against him by superior officers, for which the later justice done to him hardly compensated. Col. Briggs, who was acknowledged to be an able and well-informed officer, had his share of trouble. In 1829, great difficulties arose in the government of the Mysore, owing partly to the mixture of native rule, and partly to the province being under the direction of the Governor of Madras, who appointed Commissioners for it. Owing to these disorders, Lord W. Bentinck, the Governor-General, determined to separate the Mysore from the Madras Presidency, and appointed Col. Briggs and another officer Commissioners, with full powers over the province. Of these Col. Briggs was the chief. This appointment displeased the Governor of Madras, who left no stone unturned to reverse it, and after a year and a half succeeded in getting Col. Briggs removed, and another officer put in his place. The difficulties and real hardships (for he was then ill) which Col. Briggs underwent during this time were communicated to his sympathising nephew and friend in England, who gave what help he could by calling the attention of the Directors of the Company to the case. Nothing could be done, however, and Mr. James Mill writes, 'From all I hear, I believe Col. Briggs' friends have reason to rejoice in his dismissal.'[1] 1829.

The session of 1829–30 began nearly as the last

[1] Dr. Briggs' ghost story, well known in the Madras Presidency ninety years ago, was one of the best authenticated incidents of the kind I ever heard. I give it here as it was told me, first by Mr. De Morgan, who heard it from his mother; afterwards by General Briggs, who had it when a young man from Sir John Malcolm. His father could not be induced to speak of it.

When my informant was a very young infant, Dr. Briggs, who was quartered with his regiment somewhere (I forget the place) in the hill country, used to hunt once or twice a week with the officers and others, whose custom it was to breakfast at each other's houses after the sport was over. On a day on which it was Dr. Briggs' turn to receive his friends, he awoke at dawn, and saw a figure standing beside his bed. He rubbed his eyes to make sure that he was awake, got up, crossed the room, and washed his face well with cold water. He then turned, and seeing the same figure, approached it and recognised a sister whom he had left in England. He uttered some exclamation, and fell down in a swoon, in which state he was found by the servant who came to call him for the hunt. He was of course unable to join his friends, who, when at breakfast on their return, rallied him on the cause of his absence. While they were talking he suddenly looked up aghast and said trembling, 'Is it possible that none of you see the woman who stands there?' They all declared there was no one. 'I tell you there is,' he said. 'She is my sister. I beg you all to make a note of this, for we shall hear of her death.' All present, sixteen in number, of whom Sir John Malcolm was one, made an entry of the occurrence and the date in their note-books, and by the first mail which could bring the news from England the sister's death at the time was announced. She had, before leaving this world, expressed a wish that she could see her brother and leave her two young sons to his care. Dr. Briggs was a man of great nerve and courage, and one to whom the idea of a spirit's appearance would, until that time, have been utterly ridiculous. The death of General Briggs some years since, at the age of ninety, makes it allowable to publish the story, which, however, he gave me for the purpose forty years ago.

1829–30. had ended, and the dissensions in the institution were publicly known. In March 1830 the Mathematical Professor's share of the difficulties became serious, and a correspondence between himself and the Warden resulted in some modifications of the functions of that officer. But the disturbances on the Medical side continued through the year, the result of a series of alternate mistakes on the part of the authorities and remonstrances on that of the Professors. One of these remonstrances is contained in a letter written by Mr. De Morgan, who had been asked by some members of the Council to lay before that body the views he had expressed in a conference with a Committee appointed to examine into some complaints preferred by the Anatomical class against their instructor, Professor Pattison:[1]—

1831. Letter to committee.

Gentlemen,—In compliance with the wish expressed by you when I had the honour of an interview with you, I lay before you the views which I entertain on a subject most essentially connected with the welfare of the University, viz., the situation which the Professors ought to hold in the establishment. This question is of the highest importance, inasmuch as upon the manner in which it shall be settled depends the order of education and merit which will be found among the Professors in future, and the estimation in which they will be held by the public.

In order to induce men of character to fill the chairs of the University, these latter must be rendered highly independent and respectable. No man who feels (rightly) for himself will face a class of pupils as long as there is anything in the character in which he appears before them to excite any feelings but those of the most entire respect. The pupils all know that there is a body in the University superior to the Professors; they should also know that this body respects the Professors, and that the fundamental laws of the institution will protect the Professor as long as he discharges his duty, as certainly as they will lead to his ejectment in case of misconduct or negligence. Unless the pupils are well assured of this they will look upon the situation of Professor as of very ambiguous respectability, and they will

[1] I have avoided entering into details, leaving the principles at issue to be inferred from Mr. De Morgan's letter.

only be wrong inasmuch as there will be no ambiguity at all in the case. 1831.

With the public the situation will be altogether as bad. Wherever the Professor goes, he will meet no one in a similar situation to his own—that is, no one who has put his character and prospects into the hands of a number of private individuals. The clergyman, the lawyer, the physician, the tutor or Professor in the ancient Universities, will all look down upon him, for they are all secured in the possession of their characters. Nothing but the public voice, or the law of the land, can touch them, and a security as good must be given to the Professor of the London University before he can pretend to mix in their society as their equal.

If these were the sentiments of one individual only, they would merit little attention; but if they be the opinions of a majority of the present Professors, or even of a large minority, the committee may be sure that they are prevalent among the class of men from which the University ought to expect to draw its Professors. The sense of the Professors on this subject can be readily ascertained, and the committee will incur a heavy moral responsibility should they, without the most attentive examination, propose a change which may place the Professors, present or future, in the situation I have described. For mark the consequences. If I am right, every man who has the feelings of a gentleman will abandon the University in disgust; the same feeling will prevent any person of considerable attainments from offering himself for the vacant chairs; and the University, in the general school at least, will sink into the most paltry of all establishments for education, if, indeed, it long continue to exist. I am not mentioning my own opinions alone; such deductions are very common at present. I hardly meet one of my friends who does not seriously advise me to resign my situation on these very grounds.

The committee has done me the honour to ask my opinion as to the principles to be laid down for the future regulation of the Professorships. I will state, in few words, my own convictions on the subject.

The University will never be other than divided against itself as long as the principle of expediency is recognised in the dismissal of Professors. There will always be some one who, in the opinion of some of his colleagues, is doing injury to the school by his manner of teaching; and there will always be

1831. attempts in progress to remove the obnoxious individual. The medical school is peculiarly subject to this evil, owing to the very frequent jealousies of one another which arise among the members of that profession. No man will feel secure in his seat; and, consequently, no man will feel it his interest to give up his time to the affairs of his class. And yet this is absolutely necessary in the general school at least, for from the moment when a class becomes numerous the preparation, arrangement, and conduct of a system of instruction is nearly the business of a life; at least, I have found it so. If a Professor is easily removable, he will endeavour to secure something else of a more certain tenure; he will turn his attention to some literary undertaking, or to private pupils, while he remains in the institution, in order that he may not be without resource if the caprice of the governing body should remove him—and this to the manifest detriment of his class, which, when it pays him well, ought to command his best exertions. In addition to this, he will always be on the watch to establish himself in some less precarious employment, which he will do even at pecuniary loss, since, especially if he have a family, it must be his first object. In this way the University will become a nursery of Professors for better conducted institutions of all descriptions, since no man, or body of men, desirous to secure a competent teacher in any branch of knowledge, will need to give themselves the trouble to examine into the pretensions of candidates as long as any one fit for their purpose is at the University of London. The consequence will be a perpetual change of system in the different classes of the University, and the eventual loss of its reputation as a place of education. These evils may be very simply avoided by making the continuance of the Professors in their chairs determinable only by death, voluntary resignation, or misconduct either in their character of Professors or as gentlemen, proved before a competent tribunal, so framed that there shall be no doubt in the public mind of the justice of their decision.

But this, it has been said, will be to give the Professors a *vested interest.* I assert that, in the proper sense of the words vested interest, it ought so to be. Who have more interest in the well-being of the University than I and my colleagues? Is it the Proprietary and the Council, on account of the capital invested by them, and their zeal for the advancement of education? In the latter we yield to none of them; and as to pecuniary risks, I, for example, have invested the whole results of an expensive education, for the original outlay of which I might buy fifty shares at the market price; and even omitting this, I have invested here my time, character, and prospects, all and every one of which is as truly an investment of capital as that made by any proprietor—with this addition, that it is my all; whereas the portion of any proprietor is a very small part of his. Can it be expected, then, that the Professor should be the only person in the institution who has no interest in it? and that he, merely on account of the important part he has to play, should be placed in a situation not so respectable as that of a domestic servant? These are truths which cannot but have the greatest weight with every person who shall hereafter think of embarking his fortunes here; and the only way to secure proper Professors *on the whole* is to respect these truths, and not to let incidental advantages, even supposing them such now, be considered of more importance than general results.

1831.

An institution such as ours is a machine meant to last for centuries, but this it cannot do if those who manage it are content to avail themselves of expediency, which is made for the day, in preference to fixed principle, which will never wear out.

I have written these sentiments because I feel no trouble too great when the end proposed is so truly useful. Personally I feel but slightly interested, for I cannot conceal from myself that the chance of resuming my duties in the University is very small. The opinions which I have here given will be the guide of my conduct, and, I have reason to believe, of that of others also. But should the result of the present proceedings be that a Professor of the University of London need not hold down his head for shame when he hears his situation mentioned, and the terms on which he holds it, no one is more ready than myself to stand or fall with this institution. This is, I fear, not an unmeaning pledge, for past events have so fixed in the minds of men an impression unfavourable to our prospects, that I fear our number of pupils will be seriously diminished in the ensuing session.

In conclusion, gentlemen, I have to thank you for the polite attention with which I was received by you when I took an opportunity of laying these sentiments before you in person, and I beg to subscribe myself,

Your obedient servant,

AUGUSTUS DE MORGAN.

90 Guilford Street, July 15, 1831.

1831. Mr. De Morgan also wrote the following letter, officially addressed to the Council through the Warden:—

SIR,—I beg leave to address the Council through you on a subject which I approach with great reluctance.

It is well known to the Council that I have often differed from them on matters connected with the management of the University, and that, when I have done so, I have never hesitated to declare my opinions in the plainest language. The Council will therefore believe me when I say that I am convinced that they and the Professors have during the last session been coming to such an understanding as would have made the supremacy of the former quite consistent with the respectability and independence of the latter. A third body has, however, interfered in the question, whose declared intentions, if carried into effect, will render it impossible for me to continue in the situation I at present hold.

Should the result of the labours of the Select Committee be the abrogation of the by-laws alluded to at the General Meeting, I respectfully inform the Council that it is my intention to seek elsewhere the subsistence and character which I had hoped to gain in the University of London alone. At the same time I feel it would not be dealing fairly with the Council if I let them remain in ignorance of my determination, considering that the deliberation of the Proprietors may possibly be pushed to a late period in the vacation, when a proper choice of a successor to my chair may be rendered difficult by the shortness of time remaining for that purpose. Having announced my intention, I am therefore in the hands of the Council; should they consider it unfair in me to offer a conditional resignation dependent on circumstances over which they have no control, I will, on intimation to that effect, offer an absolute resignation immediately. My wish is decidedly to remain in the University, if that can be done consistently with my own notions of what is due to my character. Having thus shortly stated the predicament in which I find myself placed, I leave the matter to the decision of the Council.

I have the honour to remain, sir,

Your obedient servant,

AUGUSTUS DE MORGAN.

90 Guilford Street, July 1831.

The whole was brought to a crisis a few days after by the dismissal of the Professor of Anatomy, the resolution for which concluded with these words:—'Resolved—That, in taking this step, the Council feel it due to Professor Pattison to state that nothing which has come to their knowledge with respect to his conduct has in any way tended to impeach either his general character or professional skill and knowledge.' 1831.

Immediately on hearing of this resolution, Mr. De Morgan sent in the following letter of resignation:—

To the Council.

GENTLEMEN,—I have just seen Mr. Pattison, who has informed me of his removal from his chair, and has also shown me a resolution, of which this is a copy. [*Copy of resolution as above.*] Here is distinctly laid down the principle that a Professor may be removed, and, as far as you can do it, disgraced, without any fault of his own.

This being understood, I should think it discreditable to hold a Professorship under you one moment longer.

I have, therefore, the honour to resign my Professorship, and to remain, gentlemen,

Your obedient servant,

A. DE MORGAN.

90 Guilford Street,
Sunday, July 24, 1831.

The answer came in the words, '*The Council accept your resignation.*'

In reply to a letter from my father, he wrote:—

90 Guilford Street, July 29, 1831.

Letter to William Frend.

DEAR SIR,—I have just received your kind note, which I hasten to answer.

The Council, in a session held after the meeting on Saturday, deprived Mr. Pattison of his Professorship, alleging at the same time, in vindication of themselves, I suppose, that nothing which had ever come to their knowledge had any tendency to lower their opinion either of Mr. Pattison's general character or of his professional skill and knowledge; thus laying down the principle that a Professor might be deprived of his office without any fault *of his own*, and even under a fire of encomiums from the Council.

I had long fully made up my mind not to hold any office whatever which was not absolutely my own during good behaviour—not even in the service of Government, should such a

1831. thing ever fall in my way. Immediately, therefore, on seeing the minute of Council containing the aforesaid removal, together with their most sufficient reason for the same as a rider, I addressed a letter to the Council that, under the principle there advocated, I should consider it discreditable to hold their Professorship one moment longer. The resignation was of course accepted, and I have done with them.

This step will be against my pecuniary interest should the University ultimately succeed very well, which the present proceedings of the Council will not allow any man to think who knows how much such an institution depends on public opinion. For the present moment, and up to the present time, I shall be no loser, since I know that by my own private exertions I can gain as much as, thanks to the dissensions in the University and the conduct of the Council regarding them, I have ever done in my public capacity.

With regard to an accusation and a hearing supposed by you necessary previous to the removal of a Professor, I must enlighten you on a principle discovered in the University of London by the Council, and faithfully acted on by them up to the present moment; viz., that a Professor in their institution is on the same footing with regard to them as a domestic servant to his master, with, however, the disadvantage of the former not being able to demand a month's wages or a month's warning. The proprietors, by their sense expressed at public meetings, have agreed with them, it appears to me.

I have still some interest in the University on account of some valued friends who remain behind, having what the advertisements call encumbrances. They, however, have expressed their determination to remain only one session longer; and feeling, as I do, that I never could send a ward of mine to an institution where it has been thus admitted by precedent that the student is a proper person to dictate the continuance and decide the merits of a Professor, I cannot wish the University to succeed, because I feel it ought not to succeed upon those principles.

If there be a large body of the Proprietary really interested in the moral as well as intellectual part of education, their efforts may yet save that fine institution. As a proprietor of it I would gladly lend my humble aid.

Yours most sincerely,

A. De Morgan.

SECTION III.

1831 TO 1836.

AT the time when he left the College, Mr. De Morgan was living with his family in Guilford Street, but removed in the autumn of 1831 to 5 Upper Gower Street, where he lived till our marriage in 1837. His only sister had been married the year before to Mr. Lewis Hensley, a surgeon of ability and good practice. My own family left Stoke Newington and settled at 31 Upper Bedford Place, Russell Square, in 1830. 1831.

In May 1828, shortly after his first coming to London, Mr. De Morgan had been elected a Fellow of the Astronomical Society, and in February 1830 took his place on the Council. Of the state of Science just before that period, Sir John Herschel said: 'The end of the eighteenth and the beginning of the nineteenth century were remarkable for the small amount of scientific movement going on in this country, especially in its more exact departments. . . . Mathematics were at the last gasp, and Astronomy nearly so—I mean in those members of its frame which depend upon precise measurement and systematic calculation. The chilling torpor of routine had begun to spread itself over all those branches of Science which wanted the excitement of experimental research.' State of Science.

In 1820 the Astronomical Society was founded by Mr. Baily in conjunction with Dr. Pearson, and from the time of its formation the joint efforts of many earnest intellectual men were given to raise the higher sciences

1831. from the state of depression and inactivity described by Sir John Herschel. The work, however, had not been uninterrupted, and the difficulties attending their task were increased by some injudicious persons who liked better to attack old errors and abuses than to work harmoniously with those whose only aim was to introduce better methods and measures.

This was inseparable from a condition of reconstruction —the same spirit of change and reconstruction that was at work in the political world—and the obstacles thrown in the way of reform by men whose efforts went either in the wrong direction or too far in the right direction, were not felt only in science. Of this time my husband wrote some years after: 'I first began to know the Scientific world in 1828. The forces were then mustering for what may be called the great battle of 1830. The great epidemic which produced the French Revolution, and what is yet (1866) the English Reform Bill, showed its effect on the scientific world.' The nature and extent of the scientific works begun before this time and carried out to completeness during the half-century which followed, can be but slightly mentioned. Mr. Francis Baily had effected the improvement in the 'Nautical Almanac,' and compiled the Society's 'Catalogue of Stars.' Sir John Herschel was engaged on his 'Catalogue of Double Stars,' to complete which he left England for the Cape of Good Hope nearly three years later. The Royal Observatory, Greenwich, was in full operation, under the direction of Professor, now Sir George Airy. Astronomy was rapidly approaching that height on which it now stands, and the efforts of the Astronomical Society—a body of men working with earnestness and unanimity—did much to raise it to its present state.

Mr. De Morgan was elected honorary secretary in 1831. He entered with zeal into every question brought before the Society, and his place was not a sinecure. It is not easy to say how much of the usefulness and prosperity of the Society during the years in which Mr. De Morgan filled this place was due to his incessant energy and effort, and to his steady judgment at difficult junctures. 1831.

Astronomical Society.

His work at the Society brought him into immediate contact with all its transactions and with all concerned in them, and as he never left London, and was known to be always at hand, much more than the routine duties of an honorary secretary would have fallen to his share, even if he had not voluntarily taken them upon himself. He drew up documents, wrote letters, and arranged for the meetings and the publication of memoirs. His obituary notices, written as one after another of his fellow-workers left the world, are biographical photographs, taken with a skill that makes the sunlight bring out all the finest as well as the most characteristic lines of the face.

In the year 1831, the second of Sir James South's presidency, a royal charter was granted to this Society. It was made out in the name of the President, owing to a legal formality, which would have involved greater expense to the Society if others of the Council had been included. But though no mention of differences of opinion appears on the minutes of the Society, there was certainly anything but unanimity as to the manner of receiving this grant, for Mr. De Morgan has preserved the following letter from Captain, afterwards Admiral Smyth, in answer to the requisition officially made for another Council meeting to re-discuss the question. The style of the formal letter contrasts strongly with the friendly effusions to the 'Esteemed Sec.' and 'Dear Mentor' of after times:—

In answer to the requisition for a Council to meet on Saturday next to re-discuss the subject of the charter, I regret to say that indispensable occupations prevent my attendance; but, I must add, if leisure were at my command I should still strongly object to being called away from employment on account of the whims of an individual.

I consider the point in question to have been already as well considered as the true spirit of our association requires; that any objection that has been started is more specious than valid;

1831. and that any further alteration will be merely a distinction without a difference. I firmly believe that every member of the Council has acted to the best of his ability and opportunity, and also feel that the Council, as a body, has ever shown itself more zealous about substance than about quibbling forms; but they might as well frame laws and institutions for Mars or Jupiter as for those who are predetermined to be dissatisfied.

I therefore trust, in order that the vigour of the Society may not be fettered, that the Council will take effectual steps to repel every disorderly attempt to impute motives or impugn its conduct, as well as to stifle their rancorous disputes, which can only engender an atrophy of moral work. If this is not insisted on, the meeting, which was purely instituted for the propagation of Science, will quickly degenerate into a spouting club, in which, instead of the adduction of undistorted facts, we shall be exposed to all the artillery of premisses without conclusions, and conclusions without premisses, added to the iteration of undigested thoughts in all the turgidity of ill-taste; and even were the reasoning powers among us more perfect, we should only be making much noise and little progress, leaving the good uncertain and remote, while the evil would be certain and immediate. Moreover, the disputatious system, being both irritable and irritating, is altogether as absurd for astronomers as would be the dramatising of Newton's Principia.

I therefore firmly hope that a perfect union in the cause we are embarked on will distinguish our efforts, for the straightforward course of duty is as perfectly practicable as it is desirable. I have the honour to be, sir,

Your obedient servant,

W. H. Smyth.

Professor De Morgan, Sec. Ast. Soc.

Who the individual was whose 'whims' Captain Smyth refers to I cannot say. But it is a significant fact that Sir J. South, whose Presidency had not expired when the charter was granted, was not re-elected in the new staff of officers, nor does his name appear on the Council after this time.

Mr. De Morgan's acquaintance with his colleagues on the Council of the Astronomical Society became in several cases intimate friendship. His friends were Mr. Baily,

Sir John Herschel, the Astronomer Royal, Lord Wrottesley, 1831.
Rev. Richard Sheepshanks, Admiral Manners, Mr. Galloway, and a few others. Mr. Sheepshanks and Mr. Galloway had houses in the immediate neighbourhood of Gower Street, and Mr. Baily lived at 37 Tavistock Place—a pleasant house in a garden sheltered by sycamores. This house, rendered famous by the repetition of the Cavendish experiment, had formerly belonged to Mr. Perry, of the 'Times.' In it my mother had met Porson, and heard him repeat Greek poetry.[1]

Mr. Baily was well fitted by his clear-headed steadiness of character, as well as by his excellent temper and geniality, to form the centre of a knot of friends sharing in the same pursuits. The same qualities made him an excellent host, and a better President of the Astronomical Society than if he had been a more brilliant talker. His kindly, simple bearing gained the love of those who could only look at his work with wonder. I remember feeling proud of having played a game of chess, in which I was of course beaten, with him. His house and appointments were just what they should be, made perfect to his friends by the cordiality of his reception. After his sister came to live with him, when this welcome was extended to his friends' wives and sisters, no house in London, I suppose, had held more happy parties than 37 Tavistock Place.

I find an anecdote showing his characteristic order and neatness in a letter left by my husband for the Institute of Actuaries. The proposal referred to was made in 1835, and related to the Cavendish experiment.

'That every rule must have its exceptions is true even of Baily's accuracy, though I should have thought the assertion must have failed if I had not known the contrary. Few persons, however, know that this assertion contradicts itself. For, if it be a rule that every rule has its exception, this rule must have its exception; that is,

[1] This house was left by Miss Baily to Sir J. Herschel, and until very lately was inhabited by Mr. Digby Wyatt.

1831. there must be a rule without exception. Leaving this bone for logicians to pick, I go on with my story. About 1835 the Government made an important proposal to the Astronomical Society. Mr. Baily, the President, stated that he had summoned the Council to consider a communication from the Lords of the Admiralty, which he would forthwith read. He then put his hand in his pocket, and the paper was not there. This almost excited remark, for that Mr. Baily should not remember in which pocket what he looked for was to be found, was a very unlikely thing. But the other pockets also answered in the negative, and the end of it was that Baily announced that he must have left his papers behind him. The announcement of a comet with satellites would not have created half the surprise which followed. There was nothing for it but to take a cab and get back as quick as possible, leaving the Council to decide *nem. con.*, though it could not be entered on the minutes, that they liked the President all the better for being, to absolute demonstration, a man of like failings with themselves.'

In the Supplement to the 'Penny Cyclopædia' Mr. De Morgan wrote of Mr. Baily :—

'The history of the astronomy of the nineteenth century will be incomplete without a catalogue of his labours. He was one of the founders of the Astronomical Society, and his attention to its affairs was as accurate and minute as if it had been a firm of which he was the chief clerk, with expectation of being taken into partnership.'

Sir John Herschel, the most distinguished in general estimation of these co-workers, was not so often among them at this time. He left England for the Cape of Good Hope in 1833, and was of course unable during his absence to take part in the practical business of the Society. My husband's letters to him show how little his colleagues liked to consider him absent. This correspondence began in the year 1831, when Mr. De Morgan, as secretary, addressed him with official formality, and continued till

1870, having for many years become the expression of affectionate friendship. 1831.

The Astronomer Royal and Mrs. Airy were among the most welcome of this circle of friends, who often met at the house of Mr. Sheepshanks, where the presence of his sister, a woman full of genial kindness, made all feel welcome and happy. All were fond of music, and Mrs. Airy's and her sister's ballads, sung with a spirit that gave them a character equal to Wilson's, were sometimes accompanied by Mr. De Morgan's flute, and are still among my pleasantest remembrances.

Richard Sheepshanks.

Mr. De Morgan had a strong regard for Mr. Sheepshanks. Among many descriptive remarks, he says of him in the MS. before mentioned, 'He was the man from whom I learnt more than from all others of the way to feel and acknowledge the merits of an opponent. I have known many men cheerfully and candidly admit the good points of an antagonist, but hardly another, besides Sheepshanks, who would, in the course of opposition, systematically select them, bring them forward, maintain them against those of his own side; and this always, year after year, when engaged in warm opposition as well as in jocose conversation, when in public discussion with several as well as in private conversation with a single friend.' And that which must be noticed is the vigorous and practical character of his friendship. His active and unwearied assistance was as surely to be reckoned on as a law of nature, especially if to the cause of his friend was attached the opportunity of supporting some principle, or aiding some question of science. Nor was his kindness of feeling limited to his friends. It showed itself in real and thoughtful consideration for all with whom he came in contact. Had he been a physician, his fanciful and self-tormenting patients would have thought him the worst of their ills, his milder cases of real suffering would have been cheered by his bantering kindness, while severe and dangerous malady would have felt the presence of the

1831. sympathy which money cannot buy, shown with a delicacy which benevolence itself cannot always command.

The reference to an opponent points to Sir James South, who had become before this time a general opponent of most of his scientific friends. He joined Mr. Babbage, who had accused some members of the Astronomical Society of being in a conspiracy against him, and this accusation elicited from Mr. De Morgan the following description of his own relations with three of his friends:—

The conspirators.

'The only conspirators named were MM. Airy and Sheepshanks. These two and myself lived together in intimate friendship, officers of the Astronomical Society through a long course of years, . . . we three, and each for himself, deciding that he was a rational and practicable man, and that the other two, no doubt worthy and rational, were a couple of obstinate fellows. Francis Baily thought the same of all three. I suppose we were an equi-tenacious triangle. But never a sharp word, I am sure, passed between any two of the four. Men of Science are not always quarrelsome; and, as often happens when obstinate persons are reasoners, we were generally of one line of *action*, with occasional repudiation of each other's views. In all the many pleasant laughs we have had together about the doings of the two common assailants, nothing ever emerged which gave me the least impression of the existence of *any* common purpose in the two other minds, with reference to the eccentric anomalies of the Astronomical world.'

Captain, afterwards Admiral Smyth, soon after this time came from Bedford, and took up his abode in Cheyne Walk, Chelsea. He, assisted, I have heard, by his bowl of punch, was the life of the Astronomical Club, a little meeting of chosen friends who repaired after the business of the Society to the Piazza Coffee-house. Captain Smyth was a genial companion and a quaint, pleasant writer, devoted to Astronomical science. He also gave a good deal of attention to antiquarian research, published a quarto volume on the coins and other antiquities of Hartwell House, whither he went some years after to take charge of Dr. Lee's observatory. I think that my husband's intercourse with his co-secretary, Admiral Manners, was at first chiefly official; but in after years we saw more of him, and he continued till death our cordial friend. 1831.

Mr. and Mrs. Bishop were living at South Villa, Regent's Park. Mr. Bishop was at one time President and for many years treasurer of the Royal Astronomical Society. His love of science never abated while he lived, and it led him to undertake a difficult study at an age when most men hold elementary learning out of the question. Shortly after this time he came to Mr. De Morgan for lessons in algebra, in order to read the *Mécanique Céleste*. The little observatory in the Regent's Park was rendered famous by Mr. Hind's discovery of many asteroids. George Bishop.

It was at Mr. Baily's suggestion that in the year 1827 or 1828, the state of the Nautical Almanac was made the subject of Government inquiry. This ephemeris, which was under the management of the Admiralty, had not, as to the information it afforded to navigators, kept pace with Continental works of the same character; and its defects and errors were great in comparison with theirs. The Board of Longitude had suggested improvement, but this Board was dissolved in 1827, and there seemed to be no hope that the work, upon which the navigation of the country greatly depended, should be brought to that degree of perfection which the amount of scientific knowledge in England rendered possible. A strong remonstrance from Mr. Baily drew attention to the matter, and after some discussion in various quarters, the Commissioners of the Admiralty entrusted to the Astronomical Society the task of revising and remodelling the Nautical Almanac. A committee was appointed and a Report Nautical Almanac.

1831. drawn up by Mr. Baily, who had given the subject unremitting attention. The recommendations of this Report were adopted by Government, and the Nautical Almanac, in its improved state, was the result. Lieutenant Stratford was appointed Superintendent.

1830–31. Seconds pendulum.

The pendulum experiments had been repeated by Mr. Baily in 1828, under conditions which precluded any but an almost imperceptible amount of error. Many other determinations depended on these, a most important one being the national standard of length; for, in the event of the standard yard being lost, the length of vibration of the seconds pendulum was the only source from which a new measure could be constructed. In 1832 a new scale was formed by the Astronomical Society under Mr. Baily's superintendence. This, which was rigorously tested, was compared with the imperial standard, and with another made by Bird in 1758. It was well that this work was completed, as both these scales, as well as the national standard of weight, were destroyed by fire in the Houses of Parliament in 1834.

In all these works, after 1828, Mr. De Morgan took a deep interest, but he was not an experimenter. He had a great love for scientific instruments, and in his various writings described their construction and work in such a way as to make them readily understood by any person of average intelligence. But his want of sight prevented his using them himself, and his share of the work done at this time of revival was, at least as to applied Science, that of an expounder and historian. I believe that every discovery, or determination of fact, of any importance, was made as clear to the world as the subject allowed in his articles in the 'Companion to the Almanac,' 'The Penny Cyclopædia,' and many other works.

1831–32. Useful Knowledge Society.

The institution of the London University had been an effect of that quickening of thought and action which accompanied what Mr. De Morgan called the social pot-boiling. Another result in the same direction was the formation

of the Society for the Diffusion of Useful Knowledge. It was 1831–32.
founded in 1826 by Lord Brougham, Mr. J. Hume, M.P., and others, most of whom had also taken part in the establishment of the University. The object was to spread scientific and other knowledge, by means of cheap and clearly written treatises by the best writers of the time. Partly from the character of free thought ascribed to some of its founders, partly perhaps from its designation—for there *is* much in a name, and 'Diffusion of Useful Knowledge' sounded to some undistinguishing ears like a parody of 'Promotion of Christian Knowledge'—the Society was held by some timorous lookers on to be a sort of conspiracy to subvert all law and religion; and the publication of the 'Saturday Magazine,' a markedly religious periodical, just after the appearance of the 'Penny Magazine' of the Society, showed the feeling of opposition that was in people's minds. One reason given for this rival publication was that the 'Penny Magazine,' like the other works of the Society, was too dry and scientific for general readers. As for the Magazine itself, it spread far and wide, and the 'Penny Cyclopædia,' one volume of which appeared at the end of the first year, had a great circulation, and has taken its place as a high-class book of reference. The charge of dryness is not so easy to get rid of as regards some of the tracts; but then it would not be easy to make light and popular reading of the higher branches of Mathematics, Chemistry, Hydrostatics, or the Polarisation of Light. The Society did good to its adversaries by making them give a better and sounder character to their own works of professedly religious aim. A few words from the 'Address of the Committee' in the year 1846,[1] when the Society's labours came to an end, will give an idea of the principles on

[1] This address was drawn up by Mr. De Morgan; Lord Brougham, Sir Isaac Goldsmid, and one or two others made a few slight alterations, amounting to about twenty lines, in his proof.

1831–32. which it was founded, and to which it adhered throughout.

Aims of the Society.

'At its commencement the Society determined with obvious prudence to avoid the great subjects of religion and government, on which it was impossible to touch without provoking angry discussion. At a time when the spirit which produced the effects of 1828, 1829, and 1832, was struggling with those who, not very long before, had tried to subdue it by force; when religious disqualification and political exclusion occupied the daily attention of the press, and when the friends of education were themselves divided on the best way of adjusting these and other matters of legislation, any interference with theology or politics would have endangered the existence of a union which demanded the most cordial co-operation from all who wished well to the cause. That the Society took an appearance of political colour from the fact that almost all its original supporters were of one party in politics, is true; but it is as true that if the committee had waited to commence operations until both parties had been ready to act together the work would have been yet to begin, and the good which so many of the Society's old opponents admit that it has done would have been left undone. But the committee remember with great satisfaction that this impossibility of combining different views in support of a great object extended only to politics. From the commencement the Society consisted of men of almost every religious persuasion. The harmony in which they have worked together is sufficient proof that there is nothing in difference of doctrinal creed which need prevent successful association when the object is good and the points of dispute are avoided.'

Tracts.

Mr. De Morgan, who became a member of the committee in the year 1843, was from the first a very large contributor to its publications. His work 'The Differential and Integral Calculus' formed a portion of the series of tracts. The long list of articles in the 'Penny

Cyclopædia,' amounting in all to nearly one-sixth of the whole work, were begun by him at the outset, and concluded with the last volume of the Supplement, in 1858. 1831–32. Penny Cyclopædia.

That his labours in this direction were fully appreciated is certain. He gave time, advice, and help in every way to the Society's work.[1] I find on the title-page of the Address from which the extract is made, in his own handwriting,—

'This Address was drawn up by me; even as to p. 17, I had to blow my own trumpet, because those who insisted on its being blown, and proposed to do it for me, were going to blow louder than I liked.

'A. De Morgan.

'Aug. 26, 1852.'

P. 17 contained, to the best of my recollection, his own modified version of the laudatory expressions inserted in the rough draft by the President and Vice-President, who had taken it home for inspection.

Private pupils occupied a good deal of the time which Mr. De Morgan had before spent in lecturing in University College. He was also engaged in writing for the 'Quarterly Journal of Education' of the Useful Knowledge Society, of which the first volume appeared in 1831. It was carried on for five years under the editorship of Mr. George Long, formerly Professor in University College.

The 'Companion to the Almanac' for this year con- 1831.

[1] In 1867, Mr. Coates wrote to Mr. De Morgan, in answer to his inquiry as to the place where the relics of the Society were deposited,—

'Take my word for it, that I have the liveliest recollection of the U.K.S., mingled with some pride, that for twenty years of my life I was not altogether useless to mankind. Nor have I been since, as to that matter, in spite of your innuendoes.'

'The archives, or papers of the Society, were deposited by Conolly in (I suppose the cellars of) University College; in two boxes or chests, as I have heard.

'The process was after my dynasty was closed. The common seal is in my hands, locked up in a little brass box, whereof Sir Isaac Goldsmid had one key and Lord Brougham had another. The original charter is, I suppose, in one of the two chests aforesaid.'

1831. tains an article on Life Assurance, the first of a series of twenty-five articles contributed annually by Mr. De Morgan to this work.

In this year the 'Elements of Arithmetic' was first published. The author's old pupil, Mr. Richard Hutton, says of this book:—

'Elements of Arithmetic.'

'The publication of his "Arithmetic," a book which has not unnaturally been more useful to masters than to scholars, began a new era in the history of elementary teaching in England; devoting, as all his books did, far more space and labour to the logical processes by which the various rules are demonstrated than to the more technical parts of the subject, though of these too, in their proper place, the writer was never unmindful, spending the greatest care on teaching the art of rapid and accurate computation, no less than of the true science of number. His exposition of the theory of limits, from the earliest stage at which it entered into algebraical conception, was so masterly and exhaustive, that it haunted his pupils in the logical tangle of their later lives, and helped many a man through the puzzle of Dr. Mansel's conundrum-making as to 'the Infinite,' in his 'Limits of Religious Thought.' [1]

These few lines indicate the place which this book, an early fruit of his own methods of reasoning, held in relation to the later writings, and show how, in his most elementary teaching, he laid the foundation of principles which were afterwards fully developed, and which furnished a guide to thought on subjects whose connection with them was not at first apparent.

He liked puzzles about numbers, as he liked riddles, and, when *very good*, plays upon words and puns. So all puzzles were referred to him, and gradually all attempts to do the impossible, by circle squarers and trisectors. One

[1] For a list of all Mr. De Morgan's works see Appendix. The articles on education for the U. K. Society were reprinted in a book entitled *The Schoolmaster*, edited by Charles Knight, London, 1836.

of the puzzles had a pleasant result. Mr. Charles Butler, the Roman Catholic author of 'The Revolutions of the Germanic Empire,' &c., an old friend of my father's, was not only a very learned, but a very kind and genial man. He dabbled (by his own account not very deeply) in Mathematics, and was fond of algebraical and geometrical questions. He gave me one, declaring that he had puzzled over it in vain, and never yet had found a person who could solve it. The following in Mr. De Morgan's writing will tell the rest:— 1832. Charles Butler.

'Mr. Charles Butler betted Miss Frend a coffee party that she could not find a Mathematician who could make out a certain difficulty. Miss Frend referred it to Mr. De Morgan, who solved it. This letter is for the settlement of the bet.' A wager.

'Mr. Butler presents his compliments to Mr. De Morgan. . . . He has perused with great pleasure Mr. De Morgan's solution of the question proposed to Miss Frend. It is certainly satisfactory in the highest degree. Mr. Butler's great professional employment has prevented his giving the attention he wished to the exact sciences, but he has always entertained the greatest regard for them, lamented his inability to prosecute them, and looked with a holy envy on those who have time and talents to cultivate them. The proposed coffee party has been changed into a dinner party. It is fixed for Saturday, the 18th inst., at Mr. Butler's house, 44 Great Ormond Street. Mr. Butler requests Mr. De Morgan will do him the honour to join the party.

'February 11, 1832.'

The party, a pleasant one, as the few now living who remember Mr. Butler will readily believe, dined together as appointed, and the solver of the problem was duly honoured.

Everything belonging to education commanded Mr. De Morgan's attention from the time when he began to think. Many circumstances of his own University

1832. career had shown him how much was needed in the Cambridge system to adapt the methods and processes of instruction to the wants of learners of every degree and variety of ability. His own place in the tripos of his year was an evidence of the inadequacy of the competitive system, faulty as it was then, and as it is now, to ascertain the quantity or quality of mental power. Before he left this world he saw that the method of crowding so much learning into a short time, at an age when the brain needs vital strength to bring it to maturity, was not the right way to secure future excellence. Its results, too, were beginning to be seen in nervous and other diseases; but he felt, and often said, that remonstrance as yet would be useless, and that those who saw the evil only too plainly must wait till the conviction of its reality should be forced upon all. I have not been able to get his letters

Archdeacon Thorp.

to his old tutor, Dr. Thorp, but, judging by the replies, he must have felt and expressed this belief at an early period. The answers generally announce the reception of an essay or book, or a new pupil sent by the old one to a teacher to whom he held himself indebted. In one Dr. Thorp says:—

> You will see that I have taken some pains to attend to the spirit of your wishes about your young friend. We are going upon the plan of discouraging private tuition as much as possible, for the sake both of tutors and pupils, as I hold that a lecture-room ought to supply all that is necessary; but as long as such a various crew is sent up to us as we get every year, part requiring the highest kind of scholarship and part unacquainted with the rudiments, the latter must avail themselves of some extra help to bring them up to the comprehension of such lectures as the former require.

Surely the means and appliances now at work to prepare young men for Cambridge ought to make private tuition even less wanted than it was in 1832; and if students were examined only on the real knowledge legitimately gained in the lecture-room, which, as Dr. Thorp says, ought to supply all that is necessary, what would become of cramming and coaching?[1] If any supplementary teaching is needed, it should be only to explain difficulties in the lectures, not to introduce new subjects. 1832.

At another time Dr. Thorp writes from Cambridge:—

Believe me, my dear friend, I fully sympathise in the pleasure you derive from reading classics in your own way only and for information, and would like nothing better than to keep pace with you in studying Aratus, Theon, and Euclid *in the original*; though I doubt whether young men would be made better mathematicians, or as good classics or logicians hereby, as by the study of authors more remarkable for *system*, and for the perfection of language and of art. . . . You should come down and see us oftener, to prevent your losing the knowledge of our streets. Our streets, however, are not much more changed than our *ways*. Have you seen our Trinity lecture-rooms, which we built at a cost of 4,000*l.* (besides a hole in the tutor's pocket, which a tutor in ancient times little thought of), and without asking anybody? It would make you a scholar to see the men going in crowds every morning to be taught by fourteen tutors (that is our number: I have got four, viz., myself, Martin, Law, and John Wordsworth, on my side), each of whom gives two, and some three lectures a day.

Trinity College.

I rejoice to think that we have so much in common as—1, some affection for the University; 2, something to do with preparing young men for it; and 3, some contempt for 'politics and stuff.' But, believe me, it gives me sincere pleasure to see a few friendly and familiar lines from one whom, though I have no right to claim much merit for—which was his fault, not mine—I am not a little proud to speak of as my old pupil. Ever, dear De Morgan,

'Your attached friend,

'T. THORP.'

This letter is dated 1833.

[1] During the time in which this has been written, several cases have occurred which sadly confirm my assertions. One will suffice. A young man, a very high wrangler, full of intellectual power and aspiration, was obliged to give coaching lessons to undergraduates. The exhaustion which followed his taking his degree and his subsequent hard work led him to recruit his strength with stimulants, first opium, then liquor. He drank himself to death. Had he not done so, in all probability he would have been a victim to disease in some other form, the result of exhausted vitality. This would have been less disgraceful perhaps, but equally lamentable.

1832. With his keen interest in books and their history Mr. De Morgan had a great love for collecting rare or ancient ones. His little library was begun soon after he left Cambridge, and the Theon, Aratus and Euclid, of which he so much enjoyed the reading, were among the first he bought, and are now with his other books in the library of the University of London, in Burlington House.[1] I have heard him say that he never laid out a shilling on a book which was not repaid with interest, even as a money transaction, from the use he made of the purchase. Had he been rich his collection would have been very large and valuable, but he was soon obliged to deny himself the luxury of buying, except the chance treasures which fell in his way at bookstalls. The first English book which he bought when a boy was 'The Pilgrim's Progress.' His researches in bibliography, which afterwards resulted in the 'Arithmetical Books,' the 'Budget of Paradoxes,' and many of the tracts, date from the time of his beginning to collect.

Visitors to the University Library, who take down any of these works from the shelves, will almost certainly light upon some of the numerous marginal notes and illustrations, serious or otherwise, with which their former owner embellished them. The fly-leaves and insides of the covers are decorated with pictures from periodicals, notably *Punch*, and other collectanea, always having some reference to the contents of the work, although, to those unacquainted with the peculiarities of Mr. De Morgan's mind and style, the appropriateness of some of them may not at once appear obvious.

M. Hachette.

M. Hachette, the French mathematician, whom Mr. De Morgan had visited in his stay at Paris two years

[1] This Institution must not be confused with University College, Gower Street. The University of London is for granting degrees only. It was founded in 1836. The books were bought of me, after my husband's death, and presented to the University by Lord Overstone.

before, was one of those who felt strongly on the struggle for freedom and enlightenment. The two authors had exchanged scientific brochures, and after thanking Mr. De Morgan for some sent in 1831, M. Hachette asks this question relating to the invention of the steam-engine :— 1832.

M. Hachette's queries.

Vous me ferez bien plaisir de m'éclaircir sur un fait relatif à la construction de la machine à vapeur. Je lis dans l'ouvrage de M. Partington, 'An Historical and Descriptive Account of the Steam Engine,' London, 1822, p. 7, que Sir Samuel Morland était fils du baronet de même nom, qui suivit Charles II. dans son exil; d'autres disent que ce baronet est le mécanicien qui a le premier mesuré la densité de la vapeur d'eau; cependant M. Partington est, dit-on, le bibliothécaire de l'Institution Royale; il n'a pas écrit sans preuve un fait de cette importance. Tâchez donc de savoir la vérité. Pour l'honneur de la science, et de l'humanité, je désire que le mécanicien ne soit pas celui qu'on accuse d'avoir trahi le parti constitutionnel.

I have not Mr. De Morgan's answer, but it appears not to have settled his correspondent's doubts, though they were afterwards in some measure set at rest.

Vieta's 'Algebra Nova.'

In return for the answer to his own question, M. Hachette made some inquiries for Mr. De Morgan touching a missing book, the 'Algebra Nova' of Vieta. In the letter giving what information he had gained I find the first mention of Count Guglielmo Libri, author of the 'History of Mathematics.' The last communication made by M. Hachette touching the lost work of Vieta is as follows:—

. . . J'ai reçu la lettre que vous m'avez fait l'honneur de m'écrire le 16 Juillet.

Le fait concernant le manuscrit du Harmonicon Céleste, prêté par Bouillaud au Prince Léopold de Médicis, est consigné dans les MSS. de Bouillaud, et M. Guglielmo Libri, savant géomètre, m'en a donné l'assurance. Les omissions de Montucla, ou les erreurs de Delambre, nous prouvent qu'il faut chercher dans les manuscrits, ou dans les ouvrages publiés par leur auteur, la vérité de l'histoire. . . .

Paris, Août 15, 1832.

1832. The remainder of the letter contains a comparison of dates and facts, recurring again to his object of proving that Morland the mechanician was not the Royalist. M. Hachette was probably the more satisfied in some degree by the references Mr. De Morgan gave him, as I find in the Penny Cyclopædia article, founded on those authorities, that the history of Morland's relations with the Royalists is very doubtful.

His last letter to Mr. De Morgan is dated September 30, 1832.

Mr. De Morgan had, I think, met M. Quêtelet in Paris in 1830, but M. Quêtelet had not remembered this when he wrote to Mr. De Morgan in 1833,—

Mon cher Monsieur,—Je vous remercie beaucoup pour l'obligeance que vous avez eue de m'addresser la table que vous m'avez promise, et vos ouvrages que j'ai parcourus déjà avec le plus grand plaisir. La méthode que j'ai trouvée dans vos livres élémentaires augmente encore le prix que j'attache aux suffrages honorables que vous avez bien voulu exprimer pour les miens.

Je suis très charmé que notre ami commun, M. Babbage, m'ait procuré le plaisir de faire votre connaissance : je désire beaucoup le cultiver. Je regrette de ne pouvoir aller moi-même vous exprimer mes remercîments, mais, comme je vais aujourd'hui, j'ai dû me borner à vous écrire, comptant bien sur votre indulgence.

Recevez, je vous prie, mon cher monsieur, l'expression de mes sentimens distingués.

Tout à vous,

QUÊTELET.

Amicable Assurance Office.

The place of registrar of the Amicable Assurance Office having become vacant about this time, Mr. De Morgan sent in his name as a candidate. He was of course well qualified for the situation, and it was a lucrative one, but he would not have liked the work so well as he did teaching and writing, and he had, as he afterwards told me, but one reason for wishing to succeed. Our friend Mr. Thomas Galloway, a distinguished Mathematician, and Fellow of the Astronomical Society, a man every way suited for the place, was appointed, and held it until his death in 1851. 1833.

The 'opponent' referred to in Mr. De Morgan's little sketch of Mr. Sheepshanks was Sir James South, known as the owner of the Campden Hill Observatory, and having some name as an Astronomer on account of his dexterity in using his very fine instruments. In the year 1833 a trial of a curious character, in which Sir James South was the defendant, commenced. With other scientific men Mr. De Morgan was greatly interested in this affair, and has left the following notice of it:—

'Mr. Sheepshanks's visits to Campden Hill were in discharge of his duty as scientific adviser on the side of Messrs. Troughton and Simms, who in 1833 brought an action against Sir James South to recover payment for mounting equatorially a large object-glass. While the work was going on, Sir James thought it would not do, insisted on beginning again upon a new plan, with offer of payment of money out of pocket, and, on refusal, shut Messrs. Troughton and Simms out of his observatory. The Court of course recommended arbitration; and this arbitration, which extended over 1833–1838, is the most remarkable astronomical trial which ever took place in England. The arbitrator was Mr. Maule, afterwards judge, the senior wrangler of 1810, a powerful Mathematician, and a man of uncommon sharpness of perception. The counsel for Troughton and Simms was Mr. Starkie, the senior wrangler of 1803, with Mr. Sheepshanks, who was a witness, for his scientific adviser. The counsel for Sir James South was Mr. Drinkwater Bethune, a well-known Mathematician, and a high wrangler of 1823, as sharp as Mr. Maule. Mr. Babbage was a witness and a sort of scientific adviser. The arbitrator began by insisting that Troughton and Simms should be allowed to finish the work; he also permitted certain additions to the plan proposed by Mr. Sheepshanks, on condition that they should only be paid for if they succeeded. The

1833. instrument was made serviceable in the opinion of the arbitrator, and the whole claim was awarded, additions included. Sir James did not let the instrument remain to shame the arbitrator and the opponent witnesses; he broke it up, sold the materials by auction, and placarded the walls with a bill, headed "Observatory, Kensington;" and addressed to "shycock toy makers, smoke-jack makers, mock coin makers, &c. &c.," and stating that "several hundredweight of brass, &c., being the metal of the great equatorial instrument made for the Kensington Observatory by Messrs. Troughton and Simms, were to be sold by hand on the premises; the wooden polar axis of which, by the same artists, with its botchings cobbled up by their assistants, Mr. Airy and the Rev. R. Sheepshanks, was purchased by divers vendors of old clothes, and dealers in dead cows and horses, with the exception of a fragment of mahogany specially reserved, at the request of several distinguished philosophers, on account of the great anxiety expressed by foreign astronomers to possess them, was converted into snuff-boxes as a souvenir piquant of the state of the art of Astronomical instrument making in England during the nineteenth century, will be disposed of at —— per pound."'

I do not mention these things with any wish to throw blame on one who, as after events proved, was in a state of mind which rendered eccentricity excusable. But at that time this was not known, and, as so often happens, that which would form an excuse for foolish conduct, and ought to give others the right of restraining it, was not suspected. The troubles arising from this cause among men of science, and reaching public associations, were as real as if they had been the result of wicked designs rather than of morbid impulse. The Astronomer Royal, who wished to visit Campden Hill for the inspection of Groombridge's transit circle, begged that no reference might be made during his visit to the trial then pending. Sir James insultingly accused him of having changed his opinions on that question, &c. On Mr. Airy's declining further correspondence till assured that no disrespect was meant, Sir James 'waited on a military officer of high rank in her Majesty's service,' who, however, either refused to accept the office offered to him, or quietly stopped the aggression. Sir James afterwards published the whole account as an advertisement in the *Times* of Nov. 29, 1838, in an attack upon the Admiralty. But all persons who had taken part with the prosecutors in the arbitration, or who had expressed an opinion contrary to that of Sir James, were held by him as enemies. Mr. De Morgan, as the intimate friend of Mr. Sheepshanks, and honorary secretary of the Astronomical Society, was one of these.

1834.

Sir James South.

Meeting him one day at the rooms of Lieut. Stratford, then assistant secretary, Sir James, in a loud voice, asked the latter to show him the time when and recommendation on which Mr. De Morgan had been elected a Fellow of the R.A.S., and added something about 'those gentlemen ignorami' by whom the election had been made. Mr. De Morgan took no notice of this, but afterwards addressed a temperate note to the speaker, saying that it had appeared to him that Sir James South asking in his presence for the time, &c., when he became a member of the Society was not in accordance with the sort of courtesy which parties who wish to behave distantly towards each other usually observe when they meet in private. He asked whether this was to be imputed to forgetfulness, or to a desire to convey the impression that Sir James had no wish to practise towards himself that negative courtesy with which a stranger is usually treated. He begged for an answer, that he might know how to behave towards Sir James in case they should meet again, 'since, in any case,' he says, 'I should not consider such a breach of etiquette worth any further consideration.'

Sir James South's answer is curious. It ends with— 'As to how you regulate your demeanour towards myself if we should happen to meet again, that is a point which,

1834. though it may concern yourself, is to me a matter of the completest indifference.

'After conduct such as I have recently received, and in which you have borne, I am told, no inconsiderable part, I beg to decline further correspondence with you, and must refer you for any further information you may require to my friend Captain Francis Beaufort, R.N., to whom I have confided the preservation of my character as a man, and my honour as a gentleman.

'I remain, sir, your obedient servant,

'J. SOUTH.

'Observatory, Campden Hill,
'March 15, 1834.'

Captain Beaufort being thus referred to, Mr. De Morgan wrote to that gentleman, stating the circumstances, and saying that he was by no means sure that any offence was intended.

His letter was quiet and temperate, and Captain Beaufort and Lieut. Raper, who, I have been told, were satisfied that whatever allusions to gunpowder might be made, it was certain that any waste of that article was not really contemplated by Sir James, were soon authorised by him to assure Mr. De Morgan that no offence had been meant. The affair passed off. In these days, when good sense and good feeling are generally found more effectual in keeping the peace than 'the laws of honour,' we may remember, that although duels were lamented and reprobated forty years ago, it was often more easy to fight than to avoid one. But it must not be forgotten that had the required assurance not been given by Sir J. South, it would have been almost impossible for Mr. De Morgan as Secretary of the Society to have afterwards met him.

The death of M. Hachette at Paris occurred at this time. Mr. De Morgan, who had a strong sympathy with and regard for this excellent man, had already received the news before hearing as follows from Dr. Gregory:—

You have probably heard of the death of our friend M. Hachette at Paris. It took place in January last, but I did not learn it till about a fortnight ago, in a letter from M. Quêtelet. . . . I have met with *very* few men of science whom I have so much admired and esteemed as M. Hachette. He had an ardour in the pursuit and promotion of science not to be extinguished by the shameful treatment which for years he met with; and his gentleness, kindness, single-heartedness, and generosity were particularly engaging. . . .

1834.
Letter from Olinthus Gregory.

I quite share your feelings of indignation, not only on account of the shameful treatment experienced by M. Hachette for so many years, but also on account of the chary and meagrely doled out measure of justice he has received since his death. As a man of Science he was truly estimable, and laid Science under many obligations not yet acknowledged; and as a man among French men of Science his character was altogether unique. I am glad that you so decidedly intend doing him justice.[1]

I am glad to know that you are about a work on the Differential Calculus upon the *principles to which you refer.* I have long felt that recourse to algebraical expansions in series, in establishing the principles, is exceedingly illogical, and have therefore long been perplexed to know what book to employ as a text-book. In my own class here I have principally employed Francœur in the second vol. of his *Mathématiques Pures*. . . . The anomalies which you specify are exceedingly curious, and serve still further to confirm me in my long-cherished persuasion that the fashionable process is hollow and unstable, and referable to no irrefragable principle. I wish you complete success according to your views of what the logic and metaphysics of first principles require in your important and interesting undertaking. . . . And I am,

Yours very cordially,

OLINTHUS GREGORY.

Early in the year 1836 *The Connexion of Number and Magnitude* was published. It is an attempt to explain the fifth book of Euclid. In the Preface the author says, 'The subject is one of some real difficulty, arising from the limited character of the symbols of Arithmetic considered as representatives of ratios, and the consequent introduc-

[1] In the Astronomical *Obituary Notices.*

1834. tion of incommensurable ratios, that is, of ratios which have no arithmetical representation. The whole number of students is divided into two classes: those who do not feel satisfied without rigorous definition and deduction; and those who would rather miss both than take a long road, while a shorter one can be cut at no greater expense than that of declaring that there *shall* be propositions which arithmetical demonstrations declare there *are not*. This work is intended for the former class.'

Most of his books were illustrated after his own fashion. The connexion of Number and Magnitude is shown by a gigantic father having the contents of his pocket rifled by a crowd of dwarfish children, one meaning of which I understand to be, to represent the properties of magnitude analysed by the aid of number.

The author made a great descent in his next book, as he tells a correspondent. The Useful Knowledge Society, which, notwithstanding the Rev. Dr. Folliott's low estimate of the 'learned friend' in Peacock's 'Crotchet Castle,' was a most useful instrument in raising the objects and methods of thought of both those who had, and those who had not thought before, out of a foggy region of half-knowledge into a comparatively clear and systematised state, had published 'Maps of the Stars,' for students of Astronomy, together with smaller ones for popular use, and six maps of the Earth. Mr. De Morgan wrote for the Society an explanation of all these maps.[1]

Mr. Lubbock furnished some of the materials for the 'Explanation,' &c., in the account of the selection of objects, the authorities, and the notation employed.

On the back of the title-page is written by the author, 'Ce coquin de livre a été commencé pendant l'été de 1833, et n'a été fini que dans le mois de Mai 1836.'

[1] Entitled *An Explanation of the Gnomonic Projection of the Sphere*, 'and of such points of Astronomy as are most necessary in the use of Astronomical maps; being a description of the construction and use of the smaller and larger maps of the U.K.S.; also of the six maps of the Earth.'

This book is exceedingly clear, and even popular. The treatise on and explanation of different projections of the sphere, with the reasons for rejecting Mercator's, are given in the first chapter. A great deal of bibliographical knowledge appears in the reference to early Astronomers, and not only much knowledge of Astronomy, but much of the history of it may be gained from the work. Astronomical Science in England in the sixteenth century is represented by an explanation of the constellations given by T. Hood in 1590:— 1836.

Scholar. I marvell why, seeing she (Ursa Major) hath the forme of a beare, her taile should be so long. Astronomy in the sixteenth century.

Master. I imagine that Jupiter, fearing to come too nigh unto her teeth, layde hold of her tayle, and thereby drew her up into the heaven, so that shee of her selfe being very weightye, and the distance from the earth to the heavens very great, there was great likelihood that her taile must stretch. Other reason know I none.

A passage from the book adds interest to one of the letters to Sir John Herschel:—

The figures of the constellations are of no use to the Astronomer as such; a star is sufficiently well known when its right ascension and declination are given; and if letters referring to the constellations are used, such as β in Orion, γ in Draco, &c., it is not now to direct the attention to any imaginary figure of an armed man or a dragon, but to a particular region of the heavens, which might with equal propriety have been called region A or region B. It is to the mythological antiquary that the figures are useful, as sometimes throwing light upon his pursuits. Every ancient people has written its own account of the singular fables, which are common to all mythologies, upon groups of stars in the heavens, and it might have been thought that some feeling of congruity, if taste were too much to expect, would have prevented the burlesque of mixing the utensils of modern life with the stories of the heroic age, presenting much such an appearance as the model of a locomotive steam-engine on the top of the Parthenon. But the Lacailles, the Halleys, and the Heveliuses have arranged it otherwise; the water-bearer pours a part of the stream which should wash the southern fish into a Forms of constellations.

1836. sculptor's workshop; a carpenter's rule has got between old Chiron and the altar on which he was going to sacrifice a wolf; and the lion and the hydra, whose juxtaposition has made more than one speculator imagine he has found a key to the whole allegory, are in truth two Astronomers fighting for a sextant, which Hevelius has placed at their disposal. A great deal of the southern sphere is laid out in mathematical instruments. If figures are to be drawn at all, it is, as we have said, for the historian, and not for the Astronomer; and we imagine the former will think it no loss that in our maps the heavens of Ptolemy have been restored, and in no one drawing exceeded. The names only, and boundaries of the modern constellations are given; but all the figures are those of Ptolemy, so arranged as to represent his catalogue.

The constellations are fortunately not in danger of being renamed before the origin and meaning of the old signs and symbols are well understood. Many Astronomers, however, were then watching the names of heavenly bodies newly discovered with a jealous eye, fearing more mathematical instruments or other incongruities.

Mr. Temple Chevalier wrote to my husband some time after:—'Can there not be some proper protest against the introduction of earthly names among the heavenly bodies? The heathen mythology, independently of lending itself to analogy, is exactly fitted to the purpose, by lending itself to allusions, *φώνεντα συνετοῖσι*. Such are those contained in Parthenope, Hygeia, Calliope, Irene, and others. When a planet shall be discovered at Oxford, "Isis" will be another name of the same kind. In the *Comptes Rendus* it appears that "Lutetia" was given because no one exclaimed against "Massilia." It seems high time to avoid more mud being thrown into the skies; or are we to have Lugdunum Batavorum, and other equally barbarous incursions?'

Perhaps, barbarous as it is, the plan of calling constellations and planets after continental towns and scientific instruments is less mischievous as regards antiquarian research than mixing mythological words, 'lending them-

its consisting of men of business, who could not, or would not, take any great interest, partly by the system of voting by proxy, the Council holding, as might be supposed, a great number of proxies. 1836. Letter to Sir H. Nicolas.

Shortly after the commencement of the Institution various causes of irritation arose between the Council and Professors, partly owing, in my belief, to the desire of power and influence in an individual who stood in an ill-defined position; partly to the jealousy of some members of the Council whose political bias led them to think the best way of preventing an administrative officer from going wrong was to tie him up so tight that he could neither go right nor wrong, but very much from a feeling among the Professors that their position was not safe, and in particular a suspicion, which suppose well founded, that the Council intended to divide the Professorships as soon as the income became considerable.

In the course of the years 1828 and 1829 the Professors—that is, a considerable number of them—made such representations to the Council of their unwillingness to remain in so ambiguous a position, backed with a declaration of their intention to retire, as induced that body to subject themselves to by-laws in regard to dismissal of a Professor, requiring long notice, considerable attendance, and decided majority before a Professor could be dismissed. It is to be noticed that these by-laws, though rescindible at the pleasure of the body which imposed them, were honourably adhered to in the subsequent matters, and that no technical difficulties were thrown in the way of the appeal to the Court of Proprietors.

This matter being settled for the present, though no great confidence in either body existed on the part of the other, disturbances arose in the Anatomical class, the pupils questioning the competency of their Professor. Suppose it admitted that these disturbances were excited in the first instance by insinuations of two other Professors in their lectures, and were culpably fomented by the individual already alluded to, and by certain members of the Council; suppose also that repeated investigations into the competency of the Professor in question failed in establishing anything against him, and that he was finally dismissed in consequence of the Council not being able to quell the disturbance, and of the interference of the Court of Proprietors, under the name of a Select Committee, which resolved to the effect that there could be no peace in the University while

1836. Letter to Sir H. Nicolas.

Mr. —— remained, and then denied that they had recommended his dismissal?

On this dismissal, within twenty minutes of hearing it authenticated, I retired from the University, writing the following letter to the Council.[1]

This took place in July or August 1831.

In consequence of the retirement of other Professors, and of the severe loss sustained by the classes, as I suppose, a different system of management was finally adopted. It is detailed in the printed paper enclosed, of which the parts in question are scored in black ink where they relate to the Professor and Council, and in red ink where they relate to the Professor and his pupils.[2]

My successor has, most unfortunately for the University, been lost at sea, which was communicated to me very suddenly by one of my old colleagues. My first impulse was to offer to perform the duties till Christmas, which I accordingly did, looking at the moment only to the inconvenience and probable loss which would be sustained by the institution opening without one of its most material chairs.

But on looking out into the world in this new character of a *pro tempore* substitute of my former self, I find in the first place a wish on the part of all I have spoken to (or rather, who have spoken to me) that I should return to my old post permanently, mixed, I suspect, with a strong notion that such is my desire. I am, therefore, if I do not choose finally to make any overture on this subject, or to allow any to be made, in a position to be supposed to have coquetted with this *divorcée* of mine, and unsuccessfully. This I mean to avoid by taking a very early opportunity of stating to anybody who thinks it worth while to ask the question, whether I will take it or not. I want the opinion of an unprejudiced person, who knows the world, on the following questions:—

1. Do the regulations here submitted amount to *bonâ fide* moral security that Professorships in the University of London are offices tenable during *good behaviour*, and not held *at pleasure*?

2. In addition to the practical security, supposing it to exist, do they offer that exterior show of being so held which would place the holders in that advantageous position as to respectability which a gentleman (meaning only by education and

[1] See p. 39.

[2] The document itself is much too long for insertion.

sentiments, for God knows all the rest is but leather and 1836
prunella) requires, one who believes that no independent man can hold at the pleasure of any individual or corporation, except perhaps the Crown, and then only because usage has made laws?

3. Does the regulation relating to that case provide the security which a prudent man would think requisite against the subdivision of the Professorship in the event of its becoming lucrative?

We will suppose it comparatively immaterial what shall or shall not be good behaviour, and who shall decide, presuming on the check of public opinion, which operated strongly though not effectually on a former occasion. And, on the one hand, let the affirmative of the question (1) have all the advantage of its having been found very difficult to remove a Professor, even under the old *régime*; while, on the other hand, it must have all the disadvantage of the appeal to the Court of Proprietors being utterly worthless.

Your opinion should be given on no supposition of the affirmative being desired, if possible.

Should I accept any offer (for I shall certainly not be a candidate) I should rather lose than gain for the time; and I do not consider the prospect of ultimate gain as greater than that I now have. The advantage would be the resumption of an occupation which is in itself pleasant to me, and which has some few pleasing associations. But in a thing so nearly indifferent to myself, the notion of what people in general would think would have some weight.

If your answers are such as would not please any parties concerned, I will keep this communication entirely secret, and remain,

Yours sincerely,

A. De Morgan.

5 Upper Gower Street, Oct. 10, 1836.

N.B.—The appeal to the Court of Proprietors is abolished, which must be considered as increasing the respectability of the Professorships, since, *entre nous*, a body of commercial Englishmen got together upon a point of trade (and with these gentlemen, as was sufficiently evident before, the honour and character of a Professor was avowedly, and almost *ipsissimis verbis*, made a question of trade) knows neither right from wrong, nor reason from anything else.

The answer must have been such as would please all

1836. parties, for when, immediately after, an offer of the Mathematical chair accompanied the thanks of the Council for Mr. De Morgan's considerate kindness, the offer was accepted, and the Professor once more settled in his old place. I dare not, in the face of his and my firm belief that all things are ordered for us by a wiser judgment than our own, express regret that this should have been; but the six-and-thirty years of intense labour which followed, ill paid at the time, and terminated by a disappointment which broke his heart, may well make me hesitate to record his return with satisfaction. But he loved his work, and his pupils were endeared to him by the interest they took in his teaching, and their efforts to profit by it.

SECTION IV.

Correspondence from 1831 to 1836.

To Sir J. Herschel.

5 Upper Gower Street, Oct. 16, 1832.

My dear Sir John,—I have just duly received your Catalogue, which must in course of things be the first paper ordered for press, after those already so disposed of. I shall be very much obliged to you for all you have offered on the Catalogue, the Comet, and the Herscheliana. The crumbs which fall from a rich man's table are good—astronomically, whatever they may be gastronomically. 1832.

Have you got, or do you know anything of, Bouillaud's or Bullialdi's *Astronomia Philolaica*? There is a copy in the British Museum which wants the Prolegomena, which is the very part *I* want. The matter has reference to Vieta's *Harmonicon Celeste*, which has been supposed to be lost, and which I have a faint hope might be recovered. Bouillaud is reported to say that somebody stole it from Mersenne, and certainly Vossius quotes words to that effect from the Prolegomena. But my good friend M. Hachette assures me that this is a mistake, and that Bouillaud, in his unpublished MS. at Paris, says that he himself lent Vieta's MSS. to Leopold, Duke of Tuscany. If this be true, some library at Florence may yet contain it. I am the more inclined to hope this, as Schootten, in the Preface to Vieta, gives as his reason for omitting the Harmonicon Celeste, not that no copy was to be had, but that the only one he could get appeared imperfect. Neither Montucla, Delambre, nor Kästner is to be trusted implicitly—at least with regard to Vieta. Neither of them was aware of the fact that Vieta during his life published a collection of his works, or rather I should say that the first publication of his works was in the form of a collection, and that they did not appear severally, but were

1832. afterwards severed by editors, as Ghetaldi and others. The name of this book, which was quite lost, was *Restituta Mathematica Analysis*, and it contained, among other things, the first seven books of the *Restituta Mathematica*, which all the above historians agree are lost. Perhaps this book may yet turn up somewhere. I would hope from these circumstances that people will think it worth while to look a little more into these points.

I remain, my dear Sir John,

Yours very truly,

A. De Morgan.

To Sir John Herschel.

5 Upper Gower Street, Dec. 27, 1833.

My dear Sir John,—I have a young relation going your way, and though my lucubrations generally speaking are little worth, yet as I know paper direct from England acquires a certain value by crossing the sea, I shall try to fill up this sheet with English news, or rather with what must pass for such in stagnation of better. Thank you, in the first place, for your paper on nebulæ, which I duly received, proving that you never went to look after the southern hemisphere till you had pretty well rummaged the northern. . . .

I have written a note to Mr. Baily, informing him of this opportunity, but as I have only had twelve hours' notice of it, I am not sure that you will hear from him. In any case, he is in good health, and thriving as no man better deserves to be. The same as to predicaments of the Astronomical. Your papers, namely, Catalogues and observations of Uranus, duly received, and will be read in course. I shall take care of the proofs, and Mr. Baily also. The Royal has had several meetings about their funds. It appears that they are obliged to sell out to pay arrears, and also that their estimated expenditure exceeds their income. They do not seem to know where to reduce. I do not know whether you left England before or after Captain Ross returned. He was at the Astronomical in November in high feather. To judge by his case, the northern latitudes must be good for consumptive people.

I am not aware whether you know Mr. I——, though I suppose you do. A paper by him was read at the Astronomical, containing an account of Flamsteed, &c. As Captain Ross was there, the penny-a-liners got hold of the Astronomical, and committed paragraphology. They spelt Halley, Nalley, whereupon Mr. I—— wrote to Captain Beaufort, whom he had no acquaintance with, and asked him whether the Astronomical had been attacking anybody under fictitious surnames. Captain Beaufort answered, I believe, that the paper would be published and he might see, and thereupon sent him our abstract when it appeared. Mr. I—— then said that the whole was an attack upon him, as having copied from Professor Airy in his paper on 'Physical Astronomy,' and reasoning very correctly, said it would have been much better in the Royal to have refused to print his paper in that case than to have suborned the Astronomical to attack him, &c. 1833.

From this and several other things I have heard I am very much afraid that he (I——) is decidedly wrong in his head.

Of course you have heard of your medal from the Institute. How could they be so imprudent as to risk annihilation at the hands of Captain Forman ? [1]

Health and prosperity to you and all yours. Catalogy to the nebulæ of the southern hemisphere,

I remain, dear Sir John,
Yours truly,
A. De Morgan.

To Sir J. Herschel.

My dear Sir John,—The bearer, Mr. Templeton, now going to the Cape, has offered to take this, whereupon I have advertised Mr. Baily of the same, and his letter accompanies it. We have not up to this date heard of your arrival, or even of MacClear's, which I suppose we hardly could.

I wrote you a gossiping letter by Mr. W. Bird, which I hope you got duly. Great is your loss if you missed it, for it was replete with *on dits*. I presume Mr. Baily has made you acquainted with all that has passed, which, as far as I know, amounts to very little. The anniversary of the Astronomical Society takes place on Friday (this is Wednesday), and I am just come from a preparatory meeting of Council. There is an old proverb that when the nose itches some one is talking of the wearer. I hope for your sake that the converse is not true; but a very good way to test it will be to look in your diary, if you

[1] An irrepressible paradoxer.

1834. keep one, for the state of your nasal economy on February 14th.

Your excellent friend, Captain Forman, has got a rise in the world. Sedgewick has mentioned him in the notes to a published sermon on the studies at Cambridge; not, indeed, as an individual, but as the representative of a class, 'The Formans' of the day. You know the story of Louis XIV., who noticed a merchant very much, and thereby emboldened him to ask for letters of nobility, which he got, and the King never spoke to him after, saying, 'You were the first of your class, now you are the last.' Would you rather be the first of the Formans or the last of the *savans*?

Your paper on Uranus and Co. is in course of reading at the Astronomical. The observations of Captain Foster will nearly fill Volume VII. of our Memoirs; but if any paper is added to them, it must be your Test Objects which has been read, and this. You will therefore receive them before long. I forget at this moment whether you ordered extra copies, but I have your last letter and shall look; I shall also ask Mr. Baily. I should not have troubled you with such a scrawl had I not Mr. Baily's letter to send, to which this shall be scum or dregs, according as you think it most flighty or stupid.

Yours very truly,
A. De Morgan.

5 Upper Gower Street,
Wednesday, Feby. 14, 1834.

To William Frend.

Sept. 1, 1834.

My dear Sir,—I was not surprised to find on my return to town on Friday that you had decamped, seeing that you take pleasure in the wilderness. Neither must you be astonished that I did not exceed by a single day my estimate of the time I could bear the viridity of extra-urban scenery. I suppose you will let me know how to direct to you before long. While my health is recovering from the effects of the raw atmosphere I have been breathing, I write this in preference to more serious occupation. This is no joke, I assure you; whenever I return from the country I am knocked down for some days, and could be ill with very little contrivance or external instigation, which never happens if I stay in town. And yet I have been only two days regularly in the wilds. To give you some account of my progress,

I went to stay with a clerical friend,[1] who lives six miles from any town or village, except the thing he calls his parish, and a lone house he calls his rectory. So, he having no vehicle except a four-legged apparatus called a pony, we slung my baggage across the beast, and crossed the country on foot like a gipsy migration, talking Mathematics over his head to his very great edification. Indeed he, the quadruped, looked as wise and profited as much as some of my preceding pupils have done. How people live in such lone houses I know not. Conceive me reduced to clip hedges to pass away the time till dinner, which I did with great *goût*, seeing that it is reducing trees to something like regularity, and diminishing the sum-total of foliage. From thence I went to Oxford, where I was thrown upon my resources for a whole evening. The only incident worth notice was that, having strolled out and picked up some second-hand books at a book-stall, rather Cornelius Agrippa looking sort of things, a good-looking old gentleman (a stout Church and State man, I'll swear) was so astounded that he changed his table to increase his distance, and looked at me as if he expected to see me carried away by an Avatar of the evil principle. Thence got I to Bedford, where I stayed some days with Captain Smyth, heard all the town politics, saw a jail with two men in it, father and son, charged with cutting the tails off fifteen pigs, dined with a clericus, and did various other things, not forgetting seeing a play acted by little children. Captain Smyth's observatory is the most beautiful little thing imaginable, mounting a 5-foot transit, a 3-foot circle (belonging to our Society), and an 8-foot equatorial. We had no very fine night, so that I could not know all the merits of the latter; but judging from what I saw, it must be a very capital instrument of its kind. Thence got I to Cambridge inside a coach with a lady, whose history I wormed out of her, agreeably to a talent I have for doing those things when I like, which you will admit when I tell you that in a ride of twenty-five miles I ascertained that she had married when very young an officer of 1st Light Dragoons, with him had gone to India, was stationed at Bangalore, where she travelled; how he died, she came home, and married the vicar of some place which I now forget; and, having stayed at some place, which I equally forget, was now moving, with furniture following in a waggon, and husband deposited outside the coach, to take possession of his living, first stopping to dine with a friend, whose

1834.

[1] Rev. Arthur Neate.

1834. name I forget. I also ascertained which of all my cousins in India she had danced with in her day, which was instructive to know. These, and a great many other things, did I ascertain; so you may see that if I am not communicative myself, I know how to make other people so when they do not know what I am at. At Cambridge I found Sheepshanks, Peacock, and several new buildings, the former of whom drove me to Madingley, a place I never saw in my life before, so you may judge how far my walks extended as an undergraduate. Came home as tired as a city mouse of hedgerows and cottages, and was more nearly in a good humour with the fiddler who stands opposite my window on a Saturday night than I could previously have thought possible. If the locomotives ever come to go so quick that one tree shall not be distinguishable from another, then, and not before, do I become a traveller. Mr. Stephenson (engineer) says he shall never be satisfied till two hours take us from London to Liverpool. Blessings on his heart, but either he or, better still, one of the minority who can be spared, shall *try it first*. I have not got a large organ of caution for nothing. I am delighted with the House of Lords for throwing out the hard bargain of 80 per cent. of Irish tithes to be secured upon the land. The I. P.[1] will never get so good a composition again. I perfectly agree *now* with Lord —— that the Commons would sometimes blunder if it were not for the Lords. Can you imagine Lord ——, the quondam Liberal, instigating the House of Lords to put it in the power of Commissioners to hinder any pauper's religious instructors from having access to him unless he were of Parliament principles? No letters from you, from which I conclude that your thoughts are of trees, only interrupted by the slopping of the waves, which are always fiddling at the sand till I long to give them a thump, and tell them to be easy. The prettiest thing about the sea is the straight horizon and the isochronism of the waves in deep water, but near the shore they do not keep time like my pendulum. . . . We have got our rooms (in part) given up to me, and about the end of September shall begin to stir in getting them ready. All the people are out of town except myself, and they might as well make me Secretary of State as set me painting, plastering, and whitewashing. Stratford is gone to Ramsgate with Mrs. S (as he calls her—I abominate initial letters) tratford, Baily to Edinburgh, Henderson, &c., ditto. There is not a soul left that I know of, which is

[1] *Quære* Irish People? I cannot interpret these political allusions.

a great advantage of being in town in the summer; for, saving your absence, it is a good thing to be thrown upon oneself for a month or two, to say nothing of the quantity of work one does. I was very sorry to find when I came home that Mr. Woolgar had been very uncourteously received by B—— with my note. That unfortunate man will never rest until he succeeds in getting nobody's good word. He calculates very wrong (for a calculating machine maker) if he thinks such a thrower of stones as himself can stand alone in the world. It takes all his analysis and his machine to boot to induce me to say I will ever have any communication with him again, which nothing should induce me to do except the consideration that men of real knowledge should have more allowance made for them than some charlatans I know. I make no doubt Mr. Woolgar will detail to you the reception he received. . . . 1834.

Apropos of logarithms, give my kind regards to all your circle, and believe me, dear sir,

Yours sincerely,
A. De Morgan.

5 Upper Gower Street.

To William Frend, Hastings.

Dear Sir,—I have nothing whatever in the nature of news to tell you, except that the Astronomical Society has obtained possession of its rooms[1] and moved into them, with nothing remarkable except that one of the secretaries, whom you know, had an opportunity of confirming an observation he has often made, that upholsterers, carpenters, and all concerned in furniture, are *laudatores temporis acti*, whatever may be their age. The bookcases of the Society are 'such as are not made now;' and even the old orange chests, bought for a trifle to put books in, are 'such as they doesn't make nowadays,' according to one of the workmen. However, the race of men are not degenerated, for *four* of them took up in their arms our large iron safety affair, and carried it slick right away into a van, whereas *five* men took two hours with iron rollers, &c., to get it into the chambers a year ago.

I have got the care of all the churches upon me now; that is, builders' estimates, &c., with a hitch as to prices. Mr. Baily is out of town, and workmen must be in the premises on Wednesday at latest. We are in this condition. The Royal Society,

[1] In Somerset House.

1836. which owns the upper storey, has cut a floor through what was a staircase, so that our rooms in part present a section like the following.[1] . . . . However, I suppose it will come right somehow or other. Our meeting-room will hold from ninety to a hundred comfortably. Our largest meeting hitherto has been eighty.

Give my kind regards all round, and believe me,

Yours sincerely,

A. De Morgan.

5 Upper Gower Street.

To Dean Peacock.

Dear Sir,—I send you herewith my series, corrected and revised in the newest manner. The result is so much generalised from that in the *Calculus of Functions*, that I think it may be considered as new matter.

I send also a small work with a new kind of title, being an endeavour to make the fifth book of Euclid somewhat readable. It is meant to be the first part of a book on Trigonometry. The astronomical world here has been enlightened by a starlight Knight,[2] at the Royal Institute. What Young and Faraday have there said of physics has been completely outdone. I did not hear the lectures, but am told that if I had I should have known how George III., surrounded by his Astronomers, went to Kew to see an occultation, foregoing the stag-hunt which was going on; how a cloud hid the moon, and how the pious King, without a single murmur against Providence (a point dwelt upon as remarkable), turned the telescope at the hunters, and saw the stag killed, between the two horizontal wires. The second lecture was closed by a description of the unfitness of Mathematicians to be practical Astronomers, with the exception of Bessel. Now Sir James would have lectured at Cambridge with half the pains which were taken to get Airy.

I remain, dear Sir,

Yours very sincerely,

A. De Morgan.

Upper Gower Street, April 25, 1836.

To Sir J. Herschel.

My dear Sir John,—Some months ago, when the *Calculus of Functions* which I now send you was published, I marked

[1] The reader's imagination will easily supply the omitted sketch from the context.

[2] Sir James South.

one for you, which has been lying waiting opportunities. I am glad now of the delay, as I am enabled to send with it some maps of the stars which I have been charged to present to you in the name of the Committee of the Society for the Diffusion of Useful Knowledge. They are the first maps, I believe, in which Sir William's nebulæ and your own are laid down from the Catalogue. 1836.

You will find a great deal of new speculation about an old subject of yours in the *Calculus of Functions*, and in particular a discussion which concerns you in § 252, &c.

The two books on *Algebra* and *Number and Magnitude* contain various metaphysical points, which I heartily wish were more attended to. I have not written your name in them because *non constat* that it is reasonable to expect you to bring elementary books home again to England, where other copies will be much at your service. Present them, therefore, to any person or institution whom they will be of use to.

We are getting up a picture of Mr. Baily by subscription, and the same is limited to a guinea. It has struck me that I do not remember that anybody has put your name down, and that you would not be pleased to be left out in any association which is to do honour to such a man. I shall therefore take care that the omission is remedied. The picture is to be presented to the Astronomical Society.

Your sixth Catalogue has been printed, or will be struck off shortly. The extra copies shall be forwarded to Mr. Stewart. There is very little stirring in our world. You will have heard that Captain Smyth had already asserted that the two stars γ *Virginis* were in peri-one-another, and was laughed at by somebody for his assertion, which laughter your letter has turned on his side.

I should suppose you now can almost fix the time of your return. I take it for granted you have learned the extraordinary discoveries you have made in the moon. It was a dull joke to republish the book in England, and I suspect in America it was done to raise the wind. I flatter myself I did just as clever a thing, which, however, has failed through Mr. Warren's want of understanding; at least, I have not seen it in print. I sent him anonymously the following:—

'*Sir John Herschel.*'—This distinguished Astronomer writes this to a friend from the Cape of Good Hope:—'The climate here is so bad that my mirrors tarnish immediately. I do not know

1836. what I should have done if I had not taken the precaution to bring out with me a large supply of Warren's jet blacking, prepared at his manufactory, 30 Strand, the polish of which is so exquisite that I can see the faintest stars in the haziest evenings.' With best wishes, I remain,

Yours very truly,
A. De Morgan.

5 Upper Gower Street, July 10, 1836.

N.B.—The Southern Hemisphere on the maps looks tolerably empty of nebulæ. There is ample space and room enough the characters of He——l to trace, whence it follows that you may mark the year and mark the night; that is, dot down nebulæ through all the twenty-four hours of R.A., and the year and night will be near enough as to time. Do you not find it an awfully unromantic change to get out of the land of Hercules, Draco, Cepheus, &c., into that of Pyxis Nautica, Cælum Sculptoris, &c.? If you have to make any new constellations, remember that the president and other officers of the Astronomical Society have an official claim. Præses Societatis Astronomicæ would look pretty; and α Præs., or β Secret., would not be amiss in a list of moon culminating stars.

Please to give my kind regards to Maclear and young Smyth when you see them. Maclear's paper on the opposition of Mars has reached the Astronomical Society duly, but we wait about printing it till we hear of his observations. Of course you know Bachelier, the mathematical bookseller at Paris. All his stock has been burnt, and that of others at the same time. I wanted Libri's 'History of Mathematics' in July, and find it is all gone in the flames.

SECTION V.

1836 to 1846.

Soon after Mr. De Morgan's return to the college a great affliction befell the family in the sudden death of his sister Mrs. Hensley in her confinement. Her brother had left his home in Gower Street, satisfied that she was doing well, and on his return in the afternoon inquired as he entered the house how she was going on. The servant replied that Mrs. Hensley was dead. It had been quite unexpected, and was a terrible blow to her mother, her husband, and brothers. Mrs. Hensley left three daughters and the infant son whose birth immediately preceded her own death. It was many months before her brother Augustus recovered from the shock he received in hearing so suddenly of the event. In writing to my mother of the affliction of his own, he added, 'As for me, I am stunned, and hardly know what I write.' And it was far longer before the grief caused by this, his first experience of the death of one whom he loved most affectionately, abated. 1836.

The religious doubts and difficulties created in his mind by the doctrinal teaching of his early years were not the only troubles arising from the same cause. It was natural that a mother, so anxious and true-hearted as his, should not see without pain anything like what she thought carelessness in religious matters, and that her anxiety to produce a belief like her own should be intensified by her recent sorrow. His sister had shared her anxiety. They looked upon him, of whose intellectual powers they were proud, and who had been enabled to give such loving

1836. and dutiful help to his family, as one perhaps doomed to endless torment, because, using the power of head and integrity of heart which made him so dear to them, he rejected some orthodox creed, and the belief that his Father in heaven was more cruel and unjust than any earthly father.

After his sister's death, his mother wrote to him with painful earnestness on this subject, begging him to read some writings left by his sister. His reply shows both what he thought of the question and what he felt on the sorrowful occasion which called it forth.[1]

From this letter it may be gathered that his opinions at this time were Unitarian, according to the meaning originally given by that word, and to the belief held by those who first bore the name.[2] There is now so much confusion of ideas as to this, that it is necessary for me to say what I mean in calling my husband a Unitarian. He believed that Jesus Christ, the Son of God by the gift of the Holy Spirit without measure, was, as to his nature, a man like ourselves, except in His power of receiving the Spirit of God. That His divinity was not, like that of the Father, the Source of all things, underived and self-existent. That the Father spoke through Him by the same Spirit, sending the message and the means of redemption or bringing back erring man to God. That the mission was attested by His words and miraculous works, and that He rose from the dead, and was seen to rise to Heaven, from whence He sends the Spirit to those who are able to receive it.

Mr. De Morgan never joined any religious sect, but I think he had most respect for the Unitarians, as being most honest in their expression of opinion, and having most critical learning. The writer's belief in the supremacy of reason to sift and interpret revelation, and his implicit

[1] See correspondence following this section.

[2] When a proposal was made to require the insertion in the census return of the various religious denominations, he declared that he should describe us as 'Christians unattached.'

faith, sufficient evidence having been obtained, are shown in the letter. The two are results of the same mental power, for reason enables us as well to interpret testimony as to judge of its value. This only applies to that religion which is of the head. My husband had a deep instinctive spring of faith in his own soul, but with this, the bond of union between his heavenly Father and himself, the world had nothing to do. Years after this letter was written, he was supposed to have accepted, on slight and insufficient grounds, facts pronounced unworthy of examination by less profound thinkers. It may be that the time will come when his guarded judgment of these phenomena will be in turn condemned as too cautiously expressed.[1] 1836.

1837. Church rates.

Early in 1837, a measure for the abolition of Church rates, and the application of Church property to meet the expenses for which they were levied, was proposed by the Government. Large calculations were, of course, necessary to show in what way the property could be so managed as to meet the necessities of the Church, without injustice to those dignitaries who were its present holders, and actuaries were engaged to make these calculations, both on the part of the Ministers and on that of the Opposition. Lord Ellenborough, then in office, applied to Mr. De Morgan as follows:—

Mr. Finlaison, not being authorised to communicate with Lord Ellenborough with respect to the details of the new plan for the management of Church property, has had the goodness to recommend to Lord Ellenborough that he should request the assistance of Mr. De Morgan as the ablest of actuaries in the elucidation of the subject, &c.

A very intricate calculation was gone through involving the values of leases for various terms of years, of the fines levied on change of holder, and of every part of the complicated question. Lord Ellenborough, between

[1] See Preface to *From Matter to Spirit.*

1837. whom and Mr. De Morgan several letters had passed, wrote (March 14) :--

> When the bill is printed and further information is obtained, I may have to trouble you again with further questions; but in the mean time I cannot delay offering my acknowledgment to you for the most valuable assistance you have already afforded.

This calculation occupied as many hours of several days as could be spared from his two lectures and his other work, and he had to sit up more than one night to complete it. The Bill was lost, and with it Mr. De Morgan's time and trouble.

Marriage. In the vacation of this year we were married. Mr. De Morgan's religious views are by this time well known to the reader. I had been brought up in my father's belief, but had not adhered to it without much modification. My husband's objection to the marriage ceremony was much stronger than my own, but my respect for his scruples made me willing to comply with his wish that we should not be married by the form prescribed by the Church of England. We were married at the registrar's office by the Rev. Thomas Madge, and by a form of words differing from that in the prayer book only by the omission of the very small part to which we could not assent with our whole hearts, and of the long exordium of St. Paul on the duties of husbands and wives.

Settled in Gower Street. After a short tour in Normandy we settled at our first home, 69 Gower Street. The books, which were then tolerably numerous, had been taken from 5 Upper Gower Street, a few weeks before, when his mother went to a larger house in Manchester Street, Manchester Square. Our house was so near the college that my husband could come home in the intervals between his morning and afternoon lectures, instead of remaining away from 8 A.M. till 5 P.M., as he was obliged to do afterwards when we lived at a greater distance from Gower Street.

My father was living in Tavistock Square at the time

of our marriage. My husband had long known almost all my father's circle of acquaintance. One exception—a dear and early friend of mine whom he did not know personally till shortly before our marriage—was Lady Noel Byron, whose health kept her much at home, and whom he accompanied me to see at her house near Acton. She soon became as truly his friend as she had been mine. Lady Byron was always shy with strangers, especially with those who excited her veneration. This shyness gave her an appearance of coldness, but she and my husband soon knew each other's worth, and she never lost an opportunity of showing her regard for him and trust in his judgment. He was rather surprised to find in one commonly reputed to be hard and austere, qualities of quite an opposite nature. She was impulsive and affectionate almost to a fault, but the expression of her feelings was often checked by the habitual state of repression in which the circumstances of her life had placed her. I had known her from my childhood. My father, whom she always held in the highest esteem, had taught her Mathematics, as a friend, before her marriage. My husband afterwards gave her daughter, Lady Lovelace, then Lady King, much help in her mathematical studies, which were carried farther than her mother's had been. I well remember accompanying her to see Mr. Babbage's wonderful analytical engine. While other visitors gazed at the working of this beautiful instrument with the sort of expression, and I dare say the sort of feeling, that some savages are said to have shown on first seeing a looking-glass or hearing a gun—if, indeed, they had as strong an idea of its marvellousness—Miss Byron, young as she was, understood its working, and saw the great beauty of the invention. She had read the Differential Calculus to some extent, and after her marriage she pursued the study and translated a small work of the Italian Mathematician Menabrea, in which the mathematical principles of its construction are explained.

1837.

1837. Mr. Babbage's 'Ninth Bridgwater Treatise,' at this time going through the press, contained the development of an idea suggested by the working of the engine. In the series of numbers presented by the rotation of the cylinders, a regular order, which has continued for a long time, is suddenly interrupted by the appearance of a new number. The old series is again resumed, and at another interval a number bearing relation to the first interruption makes its appearance. This process suggested to Mr. Babbage a reply to Hume's argument against miracles, founded on the experience of the world in the sequence of events. The idea of the intervention of a higher law in the processes of nature is now more familiar to the world than it was when Mr. Babbage gave his beautiful illustration. By theologians the book was condemned as heretical, as doing away with what was held to be the nature of miracle—an arbitrary suspension of the laws of nature. By some thoughtful men, who did not consider science and revelation incompatible, the suggestion was held valuable. My husband took a lively interest in the work, and the author, who was then on friendly terms with him, was a visitor at our house.

Elizabeth Fry. In this year we made the acquaintance and gained the friendship of Elizabeth Fry; of whom my husband speaks in the Budget of Paradoxes as 'one of the noblest of human beings.' Lady Noel Byron, who had heard of a scheme for a female benefit society and home, which seemed to promise great usefulness, and in which Mrs. Fry took an interest, introduced her to my husband for the sake of his advice on the calculations, and to us both, as likely to enter warmly into the design. He found the calculations utterly worthless, and loss or even ruin was prevented by the reference to him, for the projector had obtained promises of money for shares from persons who could ill afford to lose it; her vexation on the overthrow of her scheme was very plainly shown. Mrs. Fry allowed me to accompany her in a visit to this person. It was like witnessing an interview between an angel and the opposite character, and I could only compare the steady gentleness with which Mrs. Fry replied to the sharp, shrill arguments of Miss —— to sunshine clearing away a black frost. My husband, who was very sensitive on such points, was charmed with Mrs. Fry's voice and manner as much as by the simple self-forgetfulness with which she entered into this business; her own very uncomfortable share of it not being felt as an element in the question, as long as she could be useful in promoting good or preventing mischief. I can see her now as she came into our room, took off her little round Quaker cap, and laying it down, went at once into the matter. 'I have followed thy advice, and I think nothing further can be done in this case; but all harm is prevented.' In the following year I had an opportunity of seeing the effect of her most musical tones. I visited her at Stratford, taking my little baby and nurse with me, to consult her on some articles on prison discipline, which I had written for a periodical. The baby—three months old—was restless, and the nurse could not quiet her, neither could I entirely, until Mrs. Fry began to read something connected with the subject of my visit, when the infant, fixing her large eyes on the reader, lay listening till she fell asleep. 1837.

Introductory lecture.

On the occasion of the opening of the Faculty of Arts my husband was appointed to deliver the introductory lecture.

The establishment of the University of London had altered the relations of University College with the public and with education generally, and, as Mr. De Morgan said, 'the circumstances under which this College (University College) reopens its courses of instruction are more remarkable than any in which it has been placed since the commencement of its career.'

The University of London had been founded in the year 1836 by Government; and to prevent the confusion consequent on similarity of name, the institution which

1838. had been called the *London University* took the name of *University College*. There were then two Colleges in London affiliated to the newly established university, which bore that name with more propriety, its object being not to give education, but to examine and confer degrees upon the pupils of Colleges.

This lecture, given nine years after his first, shows the working of the same thought developed and extended over a wider field. He disclaimed being in any way the organ or representative of the College. The ideas were his own; and the principles he laid down upon public education might be consulted now with advantage in this present stage of opinion on academical training.

Our eldest child was born the year after our marriage. In the autumn of that year Lady Byron lent us her house, Fordhook, near Acton; and for the ten weeks of our stay my husband was able to go on with his writing more easily than he could have done at a greater distance from London. During the years 1836 and 1837 he had been engaged in writing his *Theory of Probabilities*. This is the description taken from the agreement made with the publishers of the 'Encyclopædia Metropolitana,' published in January 1838. 'A Mathematical Treatise on the Theory of Probabilities; containing such development of the application of Mathematics to the said Theory as shall to him (the Author) seem fit, and in particular such a view of the higher parts of the subject as laid down by Laplace in his work entitled *Théorie des Probabilités*, as can be contained in a reasonable compass, regard being had to the extent and character of the Mathematical portions of the said work.'[1]

Theory of Probabilities.

During the time which we spent at Fordhook, he completed the small volume entitled 'Essay on Probabi-

[1] From a pamphlet published in 1838, hereafter mentioned. The extract is given here for the same purpose for which the pamphlet was written, to show the difference between the scientific treatise and the popular *Essay on Probabilities*, in the Cabinet Cyclopædia.

1838. Essay on Probabilities.

lities, and on their application to Life Contingencies and Insurance Offices,' which appeared in Lardner's 'Cabinet Cyclopædia' in September 1838. The advertisement of the 'Essay' alarmed the editor of the 'Encyclopædia Metropolitana,' who, being unable to understand that a profound Mathematical work full of definite integration was altogether a different thing from a popular essay requiring only decimal fractions, and mainly devoted to life contingencies, accused the writer of having infringed the rights of the proprietors of the Encyclopædia, by publishing what he said 'might be deemed a second edition of the treatise,' and threatened, or implied a threat of prosecution. The author, who was more amused than annoyed by this want of perception in the publisher, explained to him very clearly the respective characters of the works, but failed to make him understand how widely they differed. He then proposed arbitration, he being willing to pay whatever damages should be judged proportionate to their loss to the supposed injured parties; or, in the event of the decision being in his favour, that a sum of money should be given by them to some charity, as amends for the trouble given and the false aspersions made. This last proposal being rejected, the author of the Treatise and Essay published a little pamphlet in explanation, which showed to all who cared to understand the question that the publisher's ignorance of its nature had led him into what my husband called 'wasting a good deal of good grumbling,' but which was in truth an unjust imputation on himself.

Dickens's serials.

The great amount of work which he did at this time, as at all times while his strength lasted, filled the day, so that I had but little of his society. We both naturally regretted this, but it could not be helped. He liked reading to me when he could get anything likely to please us both, so I heard several of Dickens's novels from beginning to end. They came out in monthly parts, and he would say, 'We shall have a Pickwick (or whatever it might

1838. be) to-morrow;' and on the first day of the publication we had read and commented on it. 'Punch' about that time was in the meridian, and Mrs. Caudle's Curtain Lectures threw terrible weapons into the hands of husbands. Accordingly, these last were read to me with a view to my improvement, the reader dwelling with special emphasis on any of Mrs. Caudle's most outrageous sayings which were supposed to be particularly suitable to the case. He said, 'Every man's mother is Mrs. Nickleby, and every man's wife is Mrs. Caudle.' We had more time in the vacations, but I was afterwards always parted from him for a few weeks in the autumn, as it was necessary to take the children out of town for health.

He seldom entered into any serious discussion, but liked to tell of any interesting fact which he had come across in his investigations either in reading or thinking; and many valuable bits of knowledge, which were afterwards expanded and published, were talked of first in this way. Matters of less importance, obscure derivations of words, and unsuspected translations, the origin of old customs, versions of nursery rhymes, and, above all, riddles, good *and bad*, were generally welcome.

Ideas about women.

I must not conceal the fact that in the earlier part of his life he held man-like and masterful views of women's powers and privileges. Women, he thought, ought to have everything provided for them, and every trouble taken off their hands; so the less they meddled with business in any form the better. But these very young notions gave way, as he saw more of life, to wiser and more practical ones. He found that women were not utterly helpless, and his love of justice, combined with his better opinion of their powers, made him quite willing to concede to them as much as he would have desired for himself, namely, full scope and opportunity for the exercise of all their faculties. This was shown by his giving lectures gratuitously in the Ladies' College for the first year after its foundation, and by the interest he felt in the success of those brave women who first attempted the study of medicine. 1838.

In society he seldom entered into discussion on abstract questions, except with those of whose comprehension he felt sure, but he would sometimes listen to the debates of others. I once saw him stand by, with a half-amused, half-interested look, while a discussion was going on between two learned professors on matter and spirit, the future life, and a Creator, in which the two last were on the losing side, without uttering a word. When I asked him what he thought of the arguments, he said, 'I don't understand them, but then I'm not a Philosopher.'[1]

[1] He has left some definition of Philosophers : 'The word "Philosopher" is one which has had meanings so different from each other, and has been in such demand for all manner of uses, that a person who should read the writings of one period with a notion of this word derived from the writings of another period would be in actual confusion about matters of fact and opinion both. Some movable words are understood as such : a *good man* sometimes means a just man, sometimes a benevolent man, sometimes a religious man, a rich man, or, as at Cambridge, an (undergraduate) man who is well up in his subject (of examination). This is pretty well understood, but nine-tenths of the educated think that the Philosopher is one kind of person, throughout all ages and countries.

'A Philosopher, in Greek, was originally a person who desired and sought after wisdom, especially the knowledge of man in the widest sense ; of his constitution, his capabilities, and his duties. But in history may be found this variety of meanings : 1. The original sense just described. 2. The votary of a school of opinions on man, or on nature, or on morals. 3. An ascetic, who denies himself the good things of the world. 4. A person whose temper is not easily put out. 5. A person who despises his fellow-creatures. 6. One who cares not what is said about him. 7. An academically educated man. 8. An atheist. 9. An infidel as to revelation. 10. An inquirer into the material phenomena of the universe. I need not say that this list does not include the *true* Philosopher, a genus of species innumerable, nor the technically adjectived Philosopher, as the moral Philosopher, the chemical Philosopher, &c., meaning a person who looks into morals, chemistry, &c., in a thoughtful and speculative way. These would be more rightly called Philosophic moralists, Philosophic chemists, &c. The dreadful bore who did the moral business in

1838. Then he repeated laughingly to himself a few words uttered by one of the speakers, and said, 'Poor ——, he does not see that if what he said were true, he would not be here to say it.' As he wrote, he 'had no objection to Metaphysics, far from it, but if a man takes a candle to look down his own throat, he must take care not to set his head on fire.' And this sense of the danger of fire, coupled with the fact that his own thoughts ran in new channels, made him unwilling to speak on Metaphysical questions except to the few who were already familiar with his ideas. Logic and Mathematics were different, being in some degree out of the reach of 'fire.' But his beliefs in mental and physical science were founded on that of a constantly producing and constantly sustaining Creator; and he was never found assenting to systems based *only* on observation of material nature. It might be that observation had not gone far enough when it resulted in such expressions as 'forces inherent in matter,' but he said such expressions were only a step in the road to atheism. The present state of scientific belief in some measure justifies this.

Conversation.

Lectures.

The morning lecture and explanation lasted from 9 A.M. till 10.30, when he came home, but only to attend to work of different kinds, or to a private pupil, of whom two or three came to him while we lived in Gower Street, and afterwards in Camden Street. The afternoon lecture was from 3 to 4.30, when he returned to dinner, and for the little rest he allowed himself before a long evening of writing, only interrupted by an hour's talk with me, or occasionally with some friend who might visit us.

Mr. Richard Hutton's sketch.

As I cannot describe my husband in his character of Professor, I thankfully give two little sketches of his mode

children's books of forty years ago was a Philosopher; he was sententious; he said, "From this we may learn, and let us all take warning," and he had a "small but well-selected library (may it perish with him), containing no poets except Young and Akenside."' . . . (*Unfinished.*)

and system of teaching, taken by pupils for whom he had a sincere regard, and who both loved and venerated their old master,—Mr. Richard Hutton and Mr. Sedley Taylor. Mr. Hutton says,— 1838. Mr. Hutton's recollections.

'Few men have had more eminent pupils than your husband, and few have done more to cultivate the intellects of those whom they taught. As you know, in Mr. De Morgan's time, the Mathematical classes of University College were quite as much classes in Logic, at least in the Logic of number and magnitude, as in Mathematics; but of my own fellow-pupils very few have, I think, since become eminent in the world. The present Master of the Rolls (Sir George Jessel) was, I believe, your husband's pupil a year or two before my time, as was the late Mr. Jacob Waley, who, after being his pupil, became his colleague at University College. Mr. Walter Bagehot, whose books on the working of our political constitution and on the early forms of national government have attracted the attention of most thoughtful men, was a fellow-student with me, and one of the chief subjects of discussion between us used to be the logical questions raised in the Mathematical classes, especially in your husband's lectures on the theory of limits, the theory of probabilities, the calculus of operations, and the interpretation of symbols applied, with a new and extended meaning, to cases which were not within the scope of their original definition. Professor Stanley Jevons, of Owen's College, Manchester, who has always prized very highly your husband's teaching, was his pupil many years after I had left the College, and no one has made better use of the time passed in those delightful classes; and every book he publishes bears witness to the help he has derived from your husband's teaching.

'One thing which made his classes lively to men who were up to his mark, was the humorous horror he used to express at our blunders, especially when we took the conventional or book view instead of the logical view.

1838. The bland "hush!" with which he would suppress a suggestion which was simply stupid, and the almost grotesque surprise he would feign when a man betrayed that, instead of the classification by logical principles, he was thinking of the old unmeaning classification by rule in the common school-books, were exceedingly humorous, and gave a life to the classes beyond the mere scope of their intellectual interests. I think all my fellow-pupils would agree that never was there a more curious mixture of interests than the prepared discussions of principle in his lectures, and the Johnsonian force and sometimes fun of *his* part in the short dialogues with his pupils which occurred from time to time.'

A pupil who came rather later than those mentioned, and in whose success his teacher greatly rejoiced, was Mr. Robert Bellamy Clifton, now Professor of Physical Science in Oxford. My husband early perceived talents in Mr. Clifton which had been ignored by former teachers, and the result justified his advice and predictions. Professor Clifton continued a valued friend through Mr. De Morgan's life, and gave me much kind assistance with the library, &c., after his death.

Mr. Sedley Taylor's recollections.

The work in the Mathematical lecture room, and the Professor's manner of doing it, are also well described by his pupil and friend, Mr. Sedley Taylor of Trinity College, Cambridge—one who, like himself, held conscience to be above all things, and gave up his position as a clergyman of the Church of England because he could not assent with his whole heart to her doctrine. The following is extracted from Mr. Taylor's notice of his old teacher in the *Cambridge University Reporter*:—

'As Professor of Pure Mathematics at University College, London, De Morgan regularly delivered four courses of lectures, each of three hours a week, and lasting throughout the academical year. He thus lectured two hours every day to his College classes, besides giving a course addressed to schoolmasters in the evening during a portion of the year. His courses embraced a systematic view of the whole field of Pure Mathematics, from the first book of Euclid and Elementary Arithmetic up to the Calculus of Variations. From two to three years were ordinarily spent by Mathematical students in attendance on his lectures. De Morgan was far from thinking the duties of his chair adequately performed by lecturing only. At the close of every lecture in each course he gave out a number of problems and examples illustrative of the subject which was then engaging the attention of the class. His students were expected to bring these to him worked out. He then looked them over, and returned them revised before the next lecture. Each example, if rightly done, was carefully marked with a tick, or if a mere inaccuracy occurred in the working it was crossed out, and the proper correction inserted. If, however, a mistake of *principle* was committed, the words 'show me' appeared on the exercise. The student so summoned was expected to present himself on the platform at the close of the lecture, when De Morgan would carefully go over the point with him privately, and endeavour to clear up whatever difficulty he experienced. The amount of labour thus involved was very considerable, as the number of students in attendance frequently exceeded one hundred.' 1838.

'De Morgan's exposition combined excellences of the most varied kinds. It was clear, vivid, and succinct—rich too with abundance of illustration always at the command of enormously wide reading and an astonishingly retentive memory. A voice of sonorous sweetness, a grand forehead, and a profile of classic beauty, intensified the impression of commanding power which an almost equally complete mastery over Mathematical truth, and over the forms of language in which he so attractively arrayed it, could not fail to make upon his auditors. Greater, however, than even these eminent qualities were the love of scientific truth for its own sake, and the utter contempt for all

1838. Mr. Sedley Taylor's recollections.

counterfeit knowledge, with which he was visibly possessed, and which he had an extraordinary power of arousing and sustaining in his pupils. The fundamental conceptions of each main department of Mathematics were dwelt upon and illustrated in such detail as to show that, in the judgment of the lecturer, a thorough comprehension and mental assimilation of great principles far outweighed in importance any mere analytical dexterity in the application of half-understood principles to particular cases. Thus, for instance, in Trigonometry, the wide generality of that subject, as the science of undulating or periodic magnitude, was brought out and insisted on from the very first. In like manner the Differential Calculus was approached through a rich conglomerate of elementary illustration, by which the notion of a differential coefficient was made thoroughly intelligible before any formal definition of its meaning had been given. The amount of time spent on any one subject was regulated exclusively by the importance which De Morgan held it to possess in a systematic view of Mathematical science. The claims which University or College examinations might be supposed to have on the studies of his pupils were never allowed to influence his programme in the slightest degree. He laboured to form sound scientific Mathematicians, and, if he succeeded in this, cared little whether his pupils could reproduce more or less of their knowledge on paper in a given time. On one occasion, when I had expressed regret that a most distinguished student of his had been beaten, in the Cambridge Mathematical Tripos, by several men believed to be his inferiors, De Morgan quietly remarked that he "never thought —— likely to do himself justice in THE GREAT WRITING RACE." All *cram* he held in the most sovereign contempt. I remember, during the last week of his course which preceded an annual College examination, his abruptly addressing his class as follows: "I notice that many of you have left off working my examples this week. I know perfectly well what you are doing;

YOU ARE CRAMMING FOR THE EXAMINATION. But I will set you such a paper as shall make ALL YOUR CRAM of no use." ' 1839.

Enforcement of punctuality.

His pupils' affection was not gained by any laxity of discipline, for he was strict; especially as to quietness and punctuality. His own morning lecture began at nine. That on Natural Philosophy followed it immediately, and punctuality in the first comers, to secure its full time, was important. Some of the pupils had fallen into the slovenly habit of coming into the theatre a few minutes after the bell had rung, and in this way lost, and prevented those present from hearing, the first sentences of the lecture. For the want of punctuality they could hardly be blamed, as an example was set by several of the Professors, whose entrance was delayed, as they said, 'to give the lads time to assemble.' Mr. De Morgan, after duly enjoining punctual attendance, gave notice that if the pupils came in after he had commenced, they would find the door locked, which threat after two or three days' trial was put into effect. A few enterprising youths kicked and knocked at the door, trying to burst it open, but on the appearance of a policeman, and a threat of 'the Council,' which might mean removal, they were brought to order.

At the end of this session nine pupils presented their Professor with a handsomely bound copy of Macaulay's Essays, with a letter begging his acceptance of it as 'A small expression of gratitude for the liberal and most efficient assistance which his course of mathematical lectures had afforded them in preparing to pass the examination for the degree of Bachelor of Arts in the University of London.' Among these nine gentlemen were three of the insurgents, and among the other six names were those of Jacob Waley and James Baldwin Browne. Some time had always been given after each lecture to clearing up difficulties, and rather more than usual was necessary after the outbreak, as those concerned in it were in greater need of help to make up for lost time.

1839. Very soon after the establishment of the London University, young men of Jewish parentage began to distinguish themselves by their rapid acquirement of knowledge, thus justifying the hopes of their co-religionists who had contributed so liberally to the foundation. During the years in which my husband was Professor, many Jews took the highest honours; and among his most attached and valued Jewish pupils, the late Mr. Numa Hartog was the last, and Mr. Jacob Waley the first. Mr. Hartog's career, unhappily cut short before he had applied his talents to the work of life, was a brilliant one. After taking all the honours that could be given in University College, he went to Cambridge, and took his degree as Senior Wrangler. But the mental work was too much for his strength, and an attack of small-pox in 1872 left him too weak to rally.

Jacob Waley. Mr. Jacob Waley, afterwards Professor of Law, was one of the first Jewish students, after my husband's return to his Professorship, of whom the College had reason to be proud. He was not only a successful student in class, but a diligent private pupil, and from the time of which I write till his death in 1873 a valued friend. His lessons at our house in Gower Street were pleasant to both teacher and pupil, and even to myself, for he would come to me when they were done, for a little talk about books, or a reading of his favourite writer Macaulay's *Lays* or *Essays*. Mr. Waley was the first M.A. of the University of London, which in 1836 was ready to confer degrees on students of its affiliated Colleges in London and elsewhere. Some of us now living may remember Lord Brougham's reference to this pupil in a speech made at a distribution of prizes at the time, and possibly too some may remember how the speaker dwelt upon the fact (which *was* a fact then, and we had heard it so often that we were tired of hearing it) that within those walls men of every religion were received, whether as teacher or student, without any reference to their beliefs or non-beliefs. The time at which these assertions were made was spoken of by my husband as 'before the Fall.' 1839.

At the time of our marriage Mr. George Long, who had resigned his Professorship of Greek in the University when Mr. De Morgan retired, was living with his family in Camden Street, Camden Town. He was editor of the *Penny Cyclopædia*, and others of the works of the Diffusion Society, and the work brought him and Mr. De Morgan much together. The two had several qualities in common,—integrity of purpose and simplicity of character, indefatigable industry, and a love of fun which brightened hard work and kept us always amused. Mrs. Long took great credit to herself for the fulfilment of her predictions on the subject of our marriage, which she declared she had foreseen from time immemorial. I believe her prophecies really dated from the year 1831, when my acquaintance with her began. It lasted as warm friendship till the year 1841, when to the great sorrow of all her friends she was taken from us.

George Long.

Among other visitors not connected with the College was Mr. Leslie Ellis, who left on my mind the impression of an almost perfect moral nature. This impression was confirmed when, some time after, his scientific studies were interrupted by an illness, which he bore for years with unexampled patience, trying to alleviate the intensity of his sufferings when possible by mental work, and when that was impossible, awaiting the end with perfect resignation. Dr. Logan, a learned Mathematician and afterwards Professor at the Catholic College of Oscott, was among our friends. When Mr. Leslie Ellis's sister left him on the occasion of her marriage with Dr. Whewell, the Master of Trinity, our friend Dr. Logan took her place near the sufferer, and attended him with unremitting friendship and affection till his death. I have none of my husband's letters to Dr. Logan, but I know that the correspondence was large. Mr. De Morgan was indebted to him for the volume of *Ploucquet* which was

1839. afterwards of such essential service in his logical controversy with Sir William Hamilton.

Rev. James Tate.

Another and an older friend not connected with the College was James Tate of Richmond, Yorkshire, who at this time was living with his family at the residence as Canon of St. Paul's, Amen Corner. His own story was nearly connected with that of my mother's family. Mr. Tate, like so many of his own scholars—'northern lights,' as they were called—was of obscure or rather poor parentage. Archdeacon Blackburne,[1] who lived at Richmond, wanted a lad to act as amanuensis, and to read to him. Mr. Christopher Wyvill, his friend and contemporary, a noted Whig reformer of the time, required a young man in the same capacity. Two lads were recommended by Mr. Temple, the head-master of Richmond School. These were James Tate and Peter Fraser, the last a poor boy, but a lineal descendant of the beheaded Lord Lovat. My great-grandfather engaged young Tate, and Mr. Wyvill took Fraser. The two lads proved well deserving their appointments. Archdeacon Blackburne became greatly attached to his young amanuensis, and found the means of sending him to Cambridge,

Pupils of Richmond School.

where he gained honours as a classic. Mr. Wyvill sent his *protégé* to the University, with nearly the same success. In due time young Tate was ordained, and afterwards appointed to a tutorship in the school at Richmond, of which at Mr. Temple's death he became head-master. Some of the most distinguished men of the beginning of this century were his pupils; many of them, like himself, owing all to their own ability and industry. Of these were Dean Peacock, Professor Adam Sedgwick, Professor Whewell, Richard Sheepshanks, and many others.

[1] Archdeacon of Cleveland. His work *The Confessional* gave him a distinguished place among the writers on Divinity of his time. Mr. Fraser afterwards married his granddaughter, and died rector of Kegworth in Leicestershire.

Many as were our friends, we had but very little visiting, my husband's time being so fully filled with his work. The last was done with exceeding order and punctuality. He has himself described Mr. Baily's habits of order; and his own, though less apparent, were equally characteristic. He had the faculty of arrangement in an unusual degree, but it showed itself more in classification than in tidiness. In looking at any undertaking for scientific or practical purposes, he could not go on till all his materials were ready and arranged. This faculty is seldom so well proportioned to the power of carrying out the work projected. Mr. Baily had order of every sort, from the classification of formulæ or facts to the perfect arrangement of his house and appointments. In Mr. De Morgan it showed itself differently. Not having the means to indulge in the luxuries enjoyed by richer and more affluent writers or experimentalists, he could not furnish his library with all the writing appliances and handsome bindings that ornament rich men's studies, and his old table and desk, and other cheap contrivances, looked shabby enough. Any one who went into his room would be struck at first by the homeliness of the whole, and the quantity of old and unbound books and packets of papers. But when it was seen how the books were arranged and the papers labelled and put into their proper places according to subjects, the adaptation of means to ends became as apparent as in the clearness and precision with which he laid down principles, and showed what was to be done before making a beginning on his work. His contrivances in the way of inkstand, penholder, and blotting-block, had none of them a new or unused look, but all showed that every contingency had been carefully provided for. After gutta-percha came into use he employed it in every possible way, moulding it into penholders, caps, covers, and all sorts of fastenings. He says, in *The Budget of Paradoxes*, 'I never could spell the word, but if *cowchoke* goes, I go too;' and being

1839.

Orderly habits.

1839. once disrespectfully told that he would fasten on a head, an arm, or a leg, if he lost one, with gutta-percha, he said, 'I should like to see you do as much.'

When the repetition of the Cavendish experiment was undertaken by Mr. Baily in the year 1837 at his house in Tavistock Place—a house thereby rendered memorable—Mr. De Morgan gave so many clear descriptions of it and its object, that Mr. Baily's work in 1838 and 1839 requires a longer notice than I have given to those Astronomical achievements with which my husband had less to do. A grant of 500*l.* had been made by Government, at the representation of the Astronomical Society, for defraying expenses. The forerunners of this effort to ascertain the mean density of the earth are mentioned by Mr. De Morgan in the Life of Maskelyne, written some time before, and will give some idea of the nature and objects of the undertaking.

Cavendish experiment.

'The labour of deducing an approximation to the earth's mean density was undertaken by Dr. Hutton. By getting the best possible estimate of the materials of which Schehallien is composed, and comparing what we must call the weight of the plumb-line *towards the mountain* with its weight towards the earth, it appeared that the mean density of the latter is about five times that of water. This, considered as a numerical approximation, alone and unsupported, would have been worth little, owing to the doubt which must have existed as to the correctness of the estimation of the mountain's density. It would prove that there was attraction in the mountain, but would give no very great probability as to the value of the earth's density as deduced. But a few years afterwards Cavendish made an experiment with the same object, and by an entirely different method. By producing oscillations in leaden balls by means of other leaden balls, and by a process of reasoning wholly free from astronomical data, he inferred that the mean density of the earth was five and a half times that of water. The experiment of Cavendish was published in 1798. It is much to be wished that it should be repeated on a larger scale, but the expense of the apparatus will probably deter individuals from the attempt.' 1839.

In a pencil note in the margin of the same page I find—

'This was, I believe, the remote cause of the repetition of the experiment. Being, a few months afterwards, in the year 1835, on the Council of the Astronomical Society, something was said about the mean density of the earth, and I happened to say, "I wish Cavendish's experiment could be repeated." Mr. Airy immediately said, "Ah, that would be a good thing." Others agreed, and a committee was appointed on the spot "to consider of the practicability," &c. The result was the repetition of the experiment.'

First suggestion.

The history, the nature of the formulæ for the calculations, and the results of the discovery, are all given by Mr. De Morgan in the articles 'Attraction,' 'Cavendish Experiment,' 'Weight of the Earth,' and others, in the *Penny Cyclopædia*, and in a sufficiently popular form in an article in the *Companion to the Almanac* for 1838.

Mr. Baily's repetition, commenced in 1838, was carried on in a small upper room twelve feet by twelve, as far removed as possible from the noise and shaking of street traffic. It was, of course, an object of interest to all scientific friends, and Mr. Baily's genial kindness in explaining his beautiful apparatus and showing his progress was one of the pleasant accompaniments of his important work. The apparatus designed and constructed by Mitchell, who did not live to use it, had been used by Cavendish, and afterwards by Mr. Baily, but so greatly improved and added to by the last experimenter that it could hardly be called the same. I saw

1839. the progress of the experiment, and my husband's visits were frequent to the little room in which the world was weighed.

The Life of Maskelyne, from which the mention of my husband's suggestion is taken, is one of a series of lives of Astronomers written by him for the *Gallery of Portraits*, published by C. Knight two or three years before this time. They are those of Bradley, Delambre, Descartes, Dollond, Euler, Halley, Harrison, W. Herschel, Lagrange, Laplace, Leibnitz, and Maskelyne. They are bound up together, and illustrated in his own way, under the title of 'Mathematical Biography, extracted from the *Gallery of Portraits*, by Augustus De Morgan, H.O.M.O. P.A.U.C.A.R.U.M. L.I.T.E.R.A.R.U.M.' The letters of his literary tail were only B.A., F.R.A.S., besides those expressing membership of one or two lesser Scientific societies. On account of the declaration of belief at that time required by the University he never took his M.A. degree.

In November our eldest son, William Frend De Morgan, was born.

We had spent five weeks at Boulogne in the summer. I hoped that, as my husband always liked the sea, a French watering-place would be less irksome to him than English country or sea-coast; but he soon got tired of it, and felt glad to get back to his work.

1840. He bore a few weeks at Blackheath next year with equanimity. He was near the Observatory, and Mr. and Mrs. Airy were good neighbours, so were Mr. (afterwards Lord) Wrottesley and Mrs. Wrottesley, the former being on the Council of the Astronomical Society, and, besides his other excellent social qualifications, being a good musician. My husband liked the steamboats, of which he made much use; but the heath, which he called *desolation*, was a trial to him. After this summer he begged me to take the children without him; and I found that this arrangement, which I disliked, was the best.

He required a letter, reporting health, &c., and sent me 1840.
one in return, every day.

On our return to Gower Street I went with my two little children to Highgate for a very short time to be near my father, who had had a stroke of paralysis.

He died early in the next year, 1841, at the age of 1841.
84. His Cambridge life and early difficulties on the subject of religion have been slightly referred to. He had taken his degree as Second Wrangler, and had afterwards had a Fellowship and a College living till scruples of conscience led him to leave the Church; and his subsequent publication of a pamphlet entitled *Peace and Union* was the cause of a prosecution by the University. He was tried and sentenced to non-residence, but he retained his Fellowship till his marriage with my mother, a granddaughter of Archdeacon Blackburne. He had been a pupil of Dr. Paley, for whom he always retained an affection; and among his own pupils were Dr. Edward Daniel Clarke, the traveller, Lord Lyndhurst, afterwards Lord Chancellor, and Mr. Malthus, in whose social tenets he entirely disclaimed any share. My father's political opinions, as set forth in *Peace and Union*, were held to be extreme eighty years ago; they are as milk for babes in comparison with the strong stimulants given by the Liberal party now.

He was after he left Cambridge a friend of Sir Francis Burdett during the reforming portion of his life, of Horne Tooke, and of other reformers. What place he would have taken in politics had he lived till now I can only conjecture. He was a good Hebraist, and was trustee for Mr. Robert Tyrwhitt's Hebrew Scholarship at Cambridge. His largest work was on popular Astronomy as it was known then. This book, entitled *Evening Amusements*, came out, a volume every year, for nineteen years; each volume showing the relative positions of all the heavenly bodies for every month in the year.

My father's ideas on Algebra were peculiar; his re-

1841. jection of the use of negative quantities in algebraic operations being probably the result of the same straightforwardness and clearness rather than great depth of intellect, which had led him, with more show of reason, to reject the doctrine of the Trinity in the form in which he had received it.

He was an upright and noble-minded man, generous and disinterested to a fault, if that can be, with a vigorous perceptive mind, but little imagination. His religion was real and practical; and his death was an event to which he had always looked forward with cheerfulness and happy anticipation. My husband wrote a short biography for the *Athenæum*, and a longer one to the Astronomical Society's *Obituary Notices*.

Mr. De Morgan was at this time, I think, consulting actuary to the Family Endowment Assurance Office, which afterwards merged with another office in the Mutual. I do not think he held this place more than two years.

Accident to Francis Baily.

During this summer an accident occurred which nearly proved fatal to our friend Mr. Francis Baily. He was crossing Wellington Street, Strand, when a man on horseback, riding furiously, knocked him down and stunned him. He was taken in an unconscious state to the Charing Cross Hospital, and was found to have a severe scalp wound and to be a good deal bruised. As soon as he could bear the removal he was taken home to Tavistock Place; and after a few weeks his recovery seemed to be complete, though he remained weak for some time. But the injury to the head left more serious results than were expected, for it is most probable that the disease of which he died three years after, and which is now believed to be often the result of a shock to the nervous system, was caused by this blow. My husband was, as were all Mr. Baily's friends, extremely anxious as to the possible consequences; but their anxiety was allayed when he got up again, received his friends, resumed his work on the

Cavendish experiment, and even took some part in the Report on the Commission for restoring the National Standards, to which he had devoted so much time and thought. This Commission was appointed in 1838. The Report was sent in in 1841. On the appointment of a new Commission in 1843 to reconstruct the Standard Scales, Mr. Baily recovered sufficiently to undertake the Standard of Length, but did not live to complete it. 1841.

Mr. Hallam's *History of the Literature of the Middle Ages* was at this time coming out, and after the publication of the first edition Mr. De Morgan sent the author some observations on the history of the Mathematicians of the period embraced in the first volume. What his criticisms were can only be guessed by Mr. Hallam's reply, as I have not his own letter.

Dec. 12, 1841.

I am much obliged by your correction of some inaccuracies in the first volume of my *History of Literature*, which will be of use to me in the new edition which I am now preparing. I am always thankful for such communications, which are at least a sign that the book is thought worthy of them.

In reply to further criticism, further correspondence took place two years later, Mr. Hallam's letter touching on the subject of Logic, which had assumed so definite and important a form in Mr. De Morgan's thoughts, and which was afterwards treated in connection with points of original discovery in his *Formal Logic*.

H. Hallam.

I shall pay all attention to them (your remarks) in any new edition, and will look again at some of the works in which your more expert eye has detected my errors. I certainly searched in vain for the triangle of forces so called in Stevinus, and I was a little more led to doubt of his using it, as Montucla says, because the only demonstration with which I am acquainted involves the third law of motion; but that, according to the general opinion, was not laid down till long afterwards. Perhaps Stevinus might assume it on metaphysical principles without experiment.

1841. The early history of algebra is a very interesting subject, and I shall be glad to return to it again with more leisure. I was forced to copy writers of credit, not always seeing the full force of what they said. But it seemed to me that a sound mathematician would find that field not exhausted. . . .

As to your observation about logic—that is, the syllogistic logic—perhaps I was a little influenced by the impression that more attention was paid to it at Oxford. That, I believe, is now much less the case. But though what I have written on it is inconveniently concise, and leaves more for the reader than an ordinary author has a right to expect, I thought I had given to that logic what it can best claim, its quality of perfect demonstration; all geometrical demonstration being, in fact, one species, or rather one application, of the fundamental principle. Nor did I distrust the usefulness, to a certain degree, of an acquaintance with syllogism, though I have not found that the best reasoners are very familiar with it. However, if I have gone too far in lowering this art or science—for it is not settled which—I am very willing to retract. Let me add that I have received much pleasure from some of your writings, such as are most familiar to me, especially that on the 'Connection of Number and Magnitude.' You need not fear going *ultra crepidam*, for your *crepida* is very extensive.

Believe me, dear sir,

Your much obliged servant,

HENRY HALLAM.

In October our second son, George Campbell, was born. We lost him at the age of twenty-six, not before his mathematical talents were developed sufficiently to entitle him to notice in his father's scientific history. Of this I must speak later. He was a lovely and seemingly healthy child, sweet-tempered, quiet, and thoughtful; and though sound and certain in all, he was not quite so quick in learning as his sister and brother.

Our society was diminished by the loss of Mr. Sheepshanks, who left London to live at Reading with his sister.

The correspondence with Dr. Whewell, which had begun soon after the pupil left Cambridge, related at first to Mathematical questions. But when Mr. De Morgan began to make the application of Mathematical principles to Logic, Dr. Whewell was naturally one of the first to whom his ideas were communicated. In many instances the letters were written on the occasion of sending tracts to the *Cambridge Philosophical Transactions.* The first of these, 'On the General Equation of Surfaces of the Second Degree,' is dated 1830; 'On the Foundation of Algebra,' No. I., read 1839; Nos. II. and III., 1843; No. IV., 'On Triple Algebra,' 'On the Structure of the Syllogism, and on the Application of the Theory of Probabilities to Questions of Argument and Authority,' 1846.

1841. Cambridge Tracts.

The work on the Differential and Integral Calculus, which had been published by the Useful Knowledge Society, appeared in 1842 in his complete work, a closely printed octavo volume of 770 pages. The series, which had commenced in the year 1836, consists of twenty-five numbers, each containing thirty-two pages, and to the book is added an appendix and two numbers of elementary illustrations which had been published by the Society before. Of the work he says in the preface:—

1842.

'The method of publication in numbers has afforded time to consult a large amount of writing on the different branches of the subject; the issue of the parts has extended over six years, during two of which circumstances with which I had nothing to do stopped all progress. The first number was preceded by a short advertisement, which I should desire to be retained as part of the work, for I have no opinion there expressed to alter or modify, nor have I found occasion to depart from the plan then contemplated.

'The principal feature of that plan was the rejection of the whole doctrine of series in the establishment of the fundamental parts both of the Differential and Integral Calculus. The method of Lagrange, founded on a very

1842. defective demonstration of the possibility of expanding $\phi\ (x+h)$ in whole powers of h, had taken deep root in elementary works; it was the sacrifice of the clear and indubitable principle of limits to a phantom, the idea that an Algebra without limits was purer than one in which that notion was introduced. But, independently of the idea of limit being absolutely necessary even to the proper conception of a convergent series, it must have been obvious enough to Lagrange himself that all application of the science to concrete magnitude, even on his own system, required the theory of limits. Some time after the publication of the first numbers of this work, four different treatises appeared in the French language, all of which rejected the doctrine of series, and adopted that of limits. I have, therefore, no occasion to argue further against the former method, which has been thus abandoned in the country which saw its birth, and will certainly lose ground in England when it is no longer maintained by a supply from abroad of elementary treatises written upon its principles.'

The doctrine of series in opposition to that of limits was practically overthrown before the completion of the work, and the new principle had engaged the attention of Mathematicians. As might be expected, a volume embodying them, important in its bearing upon metaphysical as well as mathematical thought, excited great interest in the minds of the few who could enter into the question. Of these Dr. Whewell, who had written on it in 1838, was one of the most pronounced. But with one exception the ideas of cotemporary thinkers must be gathered from the letters.[1] A full review of the subject, if it were within my power to make it, would not be in place here, and an imperfect one would be useless. But some of the bearings of the principles developed in my husband's Differential Calculus were thus referred to by Mr. John Stuart Mill,

[1] See next Section.

thirty years later, in an admirable article 'On Berkeley's Life and Writings:'[1]— 1842.

It is difficult to read without *parti pris* 'The Analyst,' and the admirable rejoinder to its assailants, entitled 'A Defence of Free-thinking in Mathematics,' and not to admit that Berkeley made out his case. It was not until later that the Differential Calculus was placed on the foundation it now stands on—the conception of a limit, which is the true basis of all reasoning respecting infinitely small quantities, and properly apprehended, frees the doctrine from Berkeley's objections. Nevertheless, so deeply did those objections go into the heart of the subject, that even after the false theory had been given up, the true one was not (so far as we are aware) worked out completely in language open to no philosophical objection by any one who preceded the late Professor De Morgan, who combined with the attainments of a mathematician those of a philosopher, logician, and psychologist. Though whoever had mastered the idea of a limit could see, in a general way, that it was adequate to the solution of all difficulties, the puzzle arising from the conception of different orders of differentials—quantities infinitely small, yet infinitely greater than other infinitely small quantities—had not (to our knowledge) been thoroughly cleared up, and the meaning that lies under those mysterious expressions brought into the full light of reason by any one before Mr. De Morgan.

Mr. J. S. Mill on Differential Calculus.

My husband died shortly before this was written. He had, as his letters show, a sincere respect and regard for the writer, though they had met only on one occasion, and he had corresponded with Mr. James Mill, his father. But though, as was truly said, his mathematical reasoning had deprived Berkeley of an argument drawn from the mystery of infinitely small numbers, his sympathies were in many ways more on the side of Bishop Berkeley than on that of Mr. Mill.[2] The works of Berkeley had been, as

Berkeley's Philosophy.

[1] *Fortnightly Review*, Nov. 1871.

[2] I am aware that the principles of Berkeley's philosophy have been found by some thinkers to lead to a pantheistic materialism. Much depends upon words, but more on the minds of those who use them, and a spiritual pantheism must be a near approach to truth. When the words spiritual, material, theistic, pantheistic, and atheistic

1843. aforesaid, among his earlier studies, and his absolute conviction, or, as he said, *consciousness*, of the fatherly care of God was directly opposed to the scepticism (I use the word as expressing *doubt*, not *dis*belief) of one whose work in many directions he valued highly. He wrote with great respect of Mr. Mill's Logic, and the essay on Liberty had his cordial admiration. The essay on Comte, too, he thought very valuable. His own ideas about this great fabricator of society may easily be conjectured. I should like to give them in his own words, but can only remember their import. Just in proportion to the strength of that part of a system which is founded on the principle of love to the neighbour is the weakness of that part of it which sets aside the Divine Disposer of events, and puts an arbitrary classification in place of the natural order of the world. I hope I have not misrepresented the principles of Comtism; I know, however, that this fairly represents my husband's interpretation of them.

Our third son, Edward, was born about Midsummer in this year. His father gave him his second name, Lindsey, to perpetuate that of my mother's uncle, Theophilus Lindsey, a good man, and one of the earliest English Unitarians who, like my father, seceded from the Church, and who gave up the lucrative living of Catterick, in Yorkshire, where he was much beloved, because he could not conscientiously carry on the duty in accordance with prescribed doctrines. Such secessions, united with such strong religious belief, do not often happen in these days; but we cannot judge of the motives of those who do not feel them to be necessary for conscience' sake. Many distinguished clergymen who hold the doctrines of Christianity far more loosely,

find their proper places, there will be an end to these confusions, which result from the various ways in which the great subject is looked at by speculators whose mental eyes are differently placed in relation to it. This is only saying that the true knowledge of words will be the true knowledge of things.

and give a far less literal assent to the New Testament narrative than either my great-uncle, my father, or my husband, remain in the Church with a belief that their moral influence will be of greater use than their intellectual scruples. In this case the possibility of the Church becoming too broad to hold together is not felt to be an evil, as even should it through this cause die a natural death, its work will have been well done. Mr. De Morgan felt that the profession of belief of every clergyman implies so absolute and entire an adhesion of the whole soul to the doctrine which he undertakes to preach, that should that *animus* be altered, membership, in the sight of God, has ceased with it; and the outward and visible sign can really stand for nothing when its inward and spiritual essence is gone. But he judged no one rigorously but himself, though he was happy in knowing that he had been connected with the memories of men of worth and learning, who never hesitated when their choice lay between truth as it appeared to them and any other consideration.

1843. Belief and profession.

Shortly before Edward's birth we lost our old friend Francis Baily. Early in this year his usually fine robust health had given way, and a disease of the kidneys, probably the remote result of the shock given to the brain by his accident, declared itself. He lingered some weeks, always cheerful and hopeful, but perfectly ready for whatever turn his illness might take. His friends were less prepared to lose him than he was to go. His death occurred in June, and was a loss to Science[1] which could not well at that time be filled up. Sir John Herschel wrote that he was a man *sui generis*; and the letters

1844.

[1] I have throughout this memoir used the word *science* in reference to Mathematics and Logic, and those branches of knowledge in which processes of reasoning are applied to subjects of observation. This is the older meaning of the word. It is generally, though of course not exclusively, used now to express knowledge gained by observation alone. I remember the time when, in reference to Dalton's atomic theory, it was said that chemistry had *become a Science.*

1844. which passed among his friends contain a repetition of this feeling, with an expressed determination to do by their united effort what he had done for so many years. This refers especially to the Astronomical Society, of which he had long been President, and the members of which found it difficult to appoint a successor. Mr. Airy, Sir J. Herschel, Mr. Sheepshanks, and Mr. De Morgan composed his epitaph. It was, I think, drawn up in the first instance by Sir J. Herschel or Mr. De Morgan, and carefully revised and altered by the others.[1]

Standard scale.

Mr. Sheepshanks, who had accurate knowledge and experience of scientific instruments, undertook to complete the construction of the Standard of Length. In a letter to my husband he says:—

Baily's epitaph.

I think Airy's paper on the supports of the standard scale should be printed forthwith. . . . This naturally leads me to the final clause, the inscription *Regula mensurarum in perpetuum definita.* One would not (where Baily is concerned) even in a Latin epitaph (and the language and mode of employment are not mendacious) use exaggerations. When I undertook the scale I hoped and believed a good deal was done; but when I got from Airy a précis of facts I found that it was chiefly of a negative character, viz., that our scale had changed its form, &c., and the only positive advance (beyond preparation) was, that

[1] There is a bust of Francis Baily in the apartments of the Astronomical Society in Burlington House. If the time should ever come when observations of the form and size of the different parts of the head and face are systematically made with a view to determine the elements of character, any conclusion drawn from this bust would be a great injustice to our dear old friend. It was taken from the portrait, which is weak and inadequate, and has exaggerated these defects. While it was in progress, Mr. Baily the sculptor asked my husband and myself to see the clay model at his studio. He invited criticism, and at my suggestion added so much to the forehead that it bore a strong likeness to the subject; and Miss Baily when she saw it burst into tears, exclaiming, '*It's himself.*' But the sculptor afterwards found that the penthouse brow and large forehead were not 'ideal' enough. He said his work had been spoiled, removed all the added clay, and left the weak and characterless head which professes to be a likeness of Francis Baily.

1844.

two iron bars prepared by Colby had maintained their difference. There were, indeed, some good measurements of expansion, but these were by Simms and his nephew, and I intend repeating them with, if possible, greater nicety. If this subject is mentioned (and it should be), say that he was unhappily arrested in the act of definitely fixing the national measures.

Richard Sheepshanks's micrometer observations.

In the execution of his work, which was carried on in a cellar under the chambers of the Astronomical Society in Somerset House, Mr. Sheepshanks recorded 89,500 micrometer observations. Only those who understand the nature and object of these can estimate the enormous labour involved. He had to make frequent visits to London for this purpose. We saw him often at the time, but I have only my own memory for the statement that much discussion on the experiments and observations passed between him and my husband, and that when difficulties occurred, Mr. De Morgan was often able to assist in their solution.

Camden Street.

We moved in July from Gower Street to Camden Street, Camden Town. My husband walked to the College in time to be there every morning at 8.30, that he might look over the pupils' papers before giving his lecture. He could not come home, as before, in the middle of the day, and on this account I was sorry for the change; but in other respects it was far better, as the house was roomy and convenient for a young family, and the air I thought fresher than in Gower Street.

His readiness to serve his pupils and the College had brought him some extra work, and some pleasure in consequence. The Professor on whom the teaching of Mathematical Physics devolved proved quite unequal to his task. The pupils who came to him from the Mathematical class were already much his superiors in knowledge, and in their strait they appealed to my husband. With the approbation of the Council, he at once undertook to meet the difficulty. He gave, during the remainder of the session, extra time and instruction to these young

1844. men, who at the end of the session presented to him a handsome copy of Wilkinson's *Ancient Egyptians.* The letter accompanying the gift is signed with names of Henry Robert Reynolds, Joseph Rees, Richard Holt Hutton, and C. Howard.

In this year the Useful Knowledge Society came to an end, having completed its work. Its farewell address has been already mentioned. Mr. De Morgan's last undertaking for it was a book entitled *The Globes Celestial and Terrestrial,* written as a description of the Society's globes, published in 1844.

Malby's globe

The Society, which had already brought out the 'Maps of the Stars,' had turned its attention to the want of accurately made globes of the heavens and the earth. These were constructed by Mr. Malby, under the direction of several men of Science. The celestial globe needed most revision and improvement. Up to the year 1823 the 21-inch globe by Cary and the 18-inch globe by Bardin were the best in use; and of these two, Cary's, 'having annexed to every star its proper numerical or characteristic valuation,'[1] was judged to be the best. Both were founded on Wollaston's Catalogue of 1789; and Cary's globe contains the stars extracted from Flamsteed's Catalogue by the two Herschels.

Mr. Baily had laid down, though not with the full alteration which he thought necessary, the lines which bound the constellations. Sir J. Herschel, being appealed to by my husband on behalf of the nebulæ, wrote:—

Why globe-makers will persist in laying down nebulæ of Classes II. and III. is to me astonishing. There are but half a dozen of Messier's and my father's 1st class which can be seen with the naked eye, and the 2nd class ones are for the most part invisible with a 3 inch object-glass. The per-centage of Dunlop's nebulæ which can be seen with the naked eye is still smaller.

[1] Letter from Mr. J. W. Woollgar, of Lewes, to the *Philosophical Magazine.*

Indeed, with the 20-foot reflector, out of 629[1] of which his catalogue consists, I have succeeded in observing only 207, and of these I have great doubts of the identity of between twenty and thirty. What sort of objects Mr. D. has set down as nebulæ in the other 422 cases I have no idea. All I can say is, that out of 1,700 more or less observed by myself at the Cape, the above are all of Dunlop's which have not proved coy. You will judge by this whether or not to recommend your globe-undertakers to map down Dunlop's catalogue *in its integrity*. 1844.

As to double stars, I think Struve's great catalogue will go far to saturate a 36-inch globe.

Let me know whether you are *very much* interested in Mr. Malby's undertaking, as in that case I would send you the list of those Nos. of Dunlop's nebulæ which I either know certainly to exist in or near his places, or have found nebulæ which, by a stretch of good-natured identification, I should be disposed to admit as observed by myself.

Some large and showy globes had been made in 1823, the trustworthiness of which was not guaranteed by the name of any scientific authority. As Mr. Woollgar, in writing of these, said, 'globes are oftener purchased as articles of furniture than as philosophical instruments,' and these large globes fitted the purpose. Mr. Malby's globes could not lie under this reproach, for even as articles of furniture they were not showy enough to suit the upholsterer, while their accuracy was beyond question. But as a globe can never be even quite up to the amount of astronomical or geographical knowledge at the time of its completion, it must from time to time require additions, if not corrections, and in course of discovery will at length be superseded for practical use, as the globes made in 1844 may be at this time.

From their improved construction, great accuracy, and careful measurements, questions concerning ancient Astronomy, depending for the most part on the precession of the equinoxes, could be determined by these

[1] The figures are slightly blotted in the letter, and I have not Dunlop's Catalogue to ascertain the number.

1844. John Taylor.

globes. One who rejoiced greatly in them was Mr. John Taylor, known for his inquiry into the authorship of Junius, and for his speculations on the Great Pyramid. He had made some suggestions on the formation of the Astronomical globe, and wrote to my husband: 'I can now call up all the phenomena recorded by Aratus or Hipparchus, not to forget Homer, Hesiod, Virgil, Ovid, and Columella, and I may even venture to correct the great constructor Ptolemy, when he makes a slip in his notations. Ulugh Bey, too, and Tycho, and old crusty Flamsteed, may come in their turn to the spherical ordeal. All this I owe to you, for had it not been for your friendly interference and sanction, I might never have seen my attempt submitted fairly to the world.'

Mr. Taylor's large claim to original suggestions in this work might not perhaps be fully acknowledged by the Astronomers who helped to complete it. Notwithstanding his practical and extensive dealings with the old philosophers, and his satisfaction in correcting Ptolemy's slips, he was himself far from sound in his scientific knowledge, and subjected himself to a severe castigation from Mr. Sheepshanks for meddling with the Liverpool Observatory and its manager. He also set Astronomers right about the comet of 1848, which he declared to be the same as that of 1556. 'This,' Mr. Hind says in a letter to Mr. De Morgan, 'is the last of Mr. John Taylor's Astronomical extravagances.' The motion of the first comet was direct, that of the one observed in 1848 retrograde. Mr. Taylor's announcement was made in the *Liverpool Mercury*, and corrected, I think, by Mr. De Morgan in the *Athenæum*. But his researches on the Great Pyramid are of value. Mr. De Morgan said of him: 'He is by temperament a discoverer of hidden things, and has employed much ingenuity in discovering what we may call two crack secrets, because they have never been fairly cracked.' The other crack secret was Junius.

The moving picture of progress in any study has for its background a series of ignorant guesses and foolish conclusions. A specimen of what was taught as Astronomy in the fourteenth century is not more grotesque and is less simple than what was called teaching within our own memory, and perhaps may still be so called in some remote young ladies' school. Referring to the works of a class of authors who seem to think that the pretext of writing for the especial instruction of young ladies is more than a sufficient excuse for any amount of nonsense, Mr. De Morgan adduces the following examples. The book from which the extracts are made had reached its fourteenth edition. 1845.

Astronomy for girls fifty years back.

Among the questions on *Sagittarius* are the following:—'To what sin were the Athenians addicted? What reflection does Dr. Doddridge make on the occasion?' *Apropos* of the constellation of *Ursa Major* is this question: 'Who drove stags in his phaeton instead of horses?' and the answer is, 'Lord Orford, who died in 1791.' The concatenation is that bears can be tamed, and that Prince Radzivil drove them in his carriage at Warsaw. On *Musca* the questions are, 'What are the distinguishing characteristics of flies? In what manner, demonstrating his propensity to cruelty, did Domitian treat them? Hence what sarcasm was passed upon him? How has Sterne represented the humanity of a feigned character to a fly? How did contrary behaviour in a female (according to Darwin) break off an expected matrimonial connection?'

There were a few other books of a better sort published, but they did not reach fourteen editions, and, we may believe, seldom found their way into girls' schools.

Old Mathematical Society.

There had existed from the year 1817 a Mathematical Society, or club, which met in Crispin Street, Spitalfields. It was originally composed of working men, many of them silk weavers, and among the early members had been some men of known name. The conditions of membership were that each member should have his pipe, his pot,

1845. and his problem. A short account of this society is given in the 'Budget of Paradoxes.' It gradually declined as the harder life of the working man deprived him of leisure time, or perhaps as the pot took the place of the problem; and in this year 1845 a proposal was made to incorporate what was left of it into the Astronomical Society. Only nineteen members remained, and they were not working men. Those who were not already members of the Astronomical were received without payment of fees, and it only remained to convey the books and other property of the old society to the rooms at Somerset House. Mr. De Morgan undertook to look over and to superintend the removal of the books, which now form a small portion of the library of the Royal Astronomical Society.

Two or three years before this time, some gentlemen, interested in the history of Science, had projected a society to be called the 'Historical Society of Science.' Mr. Pettigrew, Mr. Richard Taylor, the printer, and Mr. De Morgan were among the first of these. Mr. J. O. Halliwell, the archæologist, had taken a prominent part in the scheme, and became the secretary. My husband, who had looked forward to useful results from the work of this society, found in this year that it was becoming extinct for want of attention in collecting subscriptions, and from general neglect. He immediately called the attention of the other members to this state of affairs, and the society came to an end without undue pressure on the Secretary, who was not in circumstances to meet it, but who incurred some blame from one or two of the persons concerned.

Rev. S. Maitland. When my husband was a boy, living at Taunton, the Rev. Samuel Maitland, not then in orders, was his mother's friend and neighbour. He afterwards became a friend and correspondent when the subjects of his works formed part of those over which my husband's studies extended. 'His series of essays "On the Dark Ages" was the most read of all his works. He was one of a class of whose writings it must be said that wherever they take they bite. They are imbued, but not in excess, with a kind of humour which seems almost their own. It has more likeness to the peculiar humour of Pascal than is seen in any writer of our day.'[1] Though Dr. Maitland was not a Mathematician, the subjects of mutual interest were many, and their correspondence touches upon all kinds of questions, from those of Dr. Maitland's works which involve much sound learning on the theology of the dark ages, to his latest little volume, *Superstition and Science*, in which the phenomena of spiritualism and the miracles of the Catholic Church are considered in relation to Scientific inquiry. The attention which Mr. De Morgan had given to the question of Easter was shared with Dr. Maitland. My husband had contributed an article to the *Companion to the Almanac* for 1845, giving the reasons why then, as in 1818, Easter Sunday had fallen on, instead of after, the first full moon after the Vernal Equinox. There had been much fruitless discussion on this in 1818, and to avoid a repetition of it—for the question was already agitated in Parliament—a full explanation was given, in the above-mentioned article, of the cause of deviation from the rule, and the relation of the whole subject to the Christian and Jewish calendars. In the next year, 1846, an article *On the Earliest Printed Almanacs* gave further information, and his *Book of Almanacs*, published in 1851, left no means of knowledge wanting. Dr. Maitland's letters at this period showed his interest in the Easter question. He was then librarian to the Archbishop of Canterbury, and the library at Lambeth Palace afforded him means of research in it, and experience on another question, which was valuable a few years later, when the British Museum Library Catalogue occupied the thoughts of scholars.

1845.

Date of Easter.

My husband's acquaintance with Lord Brougham,

[1] From an obituary notice by Mr. De Morgan on the Royal Society's Memoirs.

1845. Lord Brougham. who never visited us, arose out of the business of the Useful Knowledge Society and of University College. The correspondence, which lasted from 1830 to Lord Brougham's death, is chiefly on Scientific subjects, on many of which the statesman consulted the Professor.[1] One of these was the properties of curves in Optics, on which Lord Brougham had experimented and written. He was also the author of a life of Newton for the Society. Mr. De Morgan had, in this year, brought out a memoir of Newton (to be noticed further on), in Charles Knight's *British Worthies*; and many of Lord Brougham's letters refer to the claims of Newton as set forth by different writers.

When, shortly after this time, the injuries inflicted on Guglielmo Libri by an unjust accusation of the French Government aroused the indignation of most English men of Science, Lord Brougham expressed his sympathy, and tried to help M. Libri's cause by communicating with his own friends having influence in France as well in politics as in Science. He, like others, found and acknowledged the unjust bias of M. Arago wherever his national prepossession could come in. This showed itself in political antagonism (supposed, in M. Libri's case, to arise from his Italian birth and proclivities), as well in scientific questions as in the case of the simultaneous discovery of the planet Neptune by Adams and Leverrier.

Discovery of Neptune. The year 1846 was made famous by the announcement to the world of this discovery. From every point of view its history is an interesting one, but it is so familiar to most readers that I must ask pardon for reverting to its principal points, that the part taken by

[1] I regret that I have none of Mr. De Morgan's letters to Lord Brougham on the subject of Newton, or on any question of general interest. I am greatly indebted to the present Lord Brougham for his kindness in sending me a few letters, but the mass of documents is, I understand, so great at Brougham Castle as to render a thorough search exceedingly difficult.

my husband on the reception of the news by Astronomers may be understood. 1846.

This discovery was an instance of that law of progress by which we find that a truth, when the time has come for its reception, is seldom the prize of one mind only: it may be that some far-seeing solitary minds have early anticipated the knowledge for which the world is not prepared, but in its full advent a new truth has more than one recipient. It was so in the case of Neptune. In observing the place of Uranus early in this century, Astronomers had found that the observable course of the newly discovered planet did not coincide with that given by mathematical calculation. This appeared from M. Bouvard's tables of Uranus from 1781 to 1821. Other irregularities were found, but the idea of a large disturbing body was not generally entertained; and M. Poinsot, who had called the attention of the French Institute to the observation of a star, supposed to be a new planet, by Messrs. Wartmann and Cacciatore, was laughed at. It is true that in 1834 Dr. Hussey wrote to Mr. Airy that he had conjectured the possibility of some disturbing body near Uranus, and that he had found that MM. Bouvard and Haussen had corresponded on the subject. Mr. Airy, however, was doubtful of the possibility of determining the place of the planet until the nature of the irregularity should be better known. Eight years before the actual discovery Bessel gave it as his opinion to Sir John Herschel that the disturbances in question could be due only to the action of a large body beyond the orbit of Uranus. The direction of investigation was thus to a certain extent pointed out; it was no less, when attained, 'the greatest triumph of inductive Science which Astronomy has yet to record.' This greatness consisted in the fact that the exact place of the planet was obtained not by actual observation, but by mathematical calculation, founded upon the elements furnished by the action of the disturbing forces. 'Two Mathematicians

Aberrations of Uranus.

1846. had worked out the problem, the English Astronomer communicating his results in November 1845 to the highest authorities in the astronomical world, by whom they were reserved for further confirmation; the French Astronomer, with more confidence in his work, making it known to the world at once. Mr. Adams' calculations antedated M. Leverrier's by eight months, but M. Leverrier's promptitude of publication secured for him the honour of the discovery.' 'Mr. Adams,' adds Mr. De Morgan, 'furnished Mr. Challis with the means of actually securing two observations of the planet previously to any announcement by M. Leverrier.'

M. Leverrier's communication was followed immediately by recorded observations of Neptune by M. Galle, the Berlin Astronomer, who wrote in September that the planet had been seen by him as a star of the 8th magnitude not marked upon any chart.

There could be no question of M. Leverrier's discovery; it remained to be seen why, when the English discoverer was so nearly the first known to be such, he had been anticipated by the French Astronomer. Mr. Airy explained that in answer to the letter in which Mr. Adams had announced it to him, he had requested Mr. Adams to give him some further explanation. This letter of inquiry had not been answered till long after; hence the delay, deeply regretted by all. On these facts, and on the discussion which followed in the Astronomical Society, Mr. De Morgan wrote:—

No blame need be attributed to any one; but I think it will turn out that the Mathematicians of this country had not faith enough in their own Science. And, most assuredly, we may look forward to seeing the wise men who never believe until the thing is done—the sober men to whom everything that *is to be* is a figment in the brain of a visionary—the practical men who are not sure that there is a future until it comes to them in the shape of time present, all loud in their outcry, some against one, some against another, for not having done that

1846. The planet Neptune.

which, six months ago, they would have been the first to have laughed at them for doing.[1] . . .

That M. Leverrier is to all intents and purposes the discoverer of the new planet is beyond a doubt. No evidence in his favour could be stronger than that of Messrs. Adams, Challis, and Airy. It is quite within probability that it might have been discovered in November 1845, from the true elements given by Mr. Adams in October, as stated by Mr. Airy. That it was on Mr. Challis's papers before it had been seen abroad is certain. Why, then, is this remarkable discovery French, and not English? Simply because there is not sufficient faith in Mathematics among the Mathematicians of this country. I should not say this upon one instance involving only three men; I know it otherwise. Our men of science too often think it wise and practical to doubt results of pure Mathematics, and the French who run into the other extreme have a decided triumph in this instance. The results will do much good among us. Few of our philosophers are deep Mathematicians, and those who aspire to the character without laying the foundations of exact science, are apt to take a tone with respect to it to which its cultivators have deferred until their deference has acted on their own minds, and affected the rising generation. In one sense, we may rejoice at the check which this spirit has received. For a long time to come, in every instance in which it shall show itself, it will be put down by the magic word *Leverrier*.

Sir John Herschel, who declared at the British Association that the movement of the planet had been felt (on paper, mind) with a certainty hardly inferior to ocular demonstration, is precisely the person who thirteen years ago (*Cabinet Cyclopædia*, 'Astronomy,' p. 5) published what there can be no doubt was meant for a rebuke to this want of faith, and also to the confidence of those who made themselves judges of what they could not possibly understand.

The history of this discovery, and of the way in which it was received, is a notable illustration of national character.

At first, on hearing how nearly Mr. Adams had anticipated him, M. Leverrier felt some apprehension that his

[1] *Athenæum.* I have changed the editorial *we* in one or two places; he never allowed these articles to be altered by the editor.

1846. The planet Neptune.

glory might be taken by another. These apprehensions were soon quieted by the generous and graceful statements of the three persons most concerned—Mr. Adams, the Astronomer Royal, and the Cambridge Astronomer. But there appeared violent articles in some of the French newspapers, which, however, were disclaimed by MM. Arago, Leverrier, and others. There was, nevertheless, some amount of irritability displayed on the first announcement to the Institute, though the more forbearing majority concurred with M. Libri, who said: 'En attendant, il est essential de procéder avec la plus grande calme à l'examen de cette affaire. Plus on y mettra de reserve et d'urbanité, plus nous en avons l'assurance l'effet sera favorable aux Astronomes Français.'

The Academy itself suppressed any feeling of jealousy, and showed itself perfectly ready to discuss the question of the relative merits of the discoverers with fairness. But M. Arago refused to allow the Englishman's claim, saying that Mr. Adams 'was not entitled to the slightest allusion in the history of the discovery.' In reference to this access of national feeling, which was afterwards carried out by M. Arago's persistent effort to have the planet named *Leverrier*, Mr. De Morgan wrote:—

Arago.

Let M. Arago refrain. There will be one part of this matter the less subjected to his distorting mirror of national bias, in which the distortion is rendered less perceptible by brightness of style and clearness of illustration. We should be the last to deny the varied talents, deep knowledge of present science, admirable enthusiasm, and concentrated power of producing effect, which the distinguished Secretary of the Institute brings to his part. But as an historian of science, he may be held to be the Bailli of the day, his mania, however, being French and not Hindoo. And we may be satisfied that among the French themselves this Bailli will some day find his Delambre. His ideas are so confused by the state in which the fear of an English claim has put him, that he styles his own determination to call the new planet by no name but that of *Leverrier*, an undeniable proof of his own love of the sciences, and an adherence to a legitimate sentiment of nationality.

But on this feeling of nationality there were other voices raised. Mr. De Morgan himself says :— 1846. The planet Neptune.

All the elements of M. Leverrier's discovery were laid elsewhere, as well as in France; let it be enough for his satisfaction, as it is for his fame, that he worked out the problem first. We may wish that the complete honour of this great fact had fallen upon the English philosopher, but far beyond any such merely national feeling is our desire that philosophers should recognise no such distinction among themselves. The petty jealousies of earth are things too poor and mean to carry up amongst the stars. Light and unmeaning as they are, they would be found heavy incumbrances on a voyage so long as that to Uranus.

M. Biot, writing of the address in which Mr. Airy narrated the facts, says :— M. Biot.

Thus, in the first week of October, 1845, precisely eight months before M. Leverrier's first announcement, the new planet was predicted by Mr. Adams, and he alone was in the secret of its position. . . .

I do not speak here in accordance with the narrow sentiments of geographical egotism, improperly called patriotism. Minds devoted to the culture of science have, in my opinion, a common intellectual country embracing every kind of polar elevation.

Professor Strüve threw in his vote of equal justice :—

Far be it from me to have any intention of withholding our entire admiration from the eminent merit of M. Leverrier. But impartial history will in the future make honourable mention of the name of Mr. Adams, and recognise two individuals as having, independently of one another, discovered the planet beyond Uranus. In the same way it attributes the discovery of the Infinitesimal Calculus at once to Newton and to Leibnitz. . . . In Mr. Airy's report we see that, in September 1845, Mr. Adams arrived at a result, and in October he transmitted to Mr. Airy a paper containing the elements of the present planet, so nearly approximating that it might have been found in the heavens ten months before it really was. Professor Strüve.

Mr. De Morgan's belief that the failure—which is almost too strong a word—on the English side was due to

1846. the want of confidence in their science among Mathematicians, received confirmation from the report, by Mr. Challis, of observations at the Cambridge Observatory, founded on Mr. Adams's calculations. The statement was laid before the Senate in December, and in it was mentioned a memorandum made in 1841, and shown by Mr. Adams to Mr. Challis, recording the writer's intention to solve the problem as soon as he had taken his degree of B.A.

The unusual character of the question is adverted to in this first statement of Mr. Challis:—

Mr. Challis's observations.

> The usual character of perturbations is to find the disturbing action of one body on another by knowing the positions of both. In the case of Uranus, Mr. Adams's problem was the inverse one; from known disturbances of a planet in a known position, to find the place of the disturbing body at a given time. . . . It will appear by the above account that my success might have been more complete if I had trusted more implicitly to the indications of the theory. It must, however, be remembered that I was quite in a novel position; the history of Astronomy does not afford a parallel instance of observation undertaken entirely in reliance upon deductions from theoretical calculations, and those, too, of a kind before untried. . . . We may certainly assert to be fact for which there is documentary evidence, that the problem of determining from perturbations the place of the disturbing body was first solved here; that the planet was here first sought for; that places of it were here first recorded, and that approximate elements of its orbit were here first deduced from observation. And that all this, it may be said, is entirely due to the talents and labours of one individual among us, who has at once done honour to the University and maintained the scientific reputation of the country.

Both discoverers in due course received every possible distinction at home and abroad. M. Leverrier, besides other honours given to him, was elected an Associate of the Royal Astronomical Society, and, immediately after the discovery, proposed for the medal. And here a difficulty arose. It was usual to give only one gold medal at any

1847.

The planet Neptune.

The Astronomical Society's medal.

time, and in order to secure the certainty of merit in the candidate, this could not be adjudged by a smaller majority than three to one on the Council. On the present occasion the Council, which for the adjudication of the medal met always in January, was so divided in opinion on the question that the requisite majority was not obtained. It was felt that although M. Leverrier's claim was unquestionable, the acknowledgment of it in this form would be a manifest injustice to Mr. Adams, whose claim in one way was possibly greater, though it failed in the requisite element of success—that of its being publicly made known. On this arose a great difference of opinion among members as to the right steps to be taken. All were anxious that full justice should be done to both discoverers, and all were naturally desirous that the Astronomical Society should not be behindhand in its acknowledgment of the great gain to Science of the discovery, made, as it had been, by Mathematical calculation.

No decision was come to, though the discussion had been long and anxiously carried on in the Council, and the time for the award went by. But the great body of the members could not readily submit to leave things as they were without further explanation, and a special general meeting was called to consider the propriety of suspending the by-laws and of reconsidering the whole question.

There were some members of the Society who took no part in the usual work, but attended meetings on great occasions, when it might be expected that their names would give weight to their opinion. One of these was Mr. Babbage, who was known to have a strong predilection for French science, and as strong a feeling against that which had any connection with Cambridge. He also attached much importance to the distinction of a medal, and thus was led strongly to support the claim of M. Leverrier, to the exclusion of that of Mr. Adams. Not succeeding in his efforts to reverse the decision, or rather

1846. The planet Neptune.

no-decision of the Council by proposing a vote of regret upon its measures, he wrote to the *Times* giving his own views of the matter, and thus complicating the difficulties of the position. The most active members of the Council declared that they would not work on the possible condition that a vote of 'regret,' or, which was the same thing, 'censure,' should be passed upon their measures.

Mr. Sheepshanks's opinion.

Mr. Sheepshanks, on the occasion of the South and Troughton arbitration, had given offence to Mr. Babbage, and his strong desire that justice should be done to Mr. Adams only increased Mr. Babbage's displeasure at the non-award of the medal. Mr. Sheepshanks was happily more prudent in his expressions, and though amusingly sarcastic in his letters, temperate in his public behaviour. If he had not been so at this time, the concussion in the Society might have ended in a complete disruption. He was a fellow of Trinity, with a strong love for his own University, and a desire that the glory so fairly earned by one of its members should not be quite lost to Cambridge. He says, in a letter to my husband dated November 20, 1846:—

As to the medal, I will tell you my *mezzo-termine.* To give the medal in due course to Leverrier, and, if the Council think fit, after due deliberation, to grant, by means of a special meeting, a medal to Adams, who did undoubtedly discover the planet nine months before Leverrier, and it was by no fault of his that we did not catch it first. His communication to the Astronomer Royal and to the Plumian Professor on an astronomical subject is surely a publication so far as Adams is concerned—according, at least, to all rules hitherto recognised. He saves us, I think, all real difficulty by waiving his claim to the discovery; for if we were called upon to decide by Waring's rule we should be *compelled* to decide in his favour, at least, after verifying the postmarks of the letter quoted by Airy.

Now, as he has not raised this very thorny point, it seems to me that quiet and good-tempered and sensible people, who have not committed themselves to a positive opinion *before they had heard all the story*, may come to some conclusion satisfactory to all parties except the ultra-French or the anti-Cambridge.

On the part taken by his three friends, Mr. Adams, Mr. Airy, and Mr. Challis, Mr. Sheepshanks says: 'I am far better pleased with the perfect candour and simple gentlemanlike feeling of these men, than by anything I have heard of for a long time.' 1846.

The writer conceded his opinion on the medals when he found how utterly impossible it would be to bring all parties to unanimity. He soon after wrote, 'If we don't get rid of the medal, it will capsize us.' And Mr. Airy, who had expressed his feeling that if no medal was given on this occasion the Society could never give one hereafter, also yielded to the present necessities of the case. My husband, from having given close attention to the whole question from the beginning, and seen its great difficulties and complications, advised a course which was taken. After the meeting, while the matter was pending, he wrote:—

Mr. De Morgan's opinion

This question of medals is almost the only one that can come before the Council, into the discussion of which may enter that question of right and wrong on which an honest man never allows his opinions to be overruled by considerations of expediency. On the knowledge of this, a wise by-law was enacted, which requires a majority of three to one in favour of the award of a medal. The consequence is, that when opinion is much divided no decision can take place. It was an unwise thing to force back upon the consideration of those who had long and anxiously deliberated without coming to any conclusion, the discussion of a question involving so many disputed points. It would have been better if the meeting had taken the matter into its own hands, and called a special meeting, not to enlarge the powers of the Council, but to do the thing itself. The meeting, however, showed, on more points than one, a strong feeling that so large a body, and so mixed, was not a proper court for the hearing of such a case. It does not follow that the special meeting when called need of necessity adopt the conclusions of the general meeting which called it. No one Parliament, though it may send business to its successor, can dictate how that business shall be done. And if the Society will take a little advice very respectfully offered, they will allow the matter

1846. The planet Neptune.

to rest where it is, and not compromise the working utility, perhaps the very existence, of their very useful body by persisting in demanding a decision from those who have the best possible reasons to know that they cannot agree.

And afterwards :—

The Society has given to both M. Leverrier and Mr. Adams the full value of twenty medals, or rather, a prize of a higher order than any medal. All the bases of the discussions take for granted that both those gentlemen possess much more than the ordinary share of merit to which, under usual circumstances, medals are awarded. All the varieties of opinion are formed upon this nucleus, and could not have existed without it. In all but the mere gold which goes to the manufacture, the discoverers have had their medals over and over again.

But, as was natural, the Astronomical Society could not feel quite satisfied to do nothing, and though gold medals were not given in the year to the two discoverers, their merits were not long afterwards acknowledged by Testimonials from the Society to each gentleman, '*For his Researches in the Problem of Inverse Perturbations, leading to the Discovery of the Planet Neptune.*'[1]

I have said more of this discovery than may be thought to belong to my husband's work in the Society. But all that he said and wrote on the question was strongly characteristic, expressing his high estimation for all the intellectual and, I may add, moral qualities of the parties concerned, and his exceeding disregard of distinctions. But besides this I have heard that the way in which many of the impediments were surmounted was due to his counsel.

Other questions connected with the Astronomical Society had arisen during the year 1846 in which Mr. De Morgan was involved.

The old difficulty of organisation was strongly felt at

[1] Testimonials were given in the same year to the Astronomer Royal, Prof. Argelander, Mr. Bishop, Sir J. Herschel, Prof. Haussen, Mr. Hencke, Mr. Hind, Sir J. Lubbock, and Mr. Weisse.

this time, because the real working men had done their parts so long and so thoroughly in the face of many embarrassments caused by the more officious but less practical members, that their time of rest was overdue. 1846.

Offer of presidency to Mr. De Morgan.

My husband never slackened in his exertions. What others could not do he undertook; but when in this year the place of President was vacant, and there were good reasons outside of the working part of the Council and one or two turbulent spirits within which made it necessary that a useful President should be secured by those who could work with him, my husband was entreated, as he had been before, to take the chair. His reasons for refusal will be found in his letters to Sir J. Herschel and Mr. Sheepshanks. The latter, who had the organ of firmness fully developed, but who considered himself 'anything but obstinate,' replied to the programme and refusal thus:—

Dear De M.,—One evening after supper John Hind of Sydney addressed Whewell thus:—'I don't quite hear what you say, but I beg to differ entirely with you.' Now I have heard and do understand all that you say, and more that you *would* say, and *I* differ with *you*. But I have given up trying to convince people against their will, ''cos I never found no good come of it.' So, just beholding your countenance as your wife painted it (decidedly obstinate if wrong, which most people are who are resolute when right), I give up all hope that my first best mode of combination will answer. I see and have seen for a long time a little cloud or two rising. Poor S—— has risen and pelted, and is, I suppose, now exhausted. We have excellent men, who don't understand or make allowances for others (I except self and you), and we have no sufficient bond of union.

If you and others can prevail on Herschel, well and good. You certainly can do a good deal as Vice, though by no means so much as if you were President. I shall look twice before I consent to continue Secretary, not merely for the trouble which this occasions to a man disliking all work (except such as he takes a whim to do), living forty miles from town, but really because a more methodical and resident person is actually required just to keep things in order.

1846. A few days after the writer added some exhortation and remonstrance to his former missive:—

'Omne ignotum'—how can you fancy that a practical or *gazing* Astronomer is wanted for our Pres.? Airy and Main and Challis and Johnson are the only members within reach who are strictly practical Astronomers, though several have, *each in our way*, practical knowledge. But I don't remember any instance where this knowledge was required in a Pres., nor can I well conceive a case wherein it would be required.

I say, as a nailer, how dare you report the proceedings of the R. A. S., not having sufficient knowledge, as you say, for President? Answer me that, good man, 'and thou shalt be to me a very stout Apollo.'

I may just ask here, with reference to my old friend's allusion to my husband's *Athenæum* reports—did Mr. Sheepshanks credit all the reporters of the learned Society's proceedings with knowledge which would qualify them for the President's chair? But Mr. S. at length gave in.

If Sir J. H. will take the Presidency, it being understood not merely that he is not required, but really not wanted, except on anniversaries and when he can make it convenient, it is the best move we can make. Pearson would do very well if we could be sure he would never come at all. Lord Wrottesley is a capital fellow, but, considering *some* of your arguments, is scarcely the person *you* should choose. Airy will take it if *necessary*, but I don't like to propose him for several reasons, one of which is that I dislike tasking his time and health so severely.

Sir John Herschel terminated the embarrassment by consenting to the wishes of his friends. Mr. De Morgan's views of the subject will be found in his letters.

SECTION VI.

Correspondence from 1836 to 1846.

To his Mother.[1]

My dear Mother,—I have read your letter carefully; and the papers, and as much of the book as was necessary to show that it contained no argument, and was in fact addressed to those who already believe all it contains. If I can make you see clearly that our modes of arriving at what we believe to be true are so totally different that an attempt to discuss the subject together would be an impossibility, it is all I expect. If your knowledge of the New Testament had been of your own getting, unwarped by the devices of a Church of which it has always been the avowed doctrine to use every means which the age will allow to force men to agree to its own interpretations, I could go much farther, and could show you that taking every book of the New Testament to be an authority to that extent only in which it was recognised as an authority in the first three centuries, and taking the words in their most probable meaning, there is no ground of fear for any honest man who uses the best means in his power to come at truth. 1836.

But between us there is in this matter no common ground on which to argue. Nothing is more easy than to be positive and certain, or to affirm the perdition of all who cannot see any particular system of doctrines to be true: but before you declare that you must be right and I must be wrong, consider the following points, and ask yourself what part of the whole New Testament has more right to a literal interpretation than this: 'In the measure which you measure with, you shall be measured.' I take the most literal translation, and not what your misleaders are pleased to call their 'authorised version.'

1. You have a number of books bound up into one, which you call the New Testament. You never meddled with the

[1] See p. 86, *ante*.

1836. question whether *all* these books are genuine, and were really written by the persons whose names they bear. Still less with the question whether being written by different persons, at different times, to different persons, &c., they can be used in interpreting each other in the same manner as the different parts of a book written by the same person. *I* have been *obliged* to consider this.

2. These books are written in a foreign language, and more than that, in a dead language, of which every one knows that it is utterly impossible to render any phrase *exactly* into corresponding English. This question no way concerns *you*. You dwell upon a single English word in the translation just as much as if it were the original itself. To me your version is useless, as I know that those who made it were utterly incompetent to take that view of the original language which the subject requires.

3. These books have come down to us in manuscripts which differ from each other repeatedly, and in one or two instances, if not in more, there is proof, which theologians of all parties unite in admitting, that additions have been made to the writers' text. You care nothing for this; I doubt if you knew the fact. *I* have been *obliged* to know it.

4. Your expressions amount to the following:—If you do not take it for granted that King James's translators chose the right Greek, and turned it into the right English, and more than that, drew all their inferences correctly, God Almighty will punish you to all eternity.

5. Out of all that precedes you have got a complicated creed, on implicit belief in which you insist. I recommend you to follow the plan adopted by Locke, when he wanted to ascertain what the Christian religion was. He looked carefully through the Acts of the Apostles, and collected every single instance in which a Christian was made by the Apostles; for, he argued—and in so doing he upset every church which has existed since A.D. 300—if I can become as much of a Christian as the first converts of the Apostles, I shall certainly obtain the essentials of Christianity. Do this yourself. Construct a creed out of all which the Apostles required, without adding a single word, and compare it with your own. For what else was so precise an account given of so many admissions into the Church? And this not with a view to changing your own opinions, for if your creed gives you comfort I would not change a letter of it; but with a view to the follow-

ing question: Do you believe that the God of truth has so misled the world as to give it a religion the essential parts of which cannot be gathered from the manner in which those who first taught it admitted proselytes? Certainly a newly baptised Christian, if sincere, did not perish everlastingly if he died the moment after his baptism. But your Church positively declares he did unless he believed what they call the 'Catholic faith.' Now ask of yourself in sincerity, where is it set down in such a manner that a wayfaring man, though a fool, could not err, that the Apostles taught this creed, or anything like it? 1836.

Again, take another test. Certain of the Apostles wrote accounts of the life and doctrines of Jesus Christ. Matthew wrote in Hebrew, no doubt for the Jews; Luke in Greek, for the Gentiles. These books were never collected into one till centuries after Christ, nor is there any proof that the earlier Jews ever saw the Greek Gospel, or the Greeks the Jewish one. It is most obvious that each of these accounts must contain the essential parts of the Christian religion. It is also most obvious that an epistle of Paul to a town in Greece must not be joined with one written to Romans, both of which were never seen for many years in Judea (so far as can be shown), to make up a doctrine essential for the salvation of Jews. Now try again. Make up your creed out of *any one* of the Gospels, if you can. Surely two fairer tests cannot be proposed to any person who knows what reason is; and still more when it is merely a question whether one person ought to believe that another must suffer eternal punishment because he will not treat as one book a number of different books in a manner which would be laughed at if applied to Livy and Tacitus. And yet they both wrote Roman history at periods as near to each other as those at which the books of the New Testament were written.

All this has no reference to the question whether the creed could be got out of the whole New Testament together if permitted. Before God I declare that I have examined closely the history of the early Church, together with abundance of controversy on both sides, not forgetting the books of the New Testament on which they are written, and can find nothing like the creed of the Churches of Rome or England. The former does not pretend to find what you call the essential doctrines of Christianity in the New Testament, but appeals to tradition. It is easy to rail at them, but to the best of my knowledge and belief, derived from historical reading and actual observation, the

1836. Church of Rome contains as much honesty as that of England, and a vast deal more knowledge. It would not take one quarter as much evidence to make me a Catholic as to make me a Church of England man.

I should have no objection to be better acquainted with Mr. Baptist Noel, but you are grievously mistaken if you think any discussion with him could have any effect on me.

1. Because he has never studied either side of any controversy, or at least of those on which such a discussion would turn, as he himself avowed to me, though he must have been at that very time meditating a controversial work, or one meant for such, which he shortly afterwards published, and which took the point in dispute for granted on his own side in the title-page.

2. Because he would be better employed in meditating how to reply to the complete and conclusive reply which he received from a Unitarian minister, whom I blame very much for replying to so weak an attack. Your letter, my dear mother, was quite as good logic as the Rev. B. Noel's book, and indeed would make a pretty abstract of it. And yet you could see that you did not profess to be able to argue the question; but the Rev. B. Noel was *not* able to see so much. He is, nevertheless, a liberal and amiable man, but he mistook his ground altogether when he thought he was a fit match for the head of a body so learned (compared with Church of England clergy) in the history of his own creed as the Unitarian ministers. If you want an opponent for me, take some one, *if you can find him*, who has studied both sides of the question. But even then I should object to discuss with him—

1. Because I never saw or heard of any one who was made to change his opinions by discussion. 2. Because such subjects are best discussed between a man and himself in retirement, and with the *real original accounts* before him. 3. Because I see in all that is orthodox a lack of that charity which Paul considers as more essential than everything else, coupled with what virtually amounts to a claim of infallibility. 4. Because numberless unanswered arguments lie before me, which the Established clergy have left off attempting to answer. Instead of attempting to drive me, an individual with little time on his hands, to go through the oft-repeated job of cutting the flimsy web of an Athanasian Christian, move your own clergy to print their assertions, and leave those who have leisure for answering to deal with what they shall advance. I shall then be able with

little trouble to verify both sides; and as what is called Christianity by the Church of England has never failed to meet with its answer, even when it was declared felony to answer it (for such were the arguments used at one time), I have little doubt any person who shows a respectable knowledge of the history of the Church will meet with a speedy reply if he will venture into the field. 1836.

Now with regard to these matters, you may surely, my dearest mother, collect from what I have said, that there will be little wisdom in attempting to revive this subject. Appeals to my feelings, whether from the bitter stroke we have lately had, or from your distress that I cannot believe as you believe, are trying to wound the wrong parts. It is impossible you can be in earnest when you think that I who have been for the full half of my animal, and the whole of intellectual life, accustomed to consider doubtful things by help of reason alone, should be moved to any opinion because any person alive believes it so strongly that he or she is grieved that I do not believe it also. Such appeals might be made to induce a person to examine that opinion, but I have examined it, and I conscientiously believe more than most of the clergy on whom you pin your faith. Still more weak are your implied assertions that my late illness, &c., are chastisements from God. How could you know this if it were true? or how can a very slight consequence of that want of tendency to inflammation which it has pleased God should preserve me from colds, fevers, cholera, &c., and which forms a constituent part of my power to sit at work many hours without headache or pain, be considered as a chastisement? Believe me, I see nothing in it but a very slight and easy composition for the want of liability to many worse things. In the name of common sense let the Almighty manage His own world. The presumption with which modern Christians explain all that happens, and point out the intention with which it all came, is one of the strong marks by which the perversion of the system may be known. If the tower of Siloam were to fall precisely upon one hundred people, all Calvinists, it would never enter my head to suppose that they were thereby declared to be objects of God's particular displeasure. As long as my reason lasts I shall never want a better argument than that.

I have looked over dear Eliza's papers with the interest with which everything that concerns her affects me at this moment. They relate to points which have now been in discussion fifteen

1836. hundred years, and on which she had to form her own opinion with such means as she had; as have you and myself. But such matters are not with me matters of feeling, they are to be tried by reason and evidence—that is *by me*, for I do not object to any one who thinks he can find truth by another method trying what he can do. You are to judge yourself, I myself, and it is impossible that any two people can usefully discuss any subject, one of whom believes that a conviction of the truth of opinions is an argument in their favour. Yours most affectionately,

A. D. M.

To Dean Peacock.

69 Gower Street, Nov. 13, 1838.

DEAR SIR,—I would not trouble you, generally speaking, to notice an undergraduate just arrived, but I think my young friend M., the bearer of this, has merit enough to be an exception. He has gone through a good course of reading, and is something like a low wrangler in his present attainments, or rather higher, with very good chance of being made into a high one. He stood a stiff examination for our Flaherty scholarship, as we call it, at Univ. Coll., and though second came off with great honour.

I remember with great gratitude and pleasure the notice I used to receive myself from those in high station, when I was an undergraduate, having no other claims than of the same description as those which M. now has. I should have left him to find his way in his own College, not doubting that his acquirements and industry would soon make him marked; but I find he is almost entirely without acquaintance. If, then, you will introduce him to one or two of your *good men*, as we used to call them, his future competitors, you will do a service where it is well deserved, and I shall feel very much obliged. I should like him early to know some reading men out of his own College, from whom he may learn that its system is not necessarily that of the University.

I do not doubt you found a letter from me containing my proposed test of convergency and divergency.

You are, of course, interested in all that concerns Dr. Young. A publisher in London has bought or will buy the plates of the *Lectures*. He proposes to republish them (catalogue excepted) in parts, transferring the copper to lithograph. My colleague Sylvester is to put notes, which with reading he will do very well. I am now proposing it to the Soc. for the Diff. of Useful

Knowledge. Do you think favourably of the plan? My own opinion is that Young's lectures are an unsearched mine of good things, often very happily expressed. But two quarto volumes frighten most people. It is of course a work which any one would like to see republished, but should your opinion be as strong as mine the expression of it would materially forward the object. I remain, dear sir, 1838.

Yours faithfully,
A. De Morgan.

69 Gower Street, Nov. 13, 1838.

To Francis Baily.

My dear Sir,—The word *rapa* or *rapum* means a *turnip*, or any small white root. There is in Italian *rapa*, a turnip, and in English the same word is seen in *rape*-seed.

Raparium and rapina (both words are used) mean a collection of many rapæ, a turnip-store, or turnip-field. Consequently your *rapina* must mean a cluster of stars.

What a capital new word for the starry heavens—a turnip-field!

Yours truly,
A. De Morgan.

69 Gower Street, Monday Morning.

To Sir John Herschel.

69 Gower Street, Nov. 22, 1842.

My dear Sir John,—Thanks for your letter and the compliments—*non omnis moriar*—with that big book. I shall be the tenant of some old book-stall a hundred years hence, and some one, perhaps myself, Lethed and transmigrated, will give half a crown for me—as I have done for others, or perhaps for myself again in some pre-transmigration—and will put me down in a bibliographical list, with a slight mistake of the name (non *omnis* moriar), and of a hundred years or thereabouts in the date.

As to Lardner's 'Cyclopædia,' you must address direct to Longman, who is the real editor now. I think a second edition stuck full of southern plums, with some nice nebulæ, neat as imported, would be a capital dish. When shall we begin to see the southern heavens through your spectacles?

1842. I forgot my hieroglyphic of Uranus; I told you, I think, that the baronetage draws Mars instead of Uranus in your shield.

I pointed out the mistake of latitude to Charles Knight's people to-day. They told me they were *partly aware* of it already. What is meant by being *partly* aware of the pole-star not being exactly on the pole?

'In order to be sure that a general proposition is true we must be sure that all its particular applications are true.' If 'in order' means *previously* I deny the assertion. Certainly, to be sure that a general proposition is true we must be sure that all its particular cases are true, for a general proposition$=\Sigma$ (particular case).

I suppose it is meant that there is no surety without examination of all the particular cases; this I admit too, but I suppose it is farther meant that there is no surety without enumeration of the particular cases: if so, this I deny.

For example, the angles at the base of an isosceles triangle are equal. In the proof I have moral certainty that I examine *every* isosceles triangle; *any one whatsoever* is *every one.* Abstraction and rejection of all that distinguishes one particular case from another is the mode of induction, if people like to call it so, which general reasoning employs.

'Consequently it is reasoning in a circle to include the truth of the particular from that of the general.' I deny the consequence. If any one likes to say, on the preceding view, that general reasoning is induction of particular cases, which I do not in a certain sense deny, he must not say that I conclude or infer, but only that I go to a drawer in which I have laid up all the cases ready for use, for this must be the correlative meaning of inference. But shall the very man who found out that I had got my drawer full be the one to deny me liberty to take out the contents, because the etymology of the words I use to describe such out-taking rather seems to describe making the goods, as wanted, by a machine, than taking them out ready made from a receptacle? No, if he goes and alters my words of first process, he must allow me either to resist or to alter the words of second process to match.

Again, how dares any one say, 'To be sure of the general, we must be sure of all the particulars'? Epimenides said *all* the Cretans are *always liars.* Now Epimenides was a Cretan himself; how could he establish his proposition? If it were true, it was therefore false.

'Has he examined all the particular cases of his proposition?' He either knows this *a priori* or he has examined (that is, if he knows it to be true); if he knows it *a priori*, then he opens the net, and out we jump; but I will answer for it he has not examined all the possible cases. 1842.

Is there not a confused way of talking about *truth*? When we *prove a truth*, that is, give ourselves certain knowledge of its being a truth, we talk as if we had *made a truth.* We say 'one proposition *is* the consequence of another;' when it should be 'our knowledge of one proposition is the consequence of our knowledge of another.'

He who said *peace among men* forbade Metaphysics. When their cloudinesses the axioms of mental philosophy declare war against 'two straight lines can't enclose a space' they remind me of the Chinese trying to take an English battery.

Here are some *undoubted* truths :—

$$\sin\infty = 0, \qquad \cos\infty = 0;$$
$$\tan\infty = \mp\sqrt{-1}, \qquad \cot\infty = \mp\sqrt{-1}.$$

As to sec ∞ and cosec ∞ I am doubtful. They are either 0 or ∞. I suspect the former.

Yours very truly,
A. DE MORGAN.

To Sir John Herschel.

69 Gower Street, Dec. 30, 1842.

MY DEAR SIR JOHN,—Many thanks for the reduction of Schiller's observations to the latitude of London. I dare say you have applied the correction—

$$+\,(\text{English} - \text{German})$$

very skilfully, but I am so ignorant of that language that I shall not find you out if you were to err in the first place of gutturals.

When your missive arrived, I was engaged with a young Turk whom I indoctrinate in differential equations and matters arising thereout. I gave him your wafer as a sort of autograph, whereupon the following dialogue took place :—

He. Oh, Sir John Herschel! what is it he has done with the

1842. Moon at the Cape of Good Hope? I have got his book, and I have read it with the greatest delight.

I. It's a hoax.

He. A hoax! Well, I was so delighted to think it was all true. I told our Ambassador of it yesterday, all that had been seen in the Moon, and his Excellency only laughed at me and said it must be a hoax.

Depend upon it there are thousands in the condition of this young man.

Yours very truly,
A. De Morgan.

To Sir John Herschel.

Camden Street, Oct. 7, 1844.

1844. My dear Sir John,—A certain man, named Malby, who makes and edits globes, has procured the copyright and plates of certain 36-inch globes, on which he means to lay down nebulæ, to wit,—

Messier,
W. Herschel (1786, 1789, 1802),
Dunlop (1828),
J. Herschel (1833);

also double stars, &c., &c. Now it strikes me that if by waiting any reasonable fraction of a revolution of the equinoxes he would have your southern patches to dig into his copper, it would be wisdom in him to wait. Can you tell me within two or three revolutions of the Moon's node when the world is likely to have your work?

To tough jobs long periods ought to be applied, both for safety and solemnity. But if you are able to reply that you will be ready in a jiffy, or a crack, or less than no time, or a brace of shakes, or the twinkling of a bedpost, or before you can say Jack Robinson, or even the sum of them, of course he must wait.

Yours very truly,
A. De Morgan.

To Sir John Herschel.

7 Camden Street, October 1844.

My dear Sir John,—I take much interest in Malby's globes.

1. Because I am writing a 'Use of the Globes' for them, of which I send you an old proof to destroy at your leisure.

2. Because Baily's *last printed work* is the set of revised boundary lines for constellations which appears on them. 1844.

3. Because Malby is a very spirited fellow. A man proposed a feasible way of mounting the globe so as to adjust it for any era, giving in fact the pole of the equator a motion round the pole of the ecliptic. Malby immediately set to work, got the globe made with an additional contrivance for solar apogee, and I have a specimen now in my hands. They are soon to be advertised. If I were your mortal enemy I should like nothing better than to dispute with you about heliacal risings, &c., of 3,000 years ago. How I should grin to think that you were at your spherical trigonometry while I was getting within more accuracy than an heliacal rising is good for, by two or three motions of the hand, and a squint or two along wooden horizons and brazen meridians!

Malby has also a very neat planisphere with *one* revolving surface and one fixed, and on bringing the hour of the day on the edge of the revolving surface to the day of the month on the edge of the fixed, the hollow part of the revolving surface shows the visible heavens for that hour and day. These planispheres usually have three surfaces and two adjustments.

But whether I shall in a month or two claim your kind offer about Dunlop's nebulæ depends upon Malby's decision about the whole scheme. I shall recommend him to defer it altogether till your work appears. What can Dunlop have been at? Your optical power must have been incomparably greater than his.

Are there any atmospheric minutiæ which last long enough for a careless observer, who never looks for a thing a second time, to note as nebulæ?

Banquo. The air hath bubbles as the water hath,
And these are of them—whither are they vanished?
Macbeth. Into the air, and what seemed *nebulæ* melted
As breath into the wind; would they had stayed!

No more at present from

Yours very truly,
A. De Morgan.

1845.

To Sir John Herschel.

7 Camden Street, May 19, 1845.

My dear Sir John,—As I gave you rather a bad impression about the Mathematical Society from their old clay and pewter days, I feel bound to tell you that Captain Smyth, Galloway, and myself went down and examined their library and themselves, and saw a good many, and got the names and occupations of all. We found an F.R.S., an F. Ant. S., an F. Linn. S., a barrister, two silk manufacturers, a surgeon, a distiller, &c.; and we found that all had really paid attention to some branch of Science. One Mr. Perrott is, I am told, a man of note as a chemist. We certainly shall not *lower* the average knowledge of our Fellows by accepting their proposition. I went down rather against the scheme, but was perfectly changed by what I saw and heard.

Their library is a good one. The matter will soon be discussed at a Council.

Yours very truly,
A. De Morgan.

To Sir John Herschel.

7 Camden Street, May 28, 1845.

My dear Sir John,—No difficulty at all. Our by-laws permit a general meeting, called by the Council, with a week's notice, to destroy every by-law, and make a complete new set. The Charter is liberty itself in this particular.

A general meeting will be called after the next ordinary meeting. Of course, we could not depart a hair's-breadth from the statutable mode of election without.

As to the Cambridge Transactions, I have not got them, and know little of them as a whole. In addition to what you name there are Murphy's papers, which are remarkable, particularly those on definite integrals; but were I you, I should consult Hopkins, the secretary, on the details.

The transactions generally may be described as having had a tendency to bring forward discussions of principle among the members of the University. There is, you may safely say, sufficient proof in them that the ordinary system of University reading, which crams details of methods, put together in examination form, with fearful rapidity upon the young student, does not destroy the power of reflecting upon the basis of mathematical knowledge, or physical. Strongly objecting, as I do, to that system in many points, I should admit the Cambridge Transactions as a decisive fact against me if there were more of it, so that the contents could be cited as a proof of the general consequences of the system. 1845.

You should not forget the Cambridge 'Mathematical Journal.' It is done by the younger men. Four octavo vols. are published. It is full of very original communications. It is, as is natural in the doings of young mathematicians, very full of symbols. The late Dr. F. Gregory, whom you must notice most honourably—I send you a mem. of him, which please to return—gave his extensions of the Calculus of Operations what used to be the separation of the symbols of operation and (quantity) in it. He was the first editor. He was the most rising man among the juniors.

Yours very truly,
A. De Morgan.

To the Master of Trinity.

7 Camden Street, June 9, 1845.

My dear Sir,—I am much obliged to you for the misprint and the supposed misprint.[1]

'Nineteenth century' is a bad misprint; and I ought to have detected it by the absence of the words 'march of intellect' in the immediate neighbourhood, for how can the first phrase come in without the second?

As to the second, I may say with Fouchy, 'C'est pire qu'une crime, c'est une faute.' For it *was* Rheticus who published it, the work being Copernicus's. But I have phrased it as if Rheticus wrote a work of his own with the title cited, whereas I meant to say that he had published one of Copernicus's, who was like Newton, and wanted a kind of half-godfather, half-midwife, for all he published.

There is another misprint which vexed me more, because context will not help. It is the old accusative decenni*o*m for decenni*u*m in the title of the canon. This was the printer's doing, after revise returned, I fully believe, for I know that I read the title most carefully.

[1] I have not found these misprints. I suppose them to be in one of his Cambridge tracts.—Ed.

1845. I have not anything to retaliate upon the Hist. Math. Sc. at present, but if you are preparing a new edition I will look at it in the vacation *quoad* Mathematics and Mathematical Physics.

You duly acknowledged the receipt of my corrections about Milbourne and Horrocks. I remain, dear sir,

Yours very truly,
A. DE MORGAN.

To Sir John Herschel.

7 Camden Street, July 21, 1845.

MY DEAR SIR JOHN,—I hope you have found rest agreeable after your Cambridge labours.

I have had a book by me for years called 'An Original Theory or New Hypothesis of the Universe,' by Thomas Wright of Durham, London, 1750, 4to, 30 mezzotinto engravings. I had always supposed it to be Ocular and Elizabethan,[1] if you know what that means, and so put it among my curiosities of that kind. But overhauling my limbo to write an article about quiddities I began to examine this book, and I find it is at great length the true theory of the milky way as a resolvable nebula, with distribution of the universe into patches of starlight.

It was a book published by subscription, and therefore not much *répandu*. It was seeing Dr. Smith's (Harmonic Smith's) name among the subscribers which first made me suspect it was not a heaven-born genius who wrote it.

Yours very truly,
A. DE MORGAN.

To Captain Smyth.

7 Camden Street, Nov. 3, 1845.

MY DEAR CAPTAIN SMYTH,—I find there is a circumstance about 'Poliphilus' which Brunet does not mention. The author's name is Francis Colonna; if the thirty-eight initial letters of the thirty-eight chapters be written down, they make *Polium frater Franciscus Columna peramavit.*

Yours truly,
A. DE MORGAN.

[1] Those accustomed to the writer's expressions will see in this a paraphrase of the slang, 'My eye and Betty Martin.'

1845.

To Sir John Herschel.

7 Camden Street, Nov. 29, 1845.

MY DEAR SIR JOHN,—First as to things in general, that you may have a piece to tear off about life insurance, if you like, and send to the managers. I have undertaken the Annual Report again. Have you any subjects to suggest? Have you anything to say about your own printing? Have you actually put metal to paper in anything more than a steel pen? Do you know anything more than mankind in general about Count Cassini, who is gone, leaving you the sole hereditary Astronomer? Will the fourth Herschel start as he did, and go and be a botanist? Ask H′ what he thinks H″ will do.

I hope you are using proper means for your cold, and getting rid of it accordingly; when it comes to loss of sleep and appetite it must be dealt with sec. art. It is *not* true that a man of forty is either a fool or a physician; he may be both, or neither. But according to that rule, I am ·9854 of a physician [1] myself, and in that capacity I beg you to take care of yourself. . . .

If you want a laugh read Sheepshanks's pamphlet, if you have not read it already. A man who acknowledges his own name to be an ugly one must be a hero of moral courage. If he had lived in the Middle Ages he would have been *vir clarissimus eruditissimusque Ricardus de Ovium Cruribus*, which would not have sounded common at all. There was one Middle Age name which I could not make out. I searched and searched, you can't think how much. It was *Jacobus Humus, Scotus Theagrius*, James Hume, a Scot—of what? I tried every part of Scotland, and endeavoured to Latinise it into Theagrius. At last I happened to mention it to a Scotchman (they all know all their lairdships), and he said, 'Oh, of course, *Hume* of *Godscroft*, petty estate.

[1] I omit the correction for folly.

To Captain Smyth.

Nov. 26, 1846. 1846. Letter to Capt. Smyth.

MY DEAR CAPTAIN SMYTH,—Sheepshanks has written to me on the same subject, and I have given him at length reasons why I will not be President. I will vote for and tolerate no President but a practical astronomer. Besides which, the chair would bring

1846. me more in contact with certain parts of our glorious constitution than I could any way stand.

But I have offered either to do the work of a Vice-President, if Herschel or some other good man will be President, or to take every duty of Secretaryship except the editing of the *Monthly Notices* and *Memoirs*. I sent Sheepshanks the following programme, which I amend for your use :—

President: Herschel, absent. Quiet enough when present.

Acting Vice-President: Capt. W. H. Smyth—too fond of cigars and black-puddings, but otherwise a capital quarter-master.

Secretaries: Rutherford, for the *Memoirs*. A. De Morgan—well enough in his way, but cranky—for all miscellaneous work, from wax candles to Council minutes.

Foreign Secretary: Rev. R. Sheepshanks (composition not wax by any means. A President overdue, but won't dub up—ought to go through the court) for the *Monthly Notices*.

If Galloway would take Rutherford's place, who wants to get off, we should be in capital force. And we look forward to Hind being a famous man.

Il n'y a que moi qui a toujours raison.

Think of the above, and see if it won't work. I will work, but in my proper place. The President must be a man of brass —a micrometer-monger, a telescope-twiddler, a star-stringer, a planet-poker, and a nebula-nabber. If we give bail that we won't let him do anything if he would, we shall be able to have him,[1] I hope. We must all give what is most wanted, and his name is even more wanted than his services. We can do without his services, not without loss, but without difficulty. I see we shall not, without great difficulty, dispense with his name. Kind regards all round.

Yours truly,
A. De Morgan.

To Sir John Herschel.

My dear Sir John,—We have just been making our arrangements for the Society for the ensuing year, and one thing is that you are not to be asked to do anything, or wished to do anything, or wanted to do anything. But we want your *name*,

[1] Herschel (understood).—S.E.De M.

and when you hear the precise state of things I hope you will see that you can give it, and can see that you will give it. 1846.

Baily, as you know, was the fly-wheel of the Society; there was an inexhaustible fund of momentum in him, which he imparted as wanted, and to the amount wanted. Since his death there is nobody of whom we can say that he is always good for $D-V$; D being the duties of any office, and V the amount of them which the holder happens to discharge. And we find it necessary to put our shoulders to the wheel, and to bestir ourselves.

Certain persons wanted to make me President, which I positively refused, holding that to appoint a person who has never promoted astronomy otherwise than as promoting mathematics is indirectly doing so (there wants a word more strong than indirectly, and less strong than directly—say $\frac{1}{2}$ directly and indirectly)—would be $\frac{1}{1001}$ (direct and indirect) confession of weakness. But I have proposed a plan of administration which I am sure of as to the working part, and which I hope you will sanction as to the proemium. Seriously, it is a *dignus vindice nodus*.

President: Sir J. Herschel. No duties. If he likes to attend anniversary meetings all the better, but nobody expects even this, except quite convenient. Work all parcelled out. None left for him. *Name* the thing wanted.

Acting[1] Vice-President: Captain Smyth. Will do it with pleasure; ascertained.

Secretaries: Galloway takes the printing of the *Memoirs*. Consents if the plan holds. Has been Sec. some years.

De Morgan takes the routine management of the Society, all but *Memoirs* and *Monthly Notices*. Has had eight years of routine, *Memoirs*, and *Notices*, all three, and would not meddle with it again except for the conviction that the Society would go to pieces forthwith if something energetic were not done at once. On condition that the President is a man of practical astronomy.

Foreign Secretary: Sheepshanks. Takes the *Monthly Notices*. Has had the Secship. for the last year.

Assistant Secretary: Williams. Seems a working man, but new.

[1] Includes making up a party for the cigar divan after every dinner.

1846. On this case I hope you will consent to your part of our Cabinet. The rest of the Council we fill up in the best way we can—no great difficulty.

If you feel able to consent, I am tolerably sure that we are safe for two years, during which time we must keep our eye upon the future.

Trusting to the *nascitur non fit* principle will give the Society a *fit* of *moritur*. I remain,

Yours very truly,

A. De Morgan.

7 Camden Street, Nov. 29, 1846.

SECTION VII.

From 1846 to 1855.

1846. Theory of the Syllogism.

'In the year 1846,' Mr. De Morgan wrote, 'I had begun to collect various matters which had suggested themselves at different times, connected with the theory of the Syllogism in Logic.' In the year 1847 the *Formal Logic* was published.

The memoirs *On the Syllogism*, Nos. I., II., III., IV., and V., are Mathematical workings of the principles developed in the *Formal Logic*; and the tracts *On the Structure of the Syllogism*, and *On the Application of the Theory of Probabilities to Questions of Argument and Authority*, immediately preceded it.

The first chapter of *Formal Logic* consists, with a few alterations, of the tract entitled *First Notions of Logic preparatory to the Study of Geometry; London*, 1839. The work as a whole, and in its higher parts, is original, but the author has been careful to distinguish between what he claimed as exclusively his own and the work of others by printing in italics, in the Table of Contents, the headings of those articles which refer to his peculiar system. A reference to this table will show how large and essential a portion was claimed as entirely new. After working these points out in his own mind, the author found that he was able to explain by their means passages of Aristotle till then obscure to himself as well as to others. The two principal features of his own system were the introduction of contraries or contradictories, and the idea of definite quantity, into the syllogism. Where *all*, *none*, or *some* had been the utmost quantification employed before,

1847. Formal Logic. the ideas of *more* or *less* with their contradictories formed an essential part of the *Formal Logic*.

The tract *On the Syllogism*, &c., read at Cambridge in November 1846, excited great interest in the minds of those Logicians who were following out the application of Mathematics to Logic. Dr. Whewell had wished the writer to read it himself, and kindly begged him to visit him at Trinity Lodge for the time. This he was not able to do. I wish it had been practicable for him to mix more freely and frequently with the friends who shared his interests; but on looking back to the great variety of subjects he treated, and the work he was engaged in at the time, it is evident that this would have been impossible. The memoir *On the Dispute between Keill and Leibnitz* was printed in this year. I hoped, on first considering in what way to mention the Logical works, to be able to supply some of my own deficiencies by inserting the letters which were addressed to him both on the reception of the tract and the publication of the volume. But they are too numerous to form a part of this work. I am, however, very thankful to be able to insert here some excellent strictures on that which I dare not myself have attempted to describe—the relation of Psychology and Mathematics in my husband's mind. These remarks were kindly given to me by Mr. De Morgan's friend and former pupil, Mr. Cecil Monro:—

Such attention to Logical method is not to be confounded with mere accuracy and explicitness of statement and demonstration. These are vital qualities, indeed, of Mathematical exposition, qualities which every one sees to be characteristic of Mr. De Morgan's work; which are, in fact, characteristic of it in an extraordinary degree; but it is in a more important sense than this that he was at least as great in Logic as in Mathematics. Even in his most strictly Mathematical writings the examination of mental processes is visibly an end in itself, as distinguished from the exhibition of mental products. In its psychological aspect his end is pursued through historical and even bibliographical inquiries, which, independently of their value as informa-

1847.

Letter of Mr. Cecil Monro.

tion, are perhaps still more interesting as episodes in that pursuit. Upon its Logical side the end may be said to have been attained, and with a completeness upon which I believe high authorities would not as yet be forward to pronounce, in a series of works of which the well-known *Formal Logic* is but a part. If, as is certainly true, his other works could only have been written by a Logician, these last could only have been written by a Mathematician. Everybody knows that Mathematics are an admirable model of exactitude; but not everybody knows that they also furnish an admirable type of generalisation. Indeed, generality is often confounded with vagueness, and therefore treated as incompatible with exactitude.

The fact is known that, having very thoroughly worked at the generalisations of Mathematics in theory and practice, Mr. De Morgan was enabled to establish with perfect precision the most highly generalised conception of Logic, perhaps, which it is possible to entertain. It is no new doctrine that Logic deals with the necessary *laws of action* of thought, and that Mathematics apply these laws to necessary *matter* of thought; but by showing that these laws can and must be applied with equal precision and equal necessity to all kinds of relations, and not only to those which the Aristotelian theory takes account of, he so enlarged the scope and intensified the power of Logic as an instrument, that we may hope for coming generations, as he must have hoped,[1] another instalment of the kind of benefit which history shows we ourselves owe to the Aristotelian theory, not merely in the analysis of one mental operation, but in the every-day practice of them. Mathematics are, meanwhile, and perhaps will always remain, the completest and most accurate example of the generalised Logic. At any rate, in the mind of the author, Logic and Mathematics as 'the two great branches of exact science, the study of the necessary laws of thought, the study of the necessary *matter* of thought,[2] were always viewed in connection and antithesis.

C. J. MONRO.

Correspondence on logic.

Mr. De Morgan had written to Dr. Logan in September 1846 that he was 'making a vigorous onslaught on the Aristotelian syllogism, which,' he says, 'I find has

[1] Syllabus, § 96, note.

[2] *On the Syllogism*, No. v. &c. (Camb. Ph. S. vol. x., Part 11, 1863), the last page.

1847. Hamilton controversy.

wanted the interference of a Mathematician very much.' He asked Dr. Logan to tell him of any place in which the history of the syllogism is given—anything about *Barbara, Celarent*, &c., their birth, parentage, and education. He also gave him some of the results of his recent inquiry into the application of the Theory of Probabilities to questions of argument and authority. In reply to this, Dr. Logan sent him a short note on the history of Logic. On September 30 he wrote to Sir William Hamilton, the Professor of Logic in the University of Edinburgh, 'inquiring about the history of the Aristotelian theory of the syllogism.'[1] This was answered in a friendly tone, Sir William Hamilton sending Mr. De Morgan 'a copy of the requisites for a prize essay,' given out to the students at the close of last session, and offering further information if required. But in all the statements of what had been taught by the Edinburgh Professor my husband did not find any reason for believing that his own discoveries on the syllogism had been anticipated; and he was able to bring forward distinct proof that, even on the quantification of the predicate, which had been taught in a form less complete than his own by Sir William Hamilton, he had himself never had the opportunity of gaining a suggestion from the Edinburgh lectures. However, after the Cambridge tract had been in Dr. Whewell's hands, and in answer to a letter on the subject from Mr. De Morgan, Sir William says that he (Mr. De Morgan) is wholly indebted to himself for information on the subject; and in his own words, 'should you, though recognising always my prior claim, give forth that doctrine as a speculation of your own, you will be guilty—pardon the plain speaking—both of an injurious breach of confidence towards me and of false dealing towards the public.'

To this the following reply was sent:—

[1] I have always used, as far as possible, Sir William's own words from his subsequent letters on the transaction.—S. E. De M.

1847. Hamilton controversy.

Dear Sir William,—Your letter of the 13th inst., which I have read with astonishment, shows me the propriety of abstaining from correspondence upon the subject in question. When my paper appears, which I expect it will do in a few days, I shall have the honour of requesting your acceptance of some copies, that you may be able to put them into the hands of those with whom you may think proper to advise.

I will not further allude to the hasty manner in which you have expressed your suspicions of an odious charge, except to state that it does not diminish the sincere respect with which I subscribe myself

Your most obedient servant,

A. De Morgan.

These few lines show the temper in which the controversy began. I purposely refrain from attempting to record in detail the arguments used on both sides. The statement appended to the *Formal Logic* will give to inquirers a full insight into my husband's reasons for believing himself the originator of those Logical processes which he claimed as his own. So far as the discussion assumed a personal character it is his biographer's duty to record it; but the questions raised were of too technical a nature to be dealt with in a work like the present, even if I were competent to discuss them. Besides the statement in the *Formal Logic*, Mr. De Morgan made some mention of the controversy in the *Budget of Paradoxes*.

Sir William returned, unread, the copy of the *Formal Logic* which Mr. De Morgan presented to him; but in the year 1852 controversial warmth must have abated, for books and courteous letters were then exchanged between the Logicians.

I cannot deny that he rather enjoyed such encounters, but no one ever engaged in them with less feeling of personal animosity. It was like a game of chess—a passage of arms. But he did full justice to Sir William's splendid metaphysical powers, and says, in reference to the controversy, 'of which I suppose that the celebrity of my opponent, and the appearance of parts of it in a

1847. Sir W. Hamilton.

journal so widely circulated as the *Athenæum*, has caused many students of Logic to hear or read something.' He himself was a mathematical Logician. Sir William's philosophy might be called poetical. I have heard his pupils speak of him as 'inspired;' and the clear mental sight which enabled the one to develop the doctrine of limits in the *Differential Calculus* was more at home in the quantification of parts of the syllogism than the genius which enabled the other to bring out his admirable analysis of cause and effect. After Sir William's death Mr. De Morgan wrote an obituary notice of him in the *Athenæum*, in which his intellectual powers and great research were set forth so luminously as to excite wonder in persons not acquainted with the writer's character.[1]

---

[1] Readers of Mr. Herbert Spencer's volume on Sociology may have been puzzled by a passage (p. 412) in which the subject of this memoir appears to be accused of a misquotation, in his own interest, of some matter relating to priority of Logical discovery. It runs as follows:—

> One further point only will I name. Professor Baynes says: 'Professor De Morgan's emphatic rejection of Mr. Bentham's claim after examining the relevant chapters of his "outline" is in striking contrast to Mr. Herbert Spencer's easy-going acceptance of it.' Now, though to many readers this will seem a telling comparison, yet to those who know that Professor De Morgan was one of the parties to the controversy, and had his own claims to establish, the comparison will not seem so telling. To me, however, and to many who have remarked the perversity of Professor De Morgan's judgments, his verdict on the matter, even were he perfectly unconcerned, will go for but little. Whoever will take the trouble to refer to the *Athenæum*, November 5, 1864, p. 600, and after reading a sentence which he there quotes, will look at either the title of the chapter it is taken from or the sentence which succeeds it, will be amazed that such recklessness of misrepresentation should be shown by a conscientious man, and will be thereafter but little inclined to abide by Professor De Morgan's authority in matters like that here in question.

The reader who 'takes the trouble' to search out the passage in this *Athenæum* a quarter of a century old will not find a 'matter like that here in question.' But he will be enabled to form a juster estimate of the above passage when he learns that the victim of Professor De Morgan's inaccuracy or unconscientiousness was neither Bentham nor Hamilton, but Mr. Herbert Spencer himself.

The quotation itself occurs in a brief notice of Mr. Spencer's *Principles of Biology*, which I here reprint in full:—

> This is one of two volumes, and the two but part of a larger work: we can

I have mentioned my husband's early interest in Berkeley's philosophy, and Mr. J. S. Mill's opinion of the manner in which some of Berkeley's arguments were affected by Mr. De Morgan's enunciation of the principle 1847. Berkeley's philosophy.

therefore but announce it. Biology means the Science of Life. As to what constitutes life, we expected to have to remain in the dark. Schelling says it is 'the tendency to individuation.' Richarand says, 'Life is a collection of phenomena which succeed each other during a limited time in an organised body:' a very good definition. But is champagne alive as long as it fizzes, and a top as long as it spins? De Blainville says, 'Life is the twofold internal movement of composition and decomposition, at once general and continuous.' Mr. Spencer formerly defined life as 'the co-ordination of actions.' Mr. Lewes says, 'Life is a series of definite and successive changes, both of structure and composition, which take place within an individual without destroying its identity.' Mr. Spencer ends with '*The* definite combination of heterogeneous changes, both simultaneous and successive.' We have heard other definitions. Time was when life, *bon ton*, and *the thing*, were synonymous terms; and, according to the City lady, it consisted in—

Drinking tea, on summer afternoons,
At Bagnigge Wells, with china and gilt spoons.

All the definitions we have given apply to the life of organised material beings. Thus restricted, our definition is, that life is that state of a material being in which structure which performs functions is maintained by matter which the living being has power to draw from without, and which, when a man and an Englishman, he calls *nutriment*.

In a later edition of the *Sociology* Mr. Spencer has, in answer to remonstrance, added a note of self-justification. This note is to the purpose so far as that it will enable readers to discern the real gravamen of Mr. De Morgan's offence. Whether Mr. Spencer had just cause for annoyance in his reviewer's evident want of respect for the science of Biology (that is, so far as it undertakes to define the nature of life) may be an open question. But with respect to the misquotation, a reading of Mr. Spencer's chapter containing his definition will show that it was the result of a simple oversight. It appears that the quotation should have stood '*the* definite combination of heterogeneous changes, both simultaneous and successive, *in correspondence with external co-existences and sequences.*' I do not presume to enter here into any discussion of the definition, and therefore only cite it that those who understand it (which I do not profess to do distinctly) may know the nature and extent of the injustice done to it in the review. I abstain from expressing an opinion as to the propriety of Mr. Spencer's mode of presenting his case, preferring to leave the decision to my readers.

It is proper to add that Mr. Spencer has in private correspondence disclaimed all intention of imputing unconscientiousness. All that can be said of this is, that better ways of not imputing unconscientiousness might be suggested.

1847. of limits in the *Differential Calculus*. But his views of the psychology of Berkeley as compared with his own are more generally interesting.

Berkeley's philosophy.

In the chapter *On Objects, Ideas, and Names* in the *Formal Logic*, the writer says :—

That our minds, souls, or thinking powers, use what name we may, exist, is the thing of which of all others we are most certain, each for himself. Next to this nothing can be more certain to us—each for himself—than that other things also exist; other minds, our own bodies, the whole world of matter. But between the character of these two certainties there is a vast difference. Any one who should deny his own existence would, if serious, be held beneath argument; he does not know the meaning of his words, or he is false or mad. But if the same man should deny that anything exists except himself—that is, if he should affirm the whole creation to be a dream of his own mind—he would be absolutely unanswerable. If I (who *know* he is wrong, for *I* am certain of *my own* existence) argue with him, and reduce him to silence, it is no more than might happen in his dream. . . .

A celebrated metaphysician, Berkeley, maintained that with regard to *matter* the above is the state of the case; that our impressions of matter are only impressions communicated by the Creator without any intervening cause of communication.

Our most convincing communicable proof of the existence of other things is, not the appearance of objects, but the necessity of admitting that there are *other minds* beside our own. The external inanimate objects might be creations of our own thought, or thinking and perceptive functions. They are so sometimes in the case of insanity, in which the mind has frequently the appearance of making the whole or part of its own external world. But when we see other beings performing similar functions to those which we ourselves perform, we come so irresistibly to the conclusion that there must be other sentients like ourselves, that we should rather compare a person who doubted it to one who denied his own existence, than to one who really denied the external existence of the material world.

In his interleaved copy of *Formal Logic* is a pencil note alongside of the foregoing :—

To read Berkeley so as to give him a fair chance, some one

else should turn the page over; for, unphilosophical as it may be, the touch of the paper periodically intervening is a snake in the grass—an unphilosophical snake. It is hard to make philosophers of the fingers. 1847. Berkeley's philosophy.

No doubt it would favour Berkeley's scheme if no appeal to the external, through the senses, were possible to his reader. But surely this would be giving him *more* than a fair chance. Another interleaved note:—

Personal identity is what every one has a clear conception of until he reads physiology and metaphysics. In that process he learns that the knife sometimes (gradually) gets a new handle, and sometimes a new blade, and all his notions of identity vanish, with nothing but memory left to puzzle him. 'Mais quand je me tâte, et quand je me rappelle, il me semble que je suis moi.'

If physiology teaches that we are *automatic*, it ought to find a new name for the nightingale and chess-player, which can be wound up when they have run down.

Dr. George Boole,[1] author of *The Laws of Thought*, had introduced himself in the year 1842 to Mr. De Morgan by a letter on the *Differential and Integral Calculus*, then recently published. His character and pursuits were in many points like those of the author, who found great pleasure in his correspondence and friendship. He was a Mathematician as well as a profound and original student of Logic and Metaphysics. In 1839, the same year in which the *First Notions of Logic* appeared, he had sent his Mathematical paper, *Researches into the Theory of Analytical Transfer*, to the *Cambridge Philosophical Journal*, and in 1844 received the gold medal of the Royal Society for a paper *On a General Method in Analysis*. In the course of these speculations he was led to consider the possibility of constructing a calculus of deductive reasoning; and he found that logical symbols conform to the same fundamental laws which govern algebraical symbols, while they are subject also to a special law. 'Mental

[1] Professor of Mathematics in Cork College.

1847. Dr. Boole. science,' says his biographer, 'then became his study, Mathematics his recreation; yet it is a remarkable fact that his more important and valuable Mathematical works were produced after he had commenced his Psychological investigations.'[1] In 1847, his attention having been drawn to the subject by the publication of Mr. De Morgan's *Formal Logic*, he published the *Mathematical Analysis of Logic*, and in the following year communicated to the *Cambridge and Dublin Mathematical Journal* a paper on the *Calculus of Logic*. His great work, *An Investigation into the Laws of Thought, on which are founded the Mathematical Theories of Logic and Probabilities*, was a development of the principle laid down in the *Calculus*, its design being 'to investigate the fundamental laws of those operations of the mind by which reasoning is performed; to give expression to them in the symbolical language of a calculus, and upon this foundation to establish the Science of Logic and construct its method; to make that method itself the basis of a general method for the application of the Mathematical doctrine of probabilities; and, finally, to collect from the various elements of truth, brought to view in the course of these inquiries, some probable intimations concerning the nature and constitution of the human mind.'

I have given Dr. Boole's own statement of the design of his work at length because it conveys in few words, not only some idea of the aim of his investigations, but of the relations between the three sciences of Psychology, Mathematics, and Logic. An estimate of his mental work and its value to science was given in a few words, after his death in 1864, by my husband:—

'His first paper in the *Cambridge Mathematical Journal* contains remarkable speculations which can here be described only in general terms, as extensions of the power of algebraic language. These papers helped to

[1] Taken substantially from a notice of Dr. Boole in the *Obituary Notices of the Royal Society*, by the Rev. R. Harley.

give that remarkable impulse which algebraic language has received in the interval from that time to the present. . . . That peculiar turn for increasing the power of mathematical language, which is the most characteristic point of Dr. Boole's genius, was shown in a remarkable way in his writings on Logic. Of late years, the two great branches of exact science, Mathematics and Logic, which had long been completely separated, have found a few common cultivators. Of these Dr. Boole has produced far the most striking results. In alluding to them we do not say that the time is come in which they can even be generally appreciated, far less extensively used. But if the public acknowledgment of progress and of genius be delayed until the whole world feels the results, the last century, which had the lunar method for finding longitude, ought to have sought for the descendants of Apollonius to reward them for his work on the Conic Sections.' 1847.

'Dr. Boole's system of logic shows that the symbols of algebra, used only to represent numbers, magnitudes, and their relations, are competent to express all the transformations and deductions which take place in inference, be the subject what it may. What he has added may be likened to a new dictionary, by consultation of which sentences written in the old grammar and syntax of a system take each a new and true meaning. No one is ignorant that the common assertion, "Nothing is both new and true," is a perfect equivalent of "Everything is either old or false, or both." Dr. Boole showed that a schoolboy who works a certain transformation, such as occurs in many a simple equation, has the form, though applied to very different matter, of this logical passage from one of two equivalents to the other. Taken alone, this is a pretty conundrum, if any one so please. But when looked at in the system of which it is a part, and when further considered as the produce of a mind which applied the same power of thought with rare success over the whole of the higher Mathematics, those who so look,

1847. George Boole. and so consider, are justified in presenting it as a type of genius, and as a specimen which may give those who are not mathematicians a faint notion of an originality of speculation which, applied to the progress of science, has attained most useful results, and made a lasting name.'[1]

My husband's regard for Dr. Boole was founded not only on admiration of his originality and power, but on sympathy with the moral and religious basis of his psychology; for Dr. Boole, like Mr. De Morgan, believed that every system which rejected the existence of God as a constantly sustaining cause of all mental as well as physical phenomena, was like a consideration of the nature and growth of a tree without reference to the root. They did not often meet, for Dr. Boole's life was passed in Ireland after his appointment to the Mathematical chair in Cork; but when his visits did occur, they were a real enjoyment to both—I believe I may say to all, for I shared in the pleasure of his conversation, ranging as it did over a wide field of thought, and touching poetry and metaphysical as well as mathematical science. My husband was, I believe, instrumental in some degree in obtaining the appointment at Cork, where Sir Robert Kane, who had married our friend Mr. Baily's niece, was Principal.

Sir Frederick Pollock. My husband's friendship with Sir Frederick Pollock, then Lord Chief Baron, had begun some time before this. Sir Frederick, who had been Senior Wrangler of his year, kept up his mathematics in the midst of his legal avocations. He was a good Mathematician, among other matters interested in the properties of triangles and in magic squares, and, I believe, made some original discoveries. He often communicated these to Mr. De Morgan, who occasionally gave him a lift when any stumbling-block came in his way—at least, so he told me. He was a most agreeable companion, full of interest in all subjects of thought, and of all men I ever knew he seemed to

[1] From a MS. unpublished paper, drawn up by Mr. De Morgan.

me to have highest value for talent of all kinds; and of all well-read persons no one was so ready with an illustration or a quotation, either in prose or poetry, suited to the occasion. 1847.

1848. Introductory Lecture.

The Introductory Lecture given by the Mathematical Professor on the opening of all the classes in University College at the session of 1848, was on a question in education on which he had thought much, and on which his opinion had strengthened as his experience in teaching increased. He considered competition—the striving one against another for the highest place among boys or young men—to be among the crudities of an imperfect system, and to be as ineffectual in gaining the end either of *making* the best scholar or *showing* the best scholar, as its moral tendency is bad.

It is quite true, as he himself has said to me in talking of this subject, that the boys are generous and sharp enough to see who deserves the prize, and very little ill-will or jealousy ever comes into the competition; but they do not know, any more than their teachers, how much easier the work is for some than for others; and as the teacher cannot take this into account, injustice in one way can hardly be avoided. Hence his objection to marks in looking over examination papers. He said he could judge of the merits of the competitor from the whole work, but he could not reckon it up by marks, and he always refused to examine in this way. But he also felt and often expressed his opinion of the terrible mischief to health done by urging a young man to go a little beyond what he could accomplish with interest and success if no undue pressure were put upon him, and by the 'cramming' to answer questions set at the pleasure of the examiners, in place of the natural and well-directed effort to learn the subject which an enlightened teacher can always evoke in an intelligent pupil. He himself, a most successful teacher, to whose instructions his pupils always looked back with the consciousness that to him they owed the

1848. Education.

habits of sound thought and correct reasoning which were among their best mental possessions, believed and said that it is owing to the imperfection of the teacher, not to the want of effort in the pupil, that the boy or young man fails to make progress. There are doubtless cases in which the capabilities are naturally wanting, and when a strain to do what cannot be done is useless; but this only shows the necessity of discernment. Some of his friends, who greatly liked the lecture of this year, saying that all his arguments were unanswerable, still hesitated to give in to all his conclusions. His old friend and fellow-collegian, the Rev. Arthur Neate, among others, asked, if the stimulus of competition were taken away, what would be put in its place. I do not know what he replied to Mr. Neate, but I know what he said to me at the time. 'With such young men as those who struggle to be highest, and who suffer in the struggle, no stimulus is needed beyond their own pleasure in learning; and if a teacher cannot make them feel this, he does not deserve the name of teacher.'

Rev. Baden Powell on competitive examinations.

Among those who fully concurred with him was our friend the Rev. Baden Powell. His thoughts on the subject were suggestive. He wrote:—'Accept my thanks for a copy of your admirable Introductory Lecture. I wish it could be more widely circulated among our candidates here and at Cambridge. Perhaps there was something in this respect better in the system of our ancestors' Disputations, in lieu of examinations. I have often wished there were something like making a man read a dissertation on a subject of his own choosing, and then cross-examining him on his *own* argument. Many would be plucked from not understanding their own meaning.' And Sir John Herschel wrote, 'I was greatly delighted with your protest against the cramming system in your opening lecture.' So also Dr. Whewell: 'I see you have been kicking against examination reading. So far good. But is your College going to do anything which will really diminish it? I should like to hear how any attempt of this kind prospers. I think the object an excellent one, and have some notions of my own as to the way in which it may be forwarded.' Such declarations of opinion from such men ought to have weight with their successors.

1848. Rev. Dr. Whewell.

Mr. De Morgan, while fully agreeing with Mr. Baden Powell on the wisdom of letting the student show what he can do by his original work, did not disapprove of examinations, not competitive ones, to test the amount of knowledge really gained. At that time his arguments were unheeded. The torrent of competition carried all before it, and not seldom swept down its victims to destruction. Wiser notions are coming into men's minds, and the evil is acknowledged, though, from what I hear of the state of things at Cambridge, better methods are not yet found. Educators have to learn that the aim of education is to develop power, not to cram knowledge; and also, what is now never thought of, that mental and moral faculties come into activity at different times in different individuals, and that a talent which is inert or unnoticed in a boy of twelve, may be a brilliant element of the mind at twenty, if not forced up or crushed out by mismanagement.

Astronomical Society.

The Astronomical Society's difficulties in obtaining good and efficient chief officers were not over. It is so long ago that I can betray no confidence by telling of these embarrassments in Captain Smyth's language:—

SIR AND SEC.,—I am but a temporary V.P., it is true, and therefore have little authority to shove my oar into others' row-locks, but the *Salus Reipublicæ* shakes me to the centre. . . .

The anniversary approaches, and so do our difficulties. The last time I saw friend Galloway he swore like a Flanders soldier that he would not serve another year! This oath fills me with consternation, and my eloquence having failed, will you exert yours?

Then, again, there's the Council! Did you ever? Pray

1848. think of these things, exert your energies, and preserve us. So shall you deserve well of your contemporaries, of posterity, and of yours very truly,—W. H. SMYTH.

The writer had filled the President's chair after Sir John Herschel and since the last difficulty, but it was again vacant, and Mr. De Morgan was very earnestly begged to take it. But he refused, for, while working harder than ever, if possible, to keep up the character and usefulness of the Society, his old reasons were still in force. He always feared the love of rank and money finding its way among the honest, useful workers who had hitherto composed it. He held up the example of the Royal Society as one in some ways to be avoided, and resisted every measure that would tend to bring in the sort of influence which had fettered its scientific work during the last century. His commentary on the Royal Society's history in times past will be found in the *Budget of Paradoxes*, as well as some allusion to the fact of his never having sought for membership. Touching the charge against Sir John Hill, that his animus towards the Society was occasioned by his failure to obtain admission to its ranks, he says, 'Whether *I* could have been a Fellow, I cannot know; as the gentleman said when asked whether he could play the violin, "I never tried."' On the last point, however, Admiral Smyth gave evidence:—

Know that I dined yesterday with the Philosophical Club, where was an ominous growl about your not being in the Royal Society, and, on the entreaties of several warm friends, I undertook to state the same to you. My own regret you are fully aware of. Pray, therefore, reconsider the case, for it is declared to be no favour at all to you, but a signal one to the Society, to allow your name to be hung up. Pray grant this, and your Petitioner will ever pray, &c.

At this time a good many friends used to meet periodically at our house, and my husband enjoyed the opportunity they gave him of seeing them in an informal way. Besides several of those I have named, many of the Professors of University College and their families of course visited us. At this time the Rev. Alexander Scott, afterwards Principal of Owen's College, Manchester, filled a chair in University College. Mr. Arthur Hugh Clough was Principal of University Hall. M. Libri and his wife, Dr. Westland Marston, Miss Mulock, now Mrs. George Craik, Mrs. Follen, the abolitionist, and Mrs. Catherine Crowe, of ghostly renown, were among our guests. As my husband was connected with the *Athenæum*, his victims and co-reviewers were a lively element in these mixed assemblies, and we found in the meeting of persons of very opposite pursuits, that some who seemed the farthest removed from each other would often rejoice to meet, and were found in unexpected connection. Several friends addicted to what are called mystical studies, found their way to us, drawn partly by my own love of trying to unveil mysteries, partly by the sounder knowledge which my husband, who did not quite despise the obscure sides of early science and mediæval philosophy, could bring to the subject. Of these, I think the Rev. Jas. Smith, author of 'The Divine Drama of History,' was the most learned and the least appreciated by the world at large; for his estimate of Swedenborg as an authority on spiritual questions, and his admiration for Joanna Southcote as a 'typical woman,' were thought to throw discredit on his good sense. Swedenborg is not held utterly contemptible now—though, as Mr. Smith said then, he is least understood by his own followers. 1848.

The Ladies College in Bedford Square.

A not infrequent visitor on these evenings was Mrs. Elizabeth Reid, a widow lady of property, whose father had been an influential Nonconformist, and who had long sought for coadjutors in her design of establishing a model ladies' College. She had written to Lady Byron and to me of it fifteen years before, but her plans were not as practical as her intentions were good, and it was

1849. Ladies' College in Bedford Square.

not until she found herself among a number of sound-thinking liberal men that she gained help and advice to carry her wishes into effect. Some of the Professors of King's College—among them the Rev. F. D. Maurice and the Rev. W. Nicolay—gave her the best advice and encouragement, and when the plan was matured, their experience drawn from the Queen's College in all technicalities respecting the arrangement of classes. Mrs. Reid took the house in Bedford Square, paid the rent and much of the expense during the first years, and otherwise endowed the Institution. She was among our friends, and we were able to interest many friends of education in the undertaking. Prof. Scott kindly promised his aid as Professor of English, Mr. Francis Newman took the Professorship of Latin, and my husband gave lectures or lessons on arithmetic and algebra for one year, to give as good a start as possible to the new college, which opened at the end of 1848. Of my own work in the formation of the Ladies' College I will only say that it was the means of ensuring his interest, and thus obtaining for the place an advantage which it could not otherwise have had.

For some years there had been a growing desire that the education of girls should be brought out of the state of absolute inanity in which it existed in ladies' schools. A specimen of the instruction has been given at p. 123, and from the Astronomy we may have an idea of the other branches of science, and form a guess at the History and Language. It was seen that something a little more efficient was wanted—some system which, if not approaching in extent, should yet be equal in soundness to the teaching given to boys and young men. To meet this want a few enlightened clergymen and others had some time before established the Queen's College in Harley Street. The orthodox King's College had grown up shortly after the establishment of the heterodox University College—as some persons thought, in rivalry and hostility to it—as wiser ones saw, another step in the right direction under different conditions. The Queen's College for ladies, which owed its origin and success to some of the Professors of King's College, was based on the same principles, but as being designed at first for members of the Church of England, was objected to by parents who did not wish their daughters either to join the prayers and Bible teaching, or to feel excluded from what their fellow-learners partook of. The growing number of this class showed that there was yet room for another College, founded, like University College, on principles of absolute neutrality as to doctrinal teaching. 1849.

Overwork.

These high-class day schools[1] have increased in number from that time; but I am told that those of the present day share the faults of their predecessors, for each teacher or Professor sets what task he likes irrespective of those imposed by his colleagues. The amount of out-of-school work was formerly excessive, and young girls suffered in proportion. I have known cases of illness for life, insanity, and even death from this cause, and as the finances of the school depended on the number of classes entered by pupils, young girls were often recommended to take ten, twelve, fourteen, or fifteen; and their parents, knowing no better, consented. I have heard entreaties on the poor girls' part to have their lessons at home shortened met by the answer, 'You have so many hours here and so many at home, there is *time* for all.' *Strength* for all was not thought of, and time to think over and assimilate what had been learned still less. Many hard-working girls became ill, many heedless ones quite indifferent, but, as a remedy for either evil, the idea of fitting the kind and amount of work to the kind and amount of power never entered the teachers' heads. It is too 'advanced' a notion, but when we have overtaken it the 'schoolmaster abroad' will be a beneficent genius,

[1] *Colleges*, such as Newnham or Girton, are of course not included in these remarks. I believe they are more wisely managed.

1849. scattering blessings in his path, instead of, what he has been made by over-driving, a rampant lion, seeking what he may devour.[1]

Book postage.

Although not among our frequent visitors, Sir Rowland Hill was an old acquaintance of my husband's. Before the time of which I am writing he had brought his great scheme of penny postage into operation, having grappled with and overcome gigantic obstacles. Mr. De Morgan had corresponded with him on this work, and in 1848 made many suggestions for the postage of books and MSS. Sir R. Hill, after stating the conditions of posting, and the amount of writing allowed in the address, requests him not to refer in writing to the source of his information, saying, 'You will perhaps think all this ridiculous, but there are real as well as imaginary difficulties in doing more—at present, at least.' My husband gave a good deal of help anonymously to this great reform, and, I think, suggested the book postage.

At this time he was actively interested in the questions raised by the proposed compilation of a complete catalogue of the British Museum Library. I need not enter into the discussion which this subject excited, further than to note that some expressions made use of by Mr. T. K. Hervey, then editor of the *Athenæum*, led to Mr. De Morgan's discontinuing his contributions to that journal for some years.

Guglielmo Libri.

During the agitation of the catalogue question he often visited the British Museum, and on one occasion met Count, or Professor,[2] Guglielmo Libri, who had come to England in a state of utter despair, owing to the charges of theft made against him by the French Government. Mr. De Morgan was at once favourably

[1] January, 1878. *Only yesterday* a friend told me that while walking in the street, violent and frightful screams startled her, and on inquiring at the house from whence they came, she was told that a young lady was dangerously ill of brain fever, *having just passed a College examination.*

[2] He preferred the latter title.

1849. M. Libri and the French Government.

impressed by M. Libri, but from his agony of mind and imperfect knowledge of English, it was a difficult matter to get at particulars of his case. Gradually facts were brought forward and documents produced. The one most patent fact, attested by M. Guizot, then Prime Minister, that M. Libri had offered all his books and manuscripts to the French nation, on condition that they should be kept together and called by his name, was a sufficient presumption of his innocence to lead to the belief that further proof would be forthcoming; for no one would believe that books, stolen from a public library, would be openly placed there by the very man who had abstracted them. M. Libri became our attached and valued friend, always recognising a firm and able defender in my husband, whose articles in the *Athenæum* and elsewhere were the means of establishing a belief in his innocence in England. Some reference to the political relations of France and Italy will throw light upon the persecution he, an Italian, experienced from the French Government; but the political condition of France, on which he expressed himself very openly, helped to determine the events of his life.

He was born in 1800, of a noble Tuscan family, and was made Professor of Mathematics in the University of Pisa when twenty years old. Being looked upon as a Liberal by the Government, he was forbidden to remain in Italy, which he had left on a visit to Paris in 1830. He returned to Paris, where he was naturalised, and in 1833 was made a member of the Institute, holding among other appointments that of Inspector of the Royal Libraries and Mathematical Professor in the Sorbonne. His *History of the Mathematical Sciences in Italy*, in four volumes, is spoken of by Mr. De Morgan as a great work. But he did not confine himself to scientific work; he helped the cause of Louis Philippe by his writings, opposed the Jesuits both in their French and Italian schemes, and gained the enmity of

1849. Guglielmo Libri.

the opposite party, by whom he was denounced as a monarchist, and as an Austrian traitor to the cause Italy. Whether he was right or wrong in his politics is matter of opinion. He had expressed his own very freely, and thus had become an object of suspicion to the Government of the time. But there was another reason for his unpopularity in France. 'In science he would not be a Frenchman, but remained an Italian. One of his great objects was to place Italian discovery, which the French historians had not treated fairly, in its proper rank. This brought him into perpetual collision with M. Arago at the Institute, and personal enmity was the consequence. Those who know French science, and how little it attends to history and to the learning which aids history, will guess what a nuisance must have been the presence of an able scholar and a profound Mathematician, with everything that the French ignore at his fingers' ends, carrying the fire of reason and the sword of reference into their most sacred haunts; and worse still, the small shot of ridicule, against which few Frenchmen have any armour. When they were establishing showers of toads by second-hand citations from old authors, M. Libri went to the originals and got them a shower of oxen upon the same evidence; *maudit Italien.* At the same time we must do the French *savans* the justice to say that M. Libri is a warm nationalist, and that we will by no means guarantee his having been always in the right. Neither can the insinuation about stealing books be traced to the Institute. We suspect that political animosity generated this slander, and a real belief in the minds of bad men that collectors always steal, and that the charge was therefore sure to be true.'

'Every one who becomes acquainted with M. Libri soon learns that the restoration of Italian fame is always in his thoughts, and, though learned in the history of other sciences, his interest in collecting is that of a propagandist, who would gladly, if he could, furnish every library with the means of verifying Italian history. . . . He specially collected Italian books, and the thefts charged are mostly of that kind of literature. He offered his whole collection, books and manuscripts, as a present to the French nation on condition that they should be kept together and called by his name, which was refused. The offer was made to M. Naudet, of the Royal Library. When difficulties arose as to the stipulation, M. Libri complained to M. Guizot, the most influential of the Ministry in literature, always his firm friend, and a firm believer in his innocence. M. Guizot certified this fact to the editor of an English journal in 1849, and gave it in evidence to a commission sent from Paris to examine him, as we learn from his handwriting. This shows the state of things in Paris with respect to M. Libri at the time of his escape to England in the year 1848. It had been rumoured that he, who was well known as having bought rare books and as having sold a large collection, had robbed the public libraries of a number of books to the amount of several hundred thousand francs, and a note was one day put into his hands at the Institute by the editor of the *National*, threatening him with popular vengeance, and advising him to disappear if he hoped to escape. A report was drawn up by M. Boucly, the Procureur du Roi, founded upon anonymous accusations, and soon after M. Libri's escape to England—a step recommended at once by his friends in France—this report was published in the *Moniteur*. To it he replied, so completely proving his innocence, that no more was heard of the document. In a letter to M. Falloux he continued his defence, which produced no effect. His books and furniture were seized, and a commission was appointed to examine them. This commission made its report in 1850, and in 1852 the *Acte d'Accusation* was passed.'

1849.

Offer of his library to the French Government.

During the time he had been in England he had gained some steady, energetic friends, many of whom gave him sympathy and assistance. Scholars and biblio-

1849. Guglielmo Libri.

graphers were convinced of his innocence, but no defence of him in France was permitted, where he had lost all property and position. But the facts of the case became evident to all who chose to examine them. M. Guizot, Lord Brougham, Mr. De Morgan, and other judicious friends found from careful inquiry in Paris that the *Acte d'Accusation* in this case implied summary conviction. They recommended him not to return to Paris for trial, and judgment went by default, though after some years the accusation was withdrawn. Mr. De Morgan said of the accusation that it involved a new form of syllogism: 'Jack lost a dog; Tom sold a cat; therefore Tom stole Jack's dog.' And it was discovered, after all, that in several cases the editions sold by M. Libri were not those which the library was reported to have lost. In several cases the library had not lost the book at all. In several cases the lost book had been found elsewhere, and in no one case was it proved that a book once belonging to a public library was found in M. Libri's possession without proof of having been honestly come by.

M. Libri had every social quality to secure regard and friendship. He was a fine classical scholar and an original thinker, having the sparkling merry humour of his countrymen, and, like an Italian, was simple and affectionate, but hasty and irascible. He had been in youth exceedingly handsome, and at this time, when of middle age, was one of the noblest-looking men I ever saw. In 1850 he married Madame Melanie Colin, a generous, self-devoted woman, who made great efforts to procure justice for her husband. She went to Paris, consulted with his friends, and appealed to his enemies, but the anxiety and exertion were more than her strength could bear, and it was thought that her subsequent illness and death were caused by the strain upon her powers.

1850. Mr. Thos. Galloway.

The death of our friend Mr. Galloway, who had been living in Torrington Square, occurred in the following year. It was preceded by some months' illness from spasms

of the heart, which he bore with calmness and patience. 1850.
Mr. De Morgan, who had a warm regard for him, spent what time he could gain in the intervals of his lectures in his friend's sick room, and his visits were looked for as affording some alleviation in a difficult nursing, not only as to such difficulties as arose in Mr. Galloway's absence from business, but, I believe, with the patient himself, who was sometimes induced by his quiet persuasion to take a remedy for which he felt disinclined. 'I can never,' Mrs. Galloway writes, 'cease to remember with love and gratitude how tenderly your beloved husband watched his downward progress, sitting day by day by his bedside, and talking to me in a low tone in the hope that it might induce sleep, and anxiously trying to get him to take food, on the amount of which the doctors said his life depended.'

Mr. Galloway had been more than once my husband's colleague as secretary of the Royal Astronomical Society, and had in many ways done service to Science. He would in all probability have been Professor of Natural Philosophy in the University of Edinburgh if he had not been elected registrar or actuary of the Amicable Life Assurance Office in 1833, as before mentioned. He had in early life intended to enter the Church, but, like Mr. De Morgan, found the teaching of Mathematics a more congenial employment than preaching, and held for a few years the appointment of Mathematical teacher at Sandhurst. His interest in the welfare of the Astronomical Society was strong and lasting, but he was very unassuming in his estimate of the work he had given to it, and begged my husband during the last days of his life to prevent anything like eulogium on his service. This arose partly, no doubt, from his own simplicity and humility of character, partly from the consciousness that Mr. De Morgan was always anxious to do full justice to all his friends. In the little memoir written by Mr. De Morgan for the Royal Society this wish is recorded, but the biographer

1850. adds that it was scarcely possible to comply with it, for a true account of Mr. Galloway's services to Science was in itself a eulogium.

1851. Negro slavery. Those of us who can look back more than thirty years will remember the feelings excited in England on negro slavery by 'Uncle Tom's Cabin'—how it brought to a climax the sympathies and efforts of those who had long worked in the same cause, and how it stimulated those who had been inactive, because ignorant of what was going on, to consider how she or he could contribute towards diminishing the sufferings of the negroes. My husband felt intense interest in this question, and pity for the sufferers on both sides. I remember his sitting up the greater part of one night reading 'Uncle Tom's Cabin,' and it was evident that the subject pressed heavily on his mind. We found several friends, among them some active abolitionists from the United States, who liked our idea of a National Address from Great Britain to the United States of America, to be signed by every one who could think and feel upon the subject. Mr. De Morgan drew up an address such as appeared to these friends calculated to encourage a wise effort at gradual but certain emancipation. It claimed for us, the writers, a right to offer sympathy and assistance, inasmuch as our countrymen and women had, until very lately, been accomplices in the enslavement of the negro. It invited mutual consultation and counsel, and promised what help could be afforded by one nation to another in the tremendous work of getting rid of the burthen of slavery with as little injury as possible to slave-owners and slaves. One or two friends, men of worth and learning, gave some suggestions in the writing of this document, of which I have not now a copy. Had it been sent in its original form, and according to the wishes of its promoters, its influence would have been hardly a drop in the ocean; and, as it afterwards proved, the time for remonstrance and argument was nearly over. But our design was not carried into effect as we intended. Before it could take form other influential opponents of slavery heard of it, and drew up an exhortatory address from the women of England to their sisters in America. This address was in the main moderate and good; the feeling it expressed was unexceptionable, but it was couched in slightly religious terms, which gave it the appearance, as we thought, of an assumption of spiritual superiority over those addressed; and we, who had hoped for the concurrence of thoughtful and influential men, felt that our effort lost strength by being made exclusively a woman's movement. Accordingly the original promoters of the plan withdrew. I do not think the Address of the Women of England, which was well introduced and signed, did either good or harm in America. Our abolitionist friends lamented our failure, but beyond causing some slight irritation among the American ladies, who did not like its tone, and did not see in it the good feeling of the writers, it had no effect at all.

1851. Address on slavery.

In the Introductory Lecture on the opening of the session of 1848 my husband had distinctly stated some of his strong objections to competitive examinations, and their preparatory *cram*, with other parts of the educational system as it was (I wish I could add, and is no longer) carried on.

1853. University of London examinations.

He had strongly expressed his disapproval of the course proposed by the University of London on its first establishment, and refused to take part in the examinations.

At that time the enormous variety of subjects on which a young man was required to answer questions, without reference to any special ability, was stultifying and confusing even to the brain which could receive them all without damage to physical health. Apropos of this reckless and fruitless waste of mental effort, my husband wrote an illustrative 'Cambridge examination:'—

1853. University of London examinations.

Q. What is knowledge?

A. A thing to be examined in.

Q. What is the instrument of knowledge?

A. A good grinding tutor.

Q. What is the end of knowledge?

A. A place in the civil service, the army, the navy, &c. (as the case may be).

Q. What must those do who would show knowledge?

A. Get up subjects and write them out.

Q. What is getting up a subject?

A. Learning to write it out.

Q. What is writing out a subject?

A. Showing that you have got it up.

His objection to the methods pursued by the University of London will be found in his letter (p. 222) written in answer to a request.

In his strictures on the teaching of Physiology he had evidently not contemplated the possibility of the dissection of living animals for demonstration, now happily forbidden by Act of Parliament. Had the question of its expediency for the sake of Science been put to him he would have said, as he always did on such occasions, that no imaginary end could justify means which were opposed to a positive law of humanity.

And his own words on the subject of vivisection show what he thought of it. A surgeon had been describing to us some of Majendie's atrocities (since equalled by those of English and Scotch physiologists), and after our friend was gone I referred with horror to what he had said. My husband, who had been silent some time, said, 'Don't talk of it;' then, in a minute or two, pausing between the sentences, he added, 'They will learn nothing by it. It's all of a piece. There is no God in their philosophy.'

Some few years after this time he came home one day from the College evidently amazed, and told me that some pupils had applied to him to interfere in the following circumstances. A cat had been poisoned 'for Scientific purposes' before one of the classes. I asked him whether a repetition of this could not be prevented. He said, 'Certainly: it must not happen again. It was too bad. I shall speak to ——,' another Professor on the medical side, 'and he will see to it.' Accordingly he spoke to ——, who satisfied him somehow that the thing would not recur. He had little notion that the Professor appealed to was and had been performing experiments before his pupils on living dogs and cats. These were of so cruel a nature that I will not describe them. They were detailed to me by a highly respectable surgeon, who had been a student of the class referred to.

1853. Experiments on living animals.

In November a circumstance which showed an uncertain interpretation on the part of the College Council of the main principles of the foundation, made my husband look forward with abated confidence to the future of his Professorship.

Dr. Peene's legacy.

During the first years of University College, its principle of non-interference with religion had been well adhered to; indeed, we received so many assurances on the subject at distribution speeches, opening lectures, and in many other ways, that no fear was felt, and my husband worked on in the happy conviction that he was aiding the great cause which he had most at heart. But for some little time before this he observed indications that the monetary success of the classes would be held a stronger motive in deciding questions connected with the working of the College, than its fulfilment of the pledges given of thoroughness in instruction and adherence to principle. He told me of these things with some anxiety. He saw, or thought he saw, a more decided tendency to temporise to secure the monetary success of the Institution in other directions; and in the year 1853 an occurrence fraught with danger to the principle on which it had been established proved that his fears were well founded.

1853. Dr. Peene's legacy.

Dr. William Gurdon Peene, of Maidstone, left seventeen hundred pounds for the purchase of books for the library of University College. These books were to be works on foreign literature and science, and the choice of them was to be entrusted to the Professors of Greek, Latin, and Mathematics, provided these three were members of the Church of England; '*Otherwise*' (as expressed in the will), '*one or more shall complete their number by choosing qualified persons from the other Professors, private teachers, or quondam alumni resident in London. If none of the three named be members of the Church of England, I beg the Council to appoint.*'

On hearing of this bequest, and learning that some members of Council were inclined to accept it with the prescribed conditions, Mr. De Morgan wrote to the chairman of the Council as follows:—

University College, Nov. 5, 1853.

SIR,—A proposal now before the Council, and to be discussed this day, involves the application of a religious test to certain Professors, with a view to their exclusion from a certain office to be founded, in the event of their opinions not being of a certain class.

I beg you will draw the attention of the Council to the following personal statement. The matter in question may never come before the Senate; and if it did, I could not expect the Senate to convey to the Council remarks which refer entirely to my own personal position. If, when I first sought the honour of a chair in this College, I had asked what security existed for my never being excluded from anything on account of my opinions, I should have been told, and with reason, that if so many public declarations as had been made, both printed and oral, uttered with every mark of sincerity and received with every appearance of enthusiasm, were not sufficient guarantees, I should do well to reconsider my intention of acting under those whom it was clear, by my question, that I mistrusted.

Again, admitting that the College, corporately, would never institute a test or create a disqualification, if I had asked whether it would allow any one else to do so within its walls, or if, giving credit for the full determination to maintain a perfect religious equality among the *students*, I had asked whether it was possible a *Professor* might be placed under disqualification, I should have been told, and with reason again, that if the length and breadth of the declarations I have alluded to were not sufficient to contain these and any other possible cases, all the lawyers who ever varied the counts of an indictment, or reckoned up the rights which pass with a freehold, would not be able to frame anything which would satisfy so suspicious a person.

1853. Letter to the Council.

I joined the College in the full conviction that the plain English of scores of declarations would have warranted the preceding replies. To my utter surprise, on the very first occasion on which money is offered on the condition of establishing a religious test, all I hear seems to indicate that it is far from certain that the offer will be rejected. What the Council has ever done to warrant such a want of certainty I cannot imagine; for if ever any Institution in this world honoured its faith and practised its professions, University College has done so, up to this moment, in the matter of religious equality. I myself should never have imagined the necessity of stating that my connection with this College was the consequence of the good and sound and religious principle shown in its leading maxim, but for the doubt to which I have referred. No one is so humble that faith need not be kept with him. In the name of all the declarations which the College has put forth from its first institution, I claim the performance of the obligation therein undertaken to maintain every student, every Professor, every officer in perfect religious equality with the rest, from the President of the Council down to the sweeper of the floor.

This I claim with the most perfect respect for the Council, which, among many other reasons, I feel because the principle of the College has always been maintained, and, I fully believe, will still be maintained. But I think it possible that the strength of the *individual* claim of those who have trusted the College, and have spent the best years of their lives in its service, may be overlooked, and for this reason only I trouble you with these remarks.—I am, sir, your most obedient servant,

A. De Morgan.

P.S.—The only precedent which bears on the matter, within my recollection is as follows: At the opening of the College, each student was desired to state whether he was Churchman or Dissenter, and the answer was affixed to his name in the list. The motive was the most innocent in the world; it was the

1853. Peene legacy.

statistical one. Great objection was raised. It was affirmed that the College on its principle could have neither need to know nor right to inquire the religious status of any student. In deference to this objection, the force of which could not be denied, the practice was discontinued.

This letter was laid before the Council. In reply to it a copy of the resolutions passed in the following month was received. My reason for giving them here will be found in the first resolution, which contains a full acknowledgment of the principle of religious equality.

Copy of Resolutions passed by the Council on December 10, 1853.

1st. That the Council cannot but regret that the late Dr. Peene should have accompanied his valuable legacy by a direction with regard to the function of choosing the books, which can, by any construction, be supposed to infringe that principle of religious equality to which the present Council and their predecessors have invariably adhered, *as well in the appointment of Professors,*[1] the admission of students, and the award of honours, as in the general administration of the affairs of the College.

Considering, however, that the function in question is totally unconnected with the ordinary duties of the Professors, and might have been assigned by the testator to persons unconnected with the Institution, and that it is to be regarded as a trust under Dr. Peene's will, and not as a duty imposed by the authorities of the College;

Considering, also, that any Professor will have the power of declining the trust altogether if he should for any reason think proper so to do, without being required to make any profession of his religious opinions;

And, lastly, considering that the value and utility of the proposed annual addition to the library are not likely to be in any degree impaired by the terms of the bequest—

The Council have determined to accept Dr. Peene's legacy, being of opinion that in so doing they do not violate *that principle of religious equality on which the College was founded.*[1]

2nd. That, as some difference of opinion has existed on this question, the Council, being anxious to prevent any misappre-

[1] The italics are mine.—S. E. De M.

hension as to the grounds of their decision, have thought it right to record their reasons in the foregoing resolution. 1853.

The principle of Universit College.

3rd. That the Secretary be directed to communicate the foregoing resolutions, together with a copy of the extract from Dr. Peene's will, to the three Professors named in the will.

It will be seen by this that the principle of religious equality was still fully recognised twenty-six years after the foundation of the College, as having been that 'on which the College was founded,' and as having been 'invariably adhered to by the present Council and their predecessors, as well in the appointment of Professors, the admission of students, and the award of honours, as as in the general administration of the affairs of the College.'

I wish I could find, for the reason advanced for passing by a principle so distinctly acknowledged, any other word than that which my husband applied to it—'a shuffle.' The determination to accept the books on the prescribed terms confirmed his fears, and on hearing of it his first impulse was to resign his chair. He was induced to remain by the consideration that the classes were not numerous, and that he wished to see the College in a more prosperous state before quitting it altogether. I did not, for my part, endeavour to influence him in this matter. Indeed, at this time my whole thoughts were filled most painfully by the illness of our eldest child, whose danger was not at first realised by her father. I think that when he spoke to me of the condition of affairs at the College, I did not strongly urge his leaving it, for I knew that his doing so would be a trial, and that he was then unprepared for the one already hanging over us. But, with reference to the resolutions, he said, 'They have got in the thin end of the wedge; the next move will be a stronger one.' And so it proved.

The end of this year was the beginning of a long period of sorrow and suffering to us. Our eldest dear child, Alice, who had caught cold after a severe attack of

1853. Death of eldest daughter. measles, died before Christmas. I had feared the termination of the great weakness and delicacy which I had vainly tried to prevent. Her father did not realise the degree of illness till the end was near, and the blow fell heavily upon him. He was not then so used to death and sorrow as we afterwards became, and his want of sight and natural hopefulness of disposition made him unaware of the degree of danger in this and in other cases. This hopefulness left him after repeated sorrows. He always dwelt on the belief that those whom God loves are the early taken, but after we lost Alice his cheerfulness diminished, and I do not think he ever laughed so heartily, or was heard whistling and singing merry snatches of songs as he used to do when all our children were with us. I cannot write of these events. A few references to them will be found in his letters.

1854. The next year passed with scarcely any incident worthy of recording. After our loss my husband remained very much at home, seeing scarcely any one but his fellow-Professors in his daily visits to University College.

An application was made to him to examine and give certificates in the City of London School, but this he declined on grounds connected with the methods and subjects of examination.

In July he gave a lecture to the Society of Arts on a kindred subject, namely, *On the Relation of Logic and Mathematics to other Branches of Science.* This lecture, which was rich in argument and illustration, was only reported in abstract in the Society's Journal. One of its strongest positions was the insufficiency of Mathematics as a mental discipline for inducing logical habits of thought, unless in conjunction with some amount of direct Logical teaching.

1855. Mr. Sheepshanks's death. In the autumn of 1855 our dear old friend Mr. Sheepshanks died at Reading. For the last few years we had seen him but seldom, for he came to London only for the Royal Astronomical Society's meetings after the work on the Standard Scale was completed. This work had been very severe, and probably reduced his strength, which was never great. His death was a blow to many friends, to none more than to my husband, who went to Reading to the funeral—a painful duty, made less painful by his habitual manner of looking at death. He wrote afterwards to me (for I was with the children at Eastbourne),— 1855.

> I returned this evening. I saw the body of my good old friend safely into a bricked vault, specially made for him and his sister, in the cemetery a mile out of the town. There were Airy, Johnson, Simms, myself, and some others. I saw Miss Sheepshanks for a few moments. . . . S. has, of course, made her his sole heiress and executrix. She intends to give all his books and instruments where they may be most useful —perhaps to the Astronomical Society. The house is a very nice one, with a garden so full of rich coloured flowers as to make me almost admire it, with greenhouses, which I did not go into, and a little observatory.

Miss Ann Sheepshanks, who had lived with her brother since the time he left Cambridge, lost with him her great interest in life. She devoted all the energy of a vigorous and self-sacrificing nature to the perpetuation of his name and memory, and the honour due to his unostentatious but most useful efforts to promote Astronomical knowledge. There was much self-denial as well as exertion in her efforts to attain her end. She gave 10,000*l*. to the University of Cambridge for an Astronomical scholarship, to be called by his name. She presented his instruments and books to the Astronomical Society, being in return elected to an honorary fellowship, and she collected materials for a memoir, which was drawn up by Mr. De Morgan.

At this time the phenomena to which I have before slightly referred began to attract general notice, chiefly under the form of table-turning; and natural philosophers, to whose experience this and all its kindred manifestations were so completely opposed, sought for explanation in the credulity and inability to observe of the believers. Mr. Faraday combated the influx of superstition in a lecture *On Mental Training*, given in the spring of 1854, at the Royal Institution. In this lecture he affirmed a principle, which Mr. De Morgan commented on two years after in a review of the printed lecture in the *Athenæum*. Will the time ever come when the reviewer's caution will be needless?

The lecturer has laid down in the strongest and plainest terms the principle of Physics, which was the bane of what is known as *the Philosophy of the Schoolmen*. It occurs in a lecture *On Mental Training*, delivered May 6, 1854, at the Royal Institution. These are his own words:—

'The laws of nature as we understand them are the foundations of our knowledge of natural things. Before we proceed to consider any questions involving physical principles we should *set out with clear ideas of the naturally possible and impossible*.'

We stared when we read this,—'set out in physical investigations with a clear idea of the naturally possible and impossible!' We thought the world had struggled forward to the knowledge that a clear idea of this was the last acquisition of study and reflection combined with observation, not the possession of our intellect at starting. We thought that mature minds were rather inclined to believe that a knowledge of the limits of possibility and impossibility was only the mirage which constantly recedes as we approach it. We remembered the Platonic idea, as clear as the crystalline orbs it led to, that the planetary motions *must* be circular, or compounded of circular motion, and that aught else was impossible. We remembered with how clear an idea of the impossibility of the earth's motion the first opponents of Galileo started these maxims into the dispute. We doubt if in any mediæval writer the principle on which they *acted* has been so broadly laid down as by our author in the phrases above quoted. The schoolmen did indeed make laws of nature the foundation of their knowledge, and clear ideas of possibility and impossibility helped them in the structure. But they rather *did* it than *professed* it.—*Athenæum*, March 1855.

Mr. Faraday believed that a full explanation of the

movement of tables might be found in the unconscious 1855.
action of the muscles of those present, and devised an instrument which he believed adequate to detect it, and to bring to the involuntary operators the conviction that the phenomena, imagined spiritual, had been caused by themselves. I remember hearing him at an evening party at Sir John Herschel's explain the action of this instrument, the *indicator*. A number of ladies and gentlemen listened with interest and attention; the explanation seemed satisfactory, and was received with the respect due to the great fame of its author. Mr. De Morgan, who was known to be one of those whose credulity required a check, stood by with some amusement on his face. I almost wished him to tell some of those things which he had seen which made him doubt the sufficiency of the explanation. But he said it would be useless.

This occurred before the lecture was printed, but it had, I think, been delivered.

SECTION VII.

CORRESPONDENCE 1846–55.

To Rev. Dr. Whewell.

Camden Street, Oct. 21, 1846.

1846. MY DEAR SIR,—First, I am very much obliged by your kind invitation, but my lectures are imperative, and I cannot leave town in November.

Next, as to *Athenæum* police report, you have made worse guesses, unless indeed you never were mistaken in your life.[1]

Now as to the papers. The only wish I have for them to appear in one is that I may get my copies all at one time, and get them disposed of with one trouble. Whether, this condition being fulfilled, they are printed in the form of two papers or one does not matter, and I agree with you that they are distinct enough to be two, and might better be so.

I am going to publish a work on Logic, which, as I told you, will appear soon after the paper. This is sufficient reason for not developing in the paper. Indeed, the Society must know that fact, and take it into consideration in deciding on the printing. There is of course an advantage in new things going first through the usual channels in which scientific matters are propagated, and so I should like the Transactions to have them. But, *tota re perspecta*, the Society may think otherwise, particularly if there is heavy matter, typographically speaking, on hand already. Your suggestion about taking a subject I will think of, but what subjects run *very thickly* in syllogisms? They are mostly full of proof of a very few. Some of Butler's *Analogy* or a chapter of Chillingworth would perhaps be promising. The syllogistic examples in books of Logic are literally nothing more than terms of one word or so substituted in the formal syllogism—I gave some examples (one of each mood) in the *Penny Cyclopædia*, article Syllogism—which (a few

[1] The '*Athenæum* police report' was a humorous skit upon the discovery of the planet Neptune.

at least of them) are more like ordinary sentences involving a syllogism. 1846.

As to subjective and objective (I shall say ideal and objective, as *subject*-ive will not do for logic) I see your difficulty, and must consider whether I have not shown that I see it when the proof sheet comes. I have a great fear of not using the word in the sense of anybody else. The *object* itself, as far as we can think of it, is the *idea* of an object. The first step I make is the existence of my own mind; the next, that of other minds. If everything in existence be a dream of my mind, a thing of which I have ideal possibility, there are no *objects*. If you attempt to argue me into belief of your existence and beat me (not argue by beating me, which is the sort of argument by which Berkeley has been answered before now), I may not be able to answer you; but all that is no more than might happen in my dream. I might sleep, as it is, and dream that I was arguing with somebody who proved to me most satisfactorily that I was awake. But getting by the argument of analogy the existence of other minds, I then begin to know *objects*—other minds get the same as I get, from somewhere. A source of ideas to more minds than one, or to all minds under the same circumstances, would be what I should call my definition of an external object, if, unfortunately, an external object *under the same circumstances* did not imply objects already. Call it then a test of objects; material or not, is of no consequence. Hence the idea of external objects.

By the *idea* I merely mean that which is in the mind. I should distinguish a *horse in the mind* from *that which is in the mind about from whence a horse comes into the mind*; idea of mental state produced, and idea of producing external cause; idea of idea, and idea of object. When I speak objectively, I refer to my idea of the object; when ideally, to my idea of the idea.

But should not objects be divided into *external* and *internal*? What am I to call an idea, looked at as presenting me with the idea of itself? I talked of the idea of a horse; I spoke then of my mind in the state of looking at itself picturing a horse; another mind would have done.

All this, I believe, is common enough. I have put it down that you may see how far our language agrees. Now as to my paper, pray observe that my notion, if such must be inferred of the case of the words *subjective* and *objective*, refers to the case in which all they have to do with formal Logic is stated. And my paper is wholly on formal Logic. The writers on this

1846. subject, so far as their confusion on this point entitles one to say they speak one way or the other, speak ideally, and not objectively. Nay, more, they even admit contradictory propositions as ideally enunciable, and subject to contradiction like others. Thus, 'every collection of two and two is five' is properly convertible into 'some fives are collections of two and two.' Accordingly they give and take no denial except contradiction; nothing with them overturns 'every A is B,' except 'some A's are not B's.' But when we come to apply Logic to the working wants of the mind we find another kind of denial, namely, denial by non-existence; necessary non-existence, or contingent, as the case may be. When we speak objectively, there may be denial by contingent non-existence perfectly distinct from denial by contradiction. Thus *objectively* I deny that 'all unicorns are animals,' not by saying that there are unicorns which are not animals, but by saying that there are no such objects as unicorns; and so far as a unicorn *is not*, so far it cannot *be* animal, or anything else. *Ideally*, I admit, unicorns are animals; my notion is the notion of animal.

I distinguish, then, denial of the *terms* from denial of the *copula*.

A is B ideally, objectively, or (say) *x-itively*.

No! for A has no x-itive existence.

No! for B has no x-itive existence.

No! for the x-itive existence of A and B belongs to *is not*, not to *is*.

Formal Logic usually is made only to treat of the copula. To be strictly *formal* I need not introduce ideal and objective, more than English and French, black and white, x and y. Two species of existence implied as belonging to the terms brought forward would do as well. But *ideal and objective* is the important distinction in practice, and as to assertion or denial, so far as I want it, is easy.

I should now ask you to consider some phraseology.

There are seven *definite* relations of term and term. I do not call $x\ )\ y$ definite, for it consists equally well with $y\ )\ x$ and $y : x$.

1, 2. Start with *identical* and *contrary*, complete co-existence or complete mutual exclusion containing all things between them. As (man being the universe) North Briton and Scotsman, or Briton and alien.

3, 4. *Sub-identical* and *super-identical*, complete content or complete containing. Thus Londoner is a sub-identical of man, man a super-identical of Londoner (man being itself a sub-identical). The case in which the super-identical is the universe may give rise to the *extreme super-identical* and *its* extreme case of *extreme identity*. 1846.

5, 6. *Sub-contrary* and *super-contrary*. The first complete exclusion of one term from the other, both terms *not* filling up the universe as (man being the universe) Englishman and Frenchman. The second contrary overlapping, or where everything in the universe is either one or the other, and some things are both. As (terrestrial object being the universe) *man* and *irrational being*, if madness and idiocy be included under irrational.

7. *Mixed* (what ought to be the name?), where *each* term has part in common with the other part not in common, and both terms do not fill up the universe. The usual form of assertion, as:—Some animals are dark-coloured. I want the word for *mixed*, and better ones for the others, if any. *Mixed* is:—Both have part in, part out, and there are which are neither. There is no hope of a word for all this. Some word formed to contain the idea of *common part* must do, and it should be Latin like the rest.

I tried an experiment yesterday with my daughter of $8\frac{1}{3}$ years old as to the ideas of necessity, and there was a dialogue as follows:—

Q. If you let a stone go, what will happen?

A. It will fall, to be sure.

Q. Always?

A. Always.

Q. How do you know?

A. I'm sure of it.

Q. How are you sure of it? Would it be true at the North Pole, where nobody has been?

A. Oh yes, people have been to the North Pole, else how could they know about the people who live there, and their kissing with their noses?

Q. That's only *near* the North Pole. Nobody has ever been *at* the Pole.

A. Well, but there's the same ground there and the same air· Hotter or colder can't make the air heavier so as to make it keep

1846. up the stones. Besides, I've read in the *Evenings at Home* that there is something in the ground which draws the stones. I am quite sure they would fall. Now, is there anything else you want to be a little more convinced of?

Q. How many do 7 and 3 make?

A. Why, 10, to be sure.

Q. At the North Pole as well as here?

A. Yes, of course.

Q. Which are you most sure of, that the pebbles fall to the ground at the North Pole, or that 7 and 3 make 10?

A. I am quite as sure of both.

Q. Can you imagine a pebble falling upwards?

A. No, it's impossible. Perhaps the birds might take them up in their beaks, but even then they wouldn't go up of themselves. They would be held up.

Q. Well, but can't you think of their falling up?

A. Oh yes, I can fancy three thousand of them going up if you like, and talking to each other too, but it's an impossible thing, I know.

Q. Can you imagine 7 and 3 making 12 at the Pole?

A. (Decided hesitation.) No, I don't think I can. No, it can't be; there aren't enough.

Here her mother came into the room. As long as the questions were challenges from me it was all defiance and certainty, but the moment Mrs. De M. appeared she ran up to her and said, 'What do you think papa has been saying? He says the stones at the North Pole don't fall to the ground. Now isn't it *very* likely they fall just as they do here and everywhere?' But she did not mention the 7 and 3=12 question, nor appeal to her mother about it. I remain, dear sir,

Yours very truly,

A. DE MORGAN.

To Rev. Dr. Whewell.

Camden Street, Camden Town,
Oct. 26, 1846.

MY DEAR SIR,—I have intended for some days to be at you once more in enunciation, on the remaining point of your letter. But I have been hindered by the necessity of looking sharply at the proofs of an account of Newton, which will appear shortly. In this matter I am the *avvocato del diavolo*, as he is called, who is the *ex-officio* opponent at Rome of canonisation. There is only one matter in which the *facts*, in the most objective sense, come out differently with me from other people. The Biog. Brit. says (copied by Brewster) that Whiston says that Newton was so offended by being represented as an Arian, that he *therefore* refused W. admission into the Royal Society. Reference is made to the edition of W.'s memoirs of 1753, which bibliographers know to contain additions. This edition is scarce, but on consulting it, I find that the representation is an absolute falsification; for W. gives the same reason as in the edition of 1749, which has nothing to do with any *ism* at all, or *arian* either. 1846.

Sir D. Brewster has had a lucky escape. It was by mere accident I looked at the Biog. Brit., a work which I never trust in the life of Newton. He gives the same account, with the same reference, without saying he has taken it from anywhere else. Had I not happened to have found his source, I should have left him to clear himself by confessing copying without verification, or otherwise at his discretion. This failing of copying references without acknowledgment has cost me hundreds of hours uselessly employed.

Now to *enunciation*. We must define. If I carry a message out of my mind into yours, and you receive it, and know that I meant to send it, and if, moreover, I *did* mean to send it—I certainly *enunciate*, if the etymology be to give the meaning. But if logical enunciation in pure form be required, there must be subject, predicate, and copula (*is* or *is not*), all duly announced.

According to Aristotle there must be in enunciation either truth or falsehood. Thus prayer, he says, is not enunciation. I say there is truth or falsehood, may be either.

Are we on a question of definition of words, or on one of separation of things? If I shut up my window, meaning to have you believe I am out, I enunciate 'A. De Morgan is not at home;' not verbally, if by enunciation is meant what I call verbal enunciation. So if I know you to be searching for, say your hat, and I *point* to the chair on which it lies, I do not *say*, 'Your hat is on that chair,' but I convey, or mean to convey, the message to your mind. If I were to chalk an X on the great gate at Trinity, meaning to charge the management with peculation, and if others so understood it, the Judge would leave it to the jury to say whether both facts were proved, my intention and others' reception. If they were satisfied on

1846. these points, he would instruct them that the X was a libel, and would leave them to find damages accordingly.

There is no doubt that in law the enunciation of a libel is wholly independent of the symbols used. The rule of law is very distinct; writing, signs, pictures, &c., are equally libels, when intention is proved; and in the civil matter the law decides, not the jury, whether the matter is libellous.

The message intended, and received as intended, constitute with me *enunciation.* If others object to the word, I must choose another word; but this is the thing I mean. Provided always that there is in reality subject, predicate, and copula. Whether message intended but not received is enunciation, that is, whether the difference should not have been a distinctive term, is matter of convenience. If I understood Arabic, to make what the French call a *fière supposition*, and thinking *you* did, wrote you تدسی ییب or whatever it might be (if more dots are wanted pray stick 'em in), and if you did not understand it, there might well be a word to denote this imperfect message.

If I were only to raise an image or single idea, not affirmation of agreement or disagreement—as, if I were merely to call your attention by uttering the single word *book,* àpropos of nothing, I could not be said to enunciate. If you took it as my saying, 'It is my pleasure to say a word, viz. *book,*' you take an enunciation. If that were what I meant, the enunciation is perfect. But if I meant nothing but to set you wondering what I meant, there would be nothing going between us. This mere utterance would, I suppose, be the λογος σημαντικος of Aristotle, as distinguished from the αποφαντικος. What I contend for is, that that which is absolutely considered semantic may be apophantic by the understanding of the parties.

I do not see how 'A is B' is in any other way more apophantic than ∼∼∼∼∼∼ which is no enunciation to you, but for what you know may be to another. This is enunciation to me—

and to all who understand Mavor's short-hand. If prayer be not enunciation, as Aristotle says it is not, how does the other party know it is prayer? Does not the pray-er say—I pray this?

I have got some further development of my Logic in *definite* syllogisms, derived from the classification in my last; with some curious entrance of a principle corresponding to that of like signs give *plus*, unlike *minus*. Common *à fortiori* reasoning will take its place in a class of distinct syllogisms. 1846.

I remain, dear sir,

Yours very truly,

A. De Morgan.

To Sir Rowland Hill.

Camden Street, Camden Town,

May 5, 1848.

My dear Sir,—I am much obliged to you for the notice. 1848.
I believe you when you say there are difficulties, because you get over them. Still, to my untutored mind, it is wonderful the Post Office should imagine that anybody would write in a book at 6*d*. a pound to save postage.

I hope that the end of it will be that anybody may write anything, and I have reason as follows:—

There is an old book I want; for example, the first edition of Wingate's *Arithmetic*, 1630. If one of my country friends finds it, what will be in the inside of an old Arithmetic? A hundred to one, something like—

Ann Price, her booke,
God give her grace therein to looke,[1]

scrawled over the inside of the cover and the fly-leaf—that is, over more than one page. Now it does not consist with the fitness of things that Ann Price's aspirations after Arithmetic in the seventeenth century should prevent a professor of Mathematics in the nineteenth from ascertaining the exact share of Wingate in the invention of decimal fractions.

You stop the circulation of old books. However, as I said, if you say it can't be, I will believe you, provided the impossibility may be interpreted as temporary.

But for the love of order, and the Constitution, and the other things that were dusted on the 10th ult., don't compel all the old-book people to stand up for equal rights and against class privileges. You'll make Chartists of Sir H. Ellis, and Hallam, &c., &c., to say nothing of,

Yours truly,

A. De Morgan.

[1] Ann Price's (probable) handwriting imitated.—S. E. De M.

To Dean Peacock on his Marriage.

7 Camden Street, Oct. 21, 1847.

1847. MY DEAR SIR,—This morning I found two cards for me at the College, which informed me of your marriage, of which I had heard nothing. In fact, for anything I knew, you might have been as confirmed a Benedict as any Pope of that name. But owing to the practice which ladies have of not putting the name they leave as well as the one they take, I had no guess who Mrs. Peacock had been; and the theory of probabilities does nothing in the way of inferring the probable name which a bride quits, having given that which she takes. So I resolved on writing hearty congratulations and warm good wishes on the existing *à priori* (or if you will have it that priors are out of question by their vows, say *à diaconiori*) presumption that you were well able to know what was good for yourself. But it so happened that an Ely man saw the cards in my hand, and, as the phrase goes, told me all about it; and I was enabled to conclude from other evidence that I might just keep my good wishes, and put good prophecies in their place. Take them both, however. As to this practice of putting only one value of the variable on wedding cards, I object to it altogether; in fact, I denounce it, and will prove my objection good. I suppose no one will deny that the cards represent the instant of the ceremony at which the contract becomes indissoluble; for before that moment the announcement would be presumptuous, and to suppose that any time elapses after it would be to suppose that a man takes that time to consider whether he will acknowledge his marriage, which is absurd. This being granted, let A B represent the duration of the lady's life, and let M be that moment of the

ceremony at which the contract becomes indissoluble. Let the lady's name during A M be Selwyn, and during M B Peacock; then, because by common courtesy a lady is not a discontinuous fraction, it follows that what is true up to the limit is true at the limit, therefore at the moment M her name is Selwyn. But for a similar reason her name at the same moment is also Peacock; therefore at the instant M she has both names, whence both ought to appear on the wedding cards. Q.E D.

I have your books on arithmetic in safety and memory, and am only waiting to return them till I have put a copy of my

Logic into the parcel, which I hope to do in about a fortnight. As matters are, I feel no compunction at having kept them so long. I beg to offer my best compliments to Mrs. Peacock, and my apologies for introducing myself by inserting her name into a demonstration. But first principles must be carried to their full extent; and I remain, my dear sir, 1847.

Yours most sincerely,

A. DE MORGAN.

To T. K. Hervey, Esq.

Dec. 1848.

DEAR HERVEY,— . . . . A man named Lacroix, a French bibliomaniac, has been over here. He came over with strong prevention against Libri, but examined his case here, and is gone back very angry with his accusers. He is preparing a pamphlet *de son chef* in defence of Libri, of which the latter promises me an early copy, or proofs if he can get them. So far good. But if you could light on any information about Lacroix (nicknamed Jacob Bibliophile in his own country), or any one of his bibliographical publications, so much the better; for this Lacroix must be looked after. Panizzi and Libri unite in declaring that of upwards of 1,700 manuscripts, sold by Libri to Lord Ashburnham some years ago, Lacroix named them all, with a few exceptions, and described where they originally came from, merely from his knowledge of existing manuscripts and their localities, thus negativing from his own personal knowledge the charge of theft as to very nearly the whole lot. This story is so extraordinary that, if true, as I cannot doubt it must be in the main, this same Lacroix should be brought forward in England and his works noticed. I can believe such a story, for I have heard such things well attested of people who pass their lives in studying the physics of books and MSS. 1848.

Yours truly,

A. DE MORGAN.

To T. K. Hervey, Esq.

Dec. 1848.

'And serve it with Hervey's sauce.'

JERDAN.

DEAR HERVEY,—That unconquerable mania which you have for thinking your puns *as good* as mine (you say *better*, but I don't believe you think *that*—the most singular fancies are

1848. sometimes carried further than they go, out of mere bravado) is a study for the psychologist. I shall forward a statement to Sir W. Hamilton.

Pray when within two years have you mentioned Lacroix?

However, since you know him, which I didn't, you now know that he has a memory.

And what makes you say that I never read any papers but those of a mathematician? Mathematica! quoth he—*are you* a mathematician? and did I not read all the we-should-gladly-forget-them-if-you-would-let-us articles, which procured you the memorable rebuke (which you will never get over) with which I have headed this letter?

And as to *preventions*, was I not talking of a Frenchman? and if he had described himself, would he not have used the word? And did I not get the word from Panizzi, and was I not assured of an Italian borrowing a Frenchman's phrase—who *deniges* of it? Yours truly,

A. De Morgan.

To Capt. Smyth.

7 Camden Street, Dec. 19, 1848.

My dear Captain Smyth,—Pray what is the matter with you? Pray write and say you are quite well; but mind, I detest lying of all things, so be sure you speak the truth.

I took a solitary glass of porter yesterday to your recovery, for I did not choose to admit any of the profane dogs about me to the ordinance, which is quite above their appreciation.

Airy gave us a very good telegraph lecture. I mean *on* telegraph, not by telegraph. But time may come when we shall sit down in our own room and hear him lecture from Greenwich.[1]

Seriously, let me know how you are. With kind regards to all, I am yours truly,

A. De Morgan.

To Sir John Herschel.

March 18, 1849.

1849. My dear Sir John,—Sir H. N. is as correct as his authorities. Censorinus, who gives the most distinct account, says that

[1] I do not suppose the writer had the smallest conception of the wonderful literalness with which his prediction would be fulfilled. It must be remembered that the telephone was not even dreamed of thirty-three years ago.—S. E. De M.

there was leap-year every fourth year, but does not say from whence the years were reckoned. 1849.

We have no authority for saying which particular years were leap-years—either in the pre-Augustan piece of the Julian Calendar, or at the start made after the Augustan reformation.

Nevertheless, I think a little train of reasoning will bring us to the following theorem.

The Julian Calendar starts with what, by reckoning back, we should call January 1 of the year -45, on the supposition that 0 does not exist, but that we pass from $+1$ to -1 consecutively, on the supposition that *every* fourth year is leap-year.

There is much reason to suppose that Cæsar began his year on January 1 because there was a new moon on this day. Otherwise it is likely he would have commenced it on the *shortest day* preceding. He is thought to have gratified the feelings of the Romans by making his start on a new moon day, and Macrobius, in the words 'Annum civilem Cæsar habitis ad unam dimensionibus constitutum, edicto palam proposito publicavit,' is held to have alluded to this. Now the fact is that January 1, -45, back-reckoned as before noted, is found to have been a day of new moon. Dr. Smith's *Dictionary of Greek and Roman Antiquities* (a book you ought to have—there is a good article on the Calendar) says it was at 6h. 16m. P.M. My rough calculation gives 10h. 55m., which I take to be within a quarter of an hour.

Now, as our tables reckon back (old style) upon the supposition of uninterrupted leap-year every four years, I take it that, as to the interval, we may depend upon knowing the exact number of days that have elapsed.

But how are we to explain the dropping into leap-year at $+4$?

Diagram I. shows us—

J. Julian leap-year.

P. Priests' mistaken leap-years.

A. Augustan leap-year after the suspension.

At the end of the year 4, the priests' leap-years and one Augustan make 13, just what there ought to have been by our back-reckoning. If, then, $+4$ was Augustan leap-year, we are all right. I assume that the first year of the reckoning was certainly not a *corrected* year. Accordingly the first priests' leap-year was -42 J., showing the Julian intention was never carried into effect.

1849. Diagram II. shows how, according to the well-known Roman mode of counting, the edict of Augustus in −8, there shall be no leap-year for twelve years, would be accounted to make +4 leap-year. If a man had been sentenced on Monday evening to six days' imprisonment, he would have been let out on Saturday morning. This seems to me to explain how we may reckon intervals from *our* January 1 −45, but from thence to +4 intervals must be corrected, though not after. The dates −45 and −8 are well fixed by the consuls being named.

The general impression is that the first of Cæsar's years, −45, itself was his bissextile. This seems to me absurd. Cæsar did not care about equinoxes. All he wanted was to keep an average of 365¼, and correcting before the error had accrued would surely never have struck him or Sosigenes.

Moreover, the preceding theory accounts for $4n$, and shows how the new moon may be made to fall where we know it did.

Yours truly,

A. De Morgan.

To the Rev. Wm. Heald.

Dear Heald,— . . . . Talking of curious powers, tell me what you think of the following story. It quite beats me.

I have seen a good deal of mesmerism, and have tried it myself on —— for the removal of ailments which required much medicine, but which mesmerism met without medicine from the time it was employed. Of the curative powers of this agent I have no more doubt than one has of things which he has constantly seen for years. But this is not the point. I had frequently heard of the thing they call clairvoyance, and had been assured of the occurrence of it in my own house, but always considered it as a thing of which I had no evidence direct or personal, and which I could not admit till such evidence came.

One evening I dined at a house about a mile from my own —a house in which my wife had never been *at that time.* I left it at half-past ten, and was in my own house at a quarter to eleven. At my entrance my wife said to me, '*We have been after you,*' and told me that a little girl whom she mesmerised for epileptic fits (and who left her cured), and of whose clairvoyance she had told me other instances, had been desired in the mesmeric state to follow me to —— Street, to ——'s house. The thing took place at a few minutes after ten. On hearing the name of the street the girl's mother said,—

1849.

'She will never find her way there. She has never been so far away from Camden Town.'

The girl in a moment got there. 'Knock at the door,' said my wife. 'I cannot,' said the girl; 'we must go in at the gate.' (The house, a most unusual thing in London, stands *in* a garden; this my wife knew nothing of.) Having made the girl go in and knock at the door, or simulate it, or whatever the people do, the girl said she heard voices upstairs, and being told to go up, exclaimed, 'What a comical house! there are three doors,' describing them thus.[1] (This was true, and is not usual in any but large houses.) On being told to go into the room from whence voices came, she said, 'Now I see Mr. De Morgan, but he has a nice coat on, and not the long coat he wears here; and he is talking to an old gentleman, and there is another old gentleman, and there are ladies.' This was a true description of the party, except that the other gentleman was not *old*. 'And now,' she said, 'there is a lady come to them, and is beginning to talk to Mr. De Morgan and the old gentleman, and Mr. De Morgan is pointing at you and the old gentleman is looking at me.' About the time indicated I happened to be talking with my host on the subject of mesmerism, and having mentioned what my wife was doing, or said she was doing, with the little girl, he said, 'Oh, my wife must hear this,' and called her, and she came up and joined us in the manner described. The girl then proceeded to describe the room; stated that there were two pianos in it. There was one, and an ornamental sideboard not much unlike a pianoforte to the daughter of a poor charwoman. That there were two kinds of curtains, white and red, and curiously looped up (all true to the letter), and that there were wine and water and biscuits on the table. Now my wife, knowing that we had dined at half-past six, and thinking it impossible that anything but coffee could be on the table, said, 'You must mean coffee.' The girl persisted, 'Wine, water, and biscuits.' My wife, still persuaded that it must be coffee, tried in every way to lead her witness, and make her say *coffee*. But still the girl persisted, 'wine, water, and biscuits,' which was literally true, it not being what people talk of under the name of a glass of wine and a biscuit, which means sandwiches, cake, &c., but strictly wine, water, and biscuits.

[1] A little diagram is given of these doors (she counted three, but indicated more) in the letter.—S. E. DE M.

1849. Now all this taking place at twenty minutes after ten, was told to me at a quarter to eleven. When I heard that I was to have such an account given, I only said, 'Tell me all of it, and I will not say one word;' and I assure you that during the narration I took the most especial care not to utter *one syllable.* For instance, when the wine and water and biscuits came up, my wife, perfectly satisfied that it must have been coffee, told me how the girl persisted, and enlarged on it as a failure, giving parallel instances of cases in which the clairvoyants had been right in all things but one. All this I heard without any interruption. Now that the things happened to me as I have described at twenty minutes after ten, and were described to me as above at a quarter to eleven, I could make oath. The curtains I ascertained next day, for I had not noticed them. When my wife came to see the room, she instantly recognised a door, which she had forgotten in her narration.

All this is no secret. You may tell whom you like, and give my name. What do you make of it? Will the never-failing doctrine of *coincidence* explain it?

I find that there are people who think that the house in the garden, the number of doors on the landing, the two gentlemen beside myself, and ladies, the red and white curtains, the singularity of the loops, the two so-called pianos, the lady joining myself and one old gentleman apart from the rest, the wine, water, and biscuits, the truth of the whole and the absence of anything false, are all things that may reasonably enough arise by coincidence, when the daughter of a poor charwoman (twelve years old[1]) undertakes to tell a lady all about where her husband is dining, in a house where neither has ever been.

I have seen other things since, and heard many more; but this is my chief personal knowledge of the subject.

Yours very sincerely,

A. De Morgan.[2]

[1] She was turned eleven—in her twelfth year.

[2] I heard all about the house and furniture, &c., *before* the girl told me what was going on. Mr. De Morgan has represented it to Mr. Heald as occurring *after*, and it is quite possible that I told him in this order. But I never heard of this letter till after his death. —S. E. De M.

To Sir John Herschel.

41 Chalcot Villas, March 26, 1850.

MY DEAR SIR JOHN,—I never heard the polar axis approximation. Pray throw it out in type, for it is quite a funny thing that we beat the French after all. And the polar axis is the only sensible diameter the earth has got. He keeps snug and quiet, and lets all the others spin about him. I think a dialogue might be written between the polar axis and an equatorial diameter—quiescence against restlessness. 1850.

And so Logical systems are bothersome. I have got sixty-four more syllogisms symbolised, in which terms take quantity *from others.* As—

For every Z there is an X, which is not Y.
Some Y's are Z's.

Required the inference.

Symbol (· (( ).
Inference (. (.
Some X's are not Z's.

These are really *hard.* To give an instance.

'To say nothing of those who succeeded by effort, there were some who owed all to fortune, for they gained the end without any attempt whatever, if indeed it be not more correct to say that the end gained them. But for every one who was successful with or without effort, at least one could be pointed out who began, but abandoned the trial before the result was declared. And yet so strangely is desert rewarded in this world, there was not one of these faint-hearted men but was as fortunate as any of those who used their best endeavours.'

I will answer for it that if this were presented to any writer on logic without warning, he would pass it over as not self-contradictory at least. But for all that, it contains the same error as the following:—'All men are animals, and some are not.'

Yours very truly,
A. DE MORGAN.

To the Rev. Dr. Whewell.

7 Camden Street, May 25, 1850.

MY DEAR SIR,—I am much obliged to you for the continuation of the chain of events. I see you are propagating an undulation

1850. through the College—a very elastic medium. I hope the matter will not lead to a gown and town dispute—a nominalist and realist discussion; the gown being nominalist, or dealing in words, while the town is realist, or, as the French say, proceeds *par voie du fait.* I mean to approfound the matter when I get an evening clear, as far as I can. I am loth to believe the text of Aristotle to be unimportant anywhere. I suspect that he shares the fate of Euclid in modern times—to wit, that everybody believed him to be so near perfection as to be willing to give him the finishing touch—to bring him quite up to it. Ptolemy has escaped this fate; but then Ptolemy, the real original, was comparatively little read—his explainers traded on their own bottoms. Compare the number of editions of Ptolemy with those of Euclid and Aristotle.

I am sorry you are all against the Royal Commission. I think that such a Commission as would certainly be appointed—properly supported by the Universities—would much tend to open the public eye to what the Universities really do. So very little is known about them that something of the kind is much wanted. If it had been a Parliamentary Commission, it would have been another thing. You might have said '— — —.[1] We do *our* work better than you do *yours*, at any rate.'

Listen to my last brand-new definition of metaphysics:—

'The science to which ignorance goes to learn its knowledge, and knowledge to learn its ignorance. On which all men agree that it is the key, but no two upon how it is to be put into the lock.'

Yours very truly,
A. De Morgan.

[1] Fill up with the τριας τετραγραμματων, which it would not be canonical to write.

To T. K. Hervey, Esq.

July 3, 1850.

Dear Hervey,—If you read again the articles[2] which have appeared in the *Athenæum*, you will see that it is not merely that as long as no proof is offered the presumptions are in favour of M. Libri, but that he, M. Libri, has actually overturned by documentary evidence—which you, speaking editorially, *saw*—every specific accusation mentioned as capable of being brought

[2] The writer's own articles in defence of M. Libri.—S. E. De M.

against him. As far as I understand the matter—I mean the recent matter—it is this. M. Libri not appearing is declared guilty, *par contumace.* No evidence is offered in such a case; an indictment only, making *allegations*, is enough. He can still return *and stand his trial*, if he should be mad enough to trust himself in a country in which his witnesses would be prevented from appearing by intimidation. 1850.

This is the reason why you saw no evidence in the *Gazette des Tribunaux*—because none was offered, or could be. All this you will find on inquiring into the French law; and you will find that the trial and sentence *par contumace* are provisional. I believe the appeal would be the real trial if he went to France. I have never communicated with you on this matter (though I confess I rather longed to do so), because I did not feel at liberty to try to make use of the *Athenæum* in a matter in which I felt personally interested, when I had, for reasons discussed between us, felt obliged to withdraw from general contribution. This would have been making a convenience of you, as *I* should have thought, even if you did not.

You will remember that I was neither friend nor acquaintance of Libri, but strongly prepossessed against him, when, as I was going to treat the subject in the *Athenæum*, I demanded of Panizzi the proof-sheets of his forthcoming defence against the allegations of M. Boucly's report, and access, which I got, to the original documents on which he founded his refutation. Being fully satisfied as to his innocence, I cultivated his acquaintance; and since that time much collateral evidence has reached me, not only as to his innocence, but as to his being in truth a high-minded and earnest employer of first-rate talents and learning in first-rate pursuits—far above what the time-serving French *savans*[1] can imagine or appreciate. As being now proud to call myself a personal friend of his, I am hardly so well qualified to treat his case in a public journal as I was when my only knowledge of him (as to his character) referred to his means of meeting the allegations made against him.

If he should entertain the idea of demanding his trial in France, *I will do all I can to hinder such a piece of insanity.*

The idea that there has been discussion of evidence in this proceeding and conviction *par contumace* is very common, I find. I have no doubt that M. Libri will take some public steps to

[1] I mean those of them (a majority, not all) who are time-serving.

1850. inform the English public how the matter stands. All this is written without communication with him.

You will, of course, take care to be well informed as to the nature of the above proceeding. That being the case, I think you will probably find that the matter stands, to any reasonable mind, just where it did. If you 'state the position in which the case stands,' I think it most likely that you will do nothing which any friend of M. Libri can regret.

On casting my eye over your note, I marked the words, which I missed at first, 'That *proof* has been given in a court of law; on what amount of valid *evidence* I cannot say.' Now I say that you will be able to ascertain that there has been neither proof nor evidence—only indictment—allegation and judgment by default of appearance. Of course, a tender of evidence is implied in the indictment, and, for aught I know, in the recital.

Yours truly,
A. DE MORGAN.

To the Rev. Dr. Whewell.

July 12, 1850.

MY DEAR SIR,—I have got my paper on logic out of hand, and have begged the Pitt Press to retain one of my copies for you, and to send it to you; which if they neglect, I shall be much obliged by your reclaiming, as the French say.

I have to-day got Sir W. Hamilton's system for the first time in a full and acknowledged form. His pupil, Spencer Baynes, has published the essay on it which got the prize in 1846; the very essay, the requisites for which, sent to me, made the foundation of Sir W. H.'s charge of theft. It has appendices and a note by the arch-syllogist himself. I and Boole come in, without being named, for a lecture against meddling with logic by help of mathematics. Pray get this work and read it carefully.

My next thoughts about the subject will be on the relation between the laws of enunciation and the laws of thought, and particularly with reference to certain invasions of each other's province which I imagine to exist.

I shall return to an objection of yours to my assertion that prayer enunciates. (You may have forgotten it, but I have all my logic correspondence together, and have been looking over it.) You say that under such an extension a man who shuts up his window enunciates that he is not at home. I dispute your example as to matter, though not as to form, if you choose a better one. Closed windows may denote death or absence, &c. But change it thus. A man who ties a *white glove* on his knocker enunciates that a child is born in the house. I believe there is no ambiguity of meaning here. I hold that he does *enunciate.* However, this is all for consideration. I remain, dear sir, 1850.

Yours truly,

A. DE MORGAN.

To Sir John Herschel.

Oct. 15, 1850.

DEAR SIR JOHN,— . . . .

I always advance the following as the infinitely small quantity which is the most puzzling of all. All others are mental creations, but this one seems different.

Let it be granted that a target which must be hit can be conceived. It may be the whole *enceinte* of the room, ceiling, floor, walls, &c.

Let it granted that the fixing of an arrow with a mathematical point can be distinctly *conceived.* I don't ask for workmanship.

Let A be a *point* in the target. Since some point must be hit, and all are equally likely, there is *some chance* of hitting A—that is, it is not impossible to hit A, which is synonymous.

But the chance of hitting a given point is certainly less than any that can be assigned.

Therefore there does exist in the mind an idea of a quantity which, not being nothing, is less than any that can be assigned.

In geometry we do not meet the same difficulty, because we learn (*how* correctly I give no opinion on) to know the point, line, surface, and solids as different *species* of magnitude, but belief cannot be subdivided into different species. Is not an expectation of hitting A *homogeneous* with that of hitting some point within a given area?

I do not know whether you have returned. I hope all your clan are well, and you yourself not disposed to give any hints about your scientific life being terminated, as you did a while ago.

Yours very truly,

A. DE MORGAN.

From my examination room, where I shall sit two hours and a half more, without anything to do except just what I please,

1850. so don't say I write in a hurry. University College, London, October 5, 9h. 30m. A.M. $\pm$ the error of my watch, 1850, the last year of the first half of the nineteenth century, let who will call it the *first year of the second half.*

To Rev. W. Heald.

7 Camden Street, Aug. 18, 1851.

1851. DEAR HEALD,—It has become quite the regular thing for the depth of vacation to remind me—not of you, for anything that carries my thoughts back to Cambridge does that,—but of inquiring how you are getting on, of which please write speedy word, according to custom, once a year. For myself I have nothing particular to report. My wife and seven children are all at Broadstairs—as they were when I last wrote—so that the information is that they really came back in the interval. I presume you really have *not* come to town to see the Exhibition, supposing that you would surely have let me know. Are you not coming? Whether I with my short sight should know you again after a quarter of a century, *plus* a quarter of a year, is a problem I should very much like to solve. But you seem determined not to furnish the data.

It seems to me that I must have written to you just before the Pope made his onslaught, which has occupied people ever since. I remember, soon after the Catholic Emancipation Bill was carried, reminding a friend of mine, a Catholic barrister, that that Bill was an experiment—a very proper experiment—one it was disgraceful not to have tried before; but still an experiment, in trial of whether it really was practically possible that people with any foreign allegiance, call it spiritual or anything else, could permanently exercise the rights of citizenship here. The occasion was his speaking *very seriously and earnestly* of it being a matter of discussion among the Roman Catholic body whether they had not in right of the E. Bill a right to proceed in Chancery against the Colleges at Oxford and Cambridge, which were founded on condition of praying for the souls of the founders, to make them either so pray or give place to those who would. It gave me at the time (the man being neither a sanguine man nor a fool) a fixed idea that from the very time of the Emancipation Bill passing there was a settled purpose of legal invasion. And I have never since faltered in the opinion that, be it settled how it might, the time would come when, on political grounds, the question would be reopened; and I prophesy it now within a few years—that is, I foretell a discussion whether the mere circumstance of owning a foreign power in any sense and manner whatever is or is not to be an absolute disqualification from even voting for a member of Parliament. 1851.

I have just heard from Arthur Neate, who with a wife and two children is doing near Alvescot what you are doing at Leeds, saving that his two parishes put together would not soul a tenth part of the bodies in your one. His father and mother are still both alive, though both very old and failing. Of other people I know nothing, I mean of your and my contemporaries. It is long since I have seen any one. I met Farish the other day, old and deaf. I am not sure I do not remember his father looking younger. I dare say you, like myself, look not very old of your age, for we both looked older than we were at Cambridge, so that if you have a provincial synod, you will hardly look ancient enough to be one of the *patres conscripti*. But you have not a Bishop, I am afraid, who will bring your part of the world abreast of H. Exon. Peace be with him, I was going to say, but I know she won't.

. . . .

Resolve me this. If our old friend P—— were alive, would he be Puseyite or not? The only one Cambridge man that I ever annoyed by taking it for granted that he was not Puseyite when he really was a strong one, was a man of whom I could tell the following story, but I won't (that is to say, you are not to repeat it, for it might get round).

I knew him at Cambridge when he was a great friend of B——, whom you perhaps have met at Neate's. A few days after he was ordained he came to see me, and being fresh off the anvil he could not but talk a little theology. So as he got over the ground he came at last to the following sentence, which brought him up all standing, as they say at sea—you are to imagine a sudden start of recollection at the *, I having stared at †:—'But you see those Catholics made a sacrament of baptism † *. Oh, by-the-bye, so do we.' Fact, upon my honour; no exaggeration. But he is now with the Bishop of Exeter on the point.

I wish you would do this: run your eye over any part of those of St. Paul's Epistles which begin with Παυλος—the Greek, I mean—and without paying any attention to the meaning. Then do the same with the Epistle to the Hebrews, and try to balance in your own mind the question whether the latter does not deal in longer words than the former. It has always run in my head

1851. that a little expenditure of money would settle questions of authorship in this way. The best mode of explaining what I would try will be to put down the results I should *expect* as if I had tried them.

Count a large number of words in Herodotus—say all the first book—and count all the letters; divide the second numbers by the first, giving the average number of letters to a word *in that book.*

Do the same with the second book. I should expect a very close approximation. If Book I. gave 5·624 letters per word, it would not surprise me if Book II. gave 5·619. I judge by other things.

But I should not wonder if the same result applied to two books of Thucydides gave, say 5·713 and 5·728. That is to say, I should expect the slight differences between one writer and another to be well maintained against each other, and very well agreeing with themselves. If this fact were established there, if St. Paul's Epistles which begin with Παυλος gave 5·428 and the Hebrews gave 5·516, for instance, I should feel quite sure that the *Greek* of the Hebrews (passing no verdict on whether Paul wrote in Hebrew and another translated) was not from the pen of Paul.

If scholars knew the law of averages as well as mathematicians, it would be easy to raise a few hundred pounds to try this experiment on a grand scale. I would have Greek, Latin, and English tried, and I should expect to find that one man writing on two different subjects agrees more nearly with himself than two different men writing on the same subject. Some of these days spurious writings will be detected by this test. Mind, I told you so. With kind regards to all your family, I remain, dear Heald,

Yours sincerely,
A. De Morgan.

To Sir John Herschel.

7 Camden Street, Aug. 29, 1852.

1852. My dear Sir John,— . . . Induction seems to lead to the conclusion that an astronomer who is Master of the Mint gets some odd mode of chronology. The first cut a great piece off the beginning, the second will cut a great piece off the end, and doom us all to be squabashed in 1865. The next, I suppose, will cut a great piece out of the middle, which will be the most singular job of the three. What if he should say that the 8th, 9th, and 10th centuries never existed? I wish they hadn't. 1852.

When De Gasparis gets his next planet, he and Hind will be six of one and half a dozen of the other. Do you mean to say that just as we have got the place snug, drained, lighted, and electro-wired and railed, that as soon as we shall just have learnt to have an idea of behaving to each other like people whose posterity may in time be Christians, we shall have to become fossils, and megatheriums, and such like, for smarter chaps than ourselves to write books upon? I will never believe it till I see it, and then only half. Why, it is only just four hundred years since printing was invented. A book, with ordinary care, will last a thousand years. It is astonishing what good condition those of 1480 are in, even after a course of bookstalls. Surely the nature of things is to live their lives out. . . .

Yours truly,
A. De Morgan.

To Dean Peacock.

7 Camden Street, Aug. 30, 1852.

My dear Sir,— . . . All I know about Young personally is, that one evening in 1828, when I first pushed my nose into the scientific world, I was presented to Young, Davies Gilbert, and Wollaston.

Wollaston said, when I was introduced as Professor of Mathematics in the University of London, 'Are they to have a Professor of Mathematics?' I told him they had one, and that I was he. Nothing more passed. Young lifted his eye-glass, and made his bow serve the double purpose of acknowledging the introduction, and bringing his eyes to the lenses. He made me certain that he saw me, and impressed me with an idea from his manner that he was *fine*. Perhaps he was only shy —shyness takes every other form to avoid its own.

Davies Gilbert was the only one of the three who had the manners of a man of the world. I believe I never saw the two first again.

I never knew till many years afterwards that I was well acquainted with some members of Young's family. His brother, Robert Young, was a Quaker, who married, as I was told when a boy, a lady, who was not a Quakeress, and was disowned by the sect. This lady was a most intimate friend of my mother, and Robert Young is one of the earliest persons I can

1852. remember. He was a banker, and something else, I think, at Taunton; he afterwards went to Bristol, and was in some business. When I was sent to school near Bristol in 1820, I was consigned to R. Y., who especially warned me not to walk in my sleep, as there were no leads outside the window—they had been removed. The consequence was that, though I never walked in my sleep before or since that I remember, I was awakened by the wind blowing on me, and found myself before the open window, with my knee on the lower ledge. I crept back to bed, leaving the window open, and the family, being alarmed by the noise, came into my room, found me asleep and the window open; so that as their fenestral logic did not reason both ways, they forgot that the leads were not there, and searched the whole house for thieves. Long afterwards I met R. Y. in Stratford's room, negotiating about some papers of Young referring to the R. A., and there I learnt whose brother he was. John Young, I am pretty sure, was a brother, if not a cousin.

You will remember that it has been said that Somersetshire has been very deficient in great men; and the *exceptiones firmantes regulam* have been Roger Bacon and John Locke. It is time that Young should make a third.

I do not know whether you have all your information about Young's family. If you want any inquiries, I have some old friends still at Taunton, and will ascertain what you want. Milverton, Young's birthplace, is a few miles from Taunton.

I hope you will not overwork yourself; and remain, dear sir,

Yours sincerely, &c., &c.

To Rev. W. Heald.

7 Camden Street, Sept. 11, 1852.

DEAR HEALD,—I make my annual renewal of correspondence, which I have got into a habit of doing when my wife and children leave town. They have gone this year to Herne Bay—not so far from London as last year, when they were at Broadstairs. By the way, a scientific friend of mine directed to me at Broadstairs, near London, when near Ramsgate would have been nearer the mark. On my asking him what he meant, he said he remembered some very broad stairs down to the river just below London Bridge, and he had a vague idea that they were the Broadstairs. Doubtless there are very broad stairs thereabouts. This put me on asking the etymology of Broadstairs, and I find that by stairs are meant passes down the cliff—natural passes. What are you and yours doing? Do not fail to tell me all about yourself, without my drawing it all out of you by specific questions. By the way, is the Mr. Prickett I see in the papers on whom somebody has been forging, any relation of our old friend? Of myself and family, nothing particular. We're all about a year older since I last wrote to you. I have been looking over and sorting correspondence of more than twenty years, and I do not see any particular marks of growing old in your handwriting. Are you not seriously contemplating the necessity of calling yourself 50 years old if things go on as they have been doing? By my estimate of your age, you will be saying 49 next birthday. I am 46 past, but, between ourselves, I have two of my wise teeth still to cut. 1852.

I looked out in the papers to see if you were moving or seconding anybody into Convocation, or being done the same to yourself. What do you think about the revival of Convocation? Did it ever happen to you to study any of their old proceedings? Where are they all? I remember that, *à propos* of the Easter Question, I wanted the acts of the Convocation which met next after the Restoration; but, though Maitland did all he could for me in the Archbishop's library, the return was *non est inventus.* Maitland is now settled at Gloucester again; what doing I don't know. He is now well stricken in years: thirty-five years ago he had completed Cambridge, had been educated for the bar and practised, had got sick of it, had retired, had married, and sat himself down comfortably at Taunton, next door but one to his father, my mother being the intermediate. I doubt his being less than thirty-five then, so that he must be seventy, I should say, at least—and he looks it. At Taunton he used to collect books and play the fiddle, and my first acquaintance with Haydn's twelve was made through him and his sister-in-law. He also bound his books himself, and he bound the upright of his bookshelves, and lettered them 'Maitland's Works,' at which his friends used to pull with great curiosity to know what he had written; and those who did not pull thought it very odd that he should write so many thin volumes on *equidistant* subjects.

I wrote you a note to see if you knew who A. E. B. of Leeds was. I suppose you do not. He shines in a publication called *Notes and Queries*, which I take in, and find a great deal of miscellaneous in it. Did you know James Parker, the vice-chan-

1852. cellor, who is just dead? He was a great friend of Farish's. Of our old friends I know but little. Neate is thriving. His father and mother are still living, which few men of forty-five or thereabouts can say. He vegetates, I am afraid; his parishes are not very populous, and though he does everything in the way of looking after them, his grasp is not full. For aught I know your parish would make fifty of his in number of souls.

August 31.—I see to-day that Maitland has published a new ittle book, combining several tracts with mediæval pictures.

I bought an auctioneer's lot the other day for one book, and found, among the rest, Hone's Trials, which I had never read through, though, when I was a boy, I had my curiosity greatly whetted by the sharp way in which they were kept out of my sight, while I was admiring the presence of mind of the defendant, and the circumstance of a man not regularly *educated* sticking logically to one point (a great rarity), namely, that the non-prosecution of parodies in favour of ministers proved that the animus was political, and that religion was a pretext. There came into my head a long-forgotten story told me by Place, the celebrated political tailor, more than twenty years ago, which shows that Cobbett, with all his pen-assurance, had not the nerve of poor Hone. When Place and some friends went to consult with Cobbett about his defence to the action for seditious libel which was coming on (on which he was convicted and imprisoned), Place told him that if he wanted to escape conviction he had only to produce the letters which public functionaries had written to him on points of his paper—bar, judges, the Speaker of the H. of C., &c.; that if he did this he would prove that he was not considered a common libeller even by the friends of Government; and that having thus made a *locus standi* he could deal with the specific charge as a fair political comment, and compare it with others. Cobbett was hardly able to speak of this plan, so great was his agitation at the boldness of producing these letters, which would have made a great sensation, for there were very curious *private applications* for his good word. He did not dare to do it, was regularly browbeat by the judge, even in what he did venture, and was convicted. Such is the difference between *pen*-courage and tongue-courage.

Pray present my best compliments to Mrs. Heald. I am sorry I cannot say remembrances. There ought to be a prospective mode of address. It would sound very odd to say, in the case of a person whom the writer had not seen, 'Present my most sanguine anticipations;' but what would be so odd as 'compliments' if used for the first time? 1852.

Surely the time must come when the vortex of London will suck you in for a few days. In the meantime let us speculate on the question whether we should know each other if we met in the street after twenty-seven years of non-visual intercourse.

Yours most sincerely,
A. DE MORGAN.

To Rev. W. Heald.

7 Camden Street, July 1853.

* * * * * *

I remember giving you my experience in regard to clairvoyance. I will now tell you some of my experience in reference to table-turning, spirit-rapping, and so on. 1853.

Mrs. Hayden, the American *medium*, came to my house, and we had a sitting of more than two hours. She had not been there many minutes[1] before some slight ticking raps were heard *in* the table apparently. The raps answered by the alphabet (pointing to the letters on a card), one after the other (a rap or two coming at the letter), to the name of a sister of my wife, who died seventeen years ago. After some questioning, she (I speak the spirit hypotheses, though I have no theory on the subject) was asked whether I might ask a question. 'Yes,' affirmative rap. I said, 'May I ask it mentally?' 'Yes.' 'May Mrs. Hayden hold up both her hands while I do it?' 'Yes.' Mrs. H. did so, and *in my mind*, without speaking, I put a question, and suggested that the answer should be in one word, which I thought of. I then took the card, and *got that word* letter by letter—C H E S S. The question was whether she remembered a letter she once wrote to me, and what was the subject? Presently came *my father* (*ob.* 1816), and after some conversation I went on as follows:—

'Do you remember a periodical I have in my head?' 'Yes.' 'Do you remember the epithets therein applied to yourself?' 'Yes.' 'Will you give me the initials of them by the card?'

[1] This is true. About ten or fifteen minutes elapsed after we sat down before the raps came; but Mr. De Morgan has not mentioned in this letter that for a few, perhaps five minutes, we sat waiting for them. On his leaving the room they were heard at once, and went on when he returned.—S. E. DE M.

1853. 'Yes.' I then began pointing to the alphabet, with a book to conceal the card, Mrs. H. being at the opposite side of a round table (large), and a bright lamp between us. I pointed letter by letter till I came to F, which I thought should be the first initial. No rapping. The people round me said, 'You have passed it; there was a rapping at the beginning.' I went back and heard the rapping distinctly at C. This puzzled me, but in a moment I saw what it was. The sentence was begun by the rapping agency earlier than I intended. I allowed C to pass, and then got D T F O C, being the initials of the consecutive words which I remembered to have been applied to my father in an old review published in 1817, which no one in the room had ever heard of but myself. C D T F O C was all right, and when I got so far I gave it up, perfectly satisfied that something, or somebody, or some spirit, was reading *my thoughts.* This and the like went on for nearly three hours, during a great part of which Mrs. H. was busy reading the 'Key to Uncle Tom's Cabin,' which she had never seen before, and I assure you she set to it, with just as much avidity as you may suppose an American lady would who saw it for the first time, while we were amusing ourselves with the raps in our own way. All this I declare to be literally true. Since that time I have seen it in my house frequently, various persons presenting themselves. The answers are given mostly by the table, on which a hand or two is gently placed, tilting up at the letters. There is much which is confused in the answers, but every now and then comes something which surprises us. I have no theory about it, but in a year or two something curious may turn up. I am, however, satisfied of the reality of the phenomenon. A great many other persons are as cognizant of these phenomena in their own houses as myself. Make what you can of it if you are a philosopher.

Now I must shut up. Give my best regards, &c.

Yours very sincerely,

A. De Morgan.

To Professor Michael Foster.

November 15, 1853.

University of London Examinations.

Dear Sir,—You have asked me for a sketch of my chief objections to the system pursued in the University of London. This is a matter into which I have not time to enter in great detail; nor would it be necessary. I have always looked forward to the time when the graduates of the University would themselves feel that their Alma Mater will not take its proper place among the Academies of Europe until its requisitions are based upon higher views of education than appear to have prevailed at its foundation. I say at its foundation, not among its founders; for the first institution preceded by several years that revival of serious thought upon mental subjects in which we now live, and which is far from having attained its full development. 1853.

With great respect for many who have been and are members of the Senate, I do not feel the slightest diffidence in opposing my opinion to the results of their collective deliberations. No man who has thought on a subject for a quarter of a century, with daily power of testing his opinions, need fear to oppose himself to a system which has not emanated from one mind. Solomon said that in the multitude of councillors there was safety; safety, not wisdom. A numerous body always compromises, and never works out a sound principle without limiting its application by considerations drawn from the expediency of the moment; practicability is the word, freedom from present difficulty is the thing.

The plan of the Universities of the Middle Ages, to which in a great degree we owe both the thought and the operative ability of the last two centuries, rested on a simple principle, which stood ready for any amount of development which its own good consequences might make possible. All existing knowledge, the pursuit of which could discipline the mind for thought and action, was collected into one system, and declared to be available for the purpose of a University. And in this manner reason, language, and observation were cultivated together. Every means was employed for forming the future man in his relation to himself, to other men, and to the external world. The worst thing, if not the only thing, that can be said against them is, that at some periods they thrashed the chaff after the corn had been beaten out. The worst thing that can be said against their successors in England is that they have not sufficiently allowed the development of the old principle in reference to branches of knowledge which progress has converted into disciplines, and that, each in its own way, they have given undue prominence to one of the ancient disciplines.

The sciences of observation occupied rather a subordinate place, because in the disciplinatory sense they had attained but little efficacy. To which it is to be added, that the very wants

1853. of daily life, in a rude state of co-operative power, made daily life itself such a discipline of observation as we have now no idea of. Every savage has all the knowledge of his tribe in matters to be drawn from observation and applied in practice. The man of the fifteenth century, much nearer to the savage than ourselves, had a considerable share of it. The man of our day has just as little as he pleases, and no more than his individual temperament and opportunities may lead him to acquire; the temperament not being fostered by education, and the opportunities being mostly subsequent to it.

The great point, then, in which the old Universities ended by ignoring the progress of the world around them, the great point on which it might have been the privilege of a new one to show them that the world could teach them something even on the fundamentals of education, was the neglect of the discipline of observation, of language as connected with it, and of inference as immediately derived from it. And how has the University of London fulfilled its especial mission? It has granted the existence of the deficiency, proclaimed its own intention to provide a remedy, and set its alumni diligently to work to read words and to look at diagrams about the way in which other people have used their eyes and their hands. This is no exaggeration. Because observation of phenomena had been neglected, and ought to have been a part of all sound discipline, the University of London demanded of its candidates a knowledge of the manner in which those who have seen things for themselves describe them to others.

For example, a candidate for the B.A. degree is required, in addition to matters which enter the ancient disciplines, to be examined in *animal physiology*. And he may pass this examination without knowing more from his own observation of what is under the skin of any animal, than he learns from the words of a book or the lines of a drawing, which no one can understand except he be familiar with the original object. I will venture to say that a large majority of those who have passed the examination in physiology know nothing about the interior of the body from their own observation except that blood follows a cut in the finger. I appeal to the examiners whether it be not as I say, and whether the answers given do not clearly show it.

Thus, for the first time in the annals of liberal education, a University has proclaimed that mere words, as words, with no meaning attached, are a worthy discipline. In learning languages words are things; they are the things to be studied, and the student compares the unknown with the known, the strange language with that which he has spoken all his life. In the exact sciences the notions treated of are present and living realities. In the common branches of physics the student has a daily knowledge of the species of phenomena which he is to study in their systematic relations; he knows air and water, and his stick is a lever. But in the physiology of the University of London he has only words descriptive of what he neither knows nor can know by words alone; if there be any shadow of advantage to be gained, it is that species of advantage which he gains to better purpose from ordinary physics. The great thing wanted, the training of the faculty of observation in connection with language and reason, is wholly left out of sight. 1853.

Will it be replied that students cannot dissect? that they have no opportunities, that they have no skill? that without such teaching as they cannot get, and such time as they cannot give, their researches into the textures would be of as much avail as those which are made with a carving-knife upon the roast or boiled joint? I freely admit it all; but I deny the conclusion that therefore the University of London should supplant observation by reading. I say nothing is proved except that physiology is a very unfit subject for the purpose, as seems to me clearly proved on other grounds.

The proposal for reform which I should submit is that actual examination upon natural objects should be a part of the trial for the B.A. degree; and that the objects should be of the vegetable world. These are accessible to all; and the matter to be tried should be, not whether the student has this or that amount of acquirement, but whether he has gained the powers of observing for himself, and stating and reasoning of the results. There are various reasons why vegetable structure is better fitted than animal for the commencing observer; but it is enough that the newly gathered plant is always within his reach, and that the newly killed animal is not.

The next point I will mention is that of the examinations for honours. There are two systems in this country,—that of Oxford, in which the candidate for classical honours is examined against his subject; that of Cambridge, in which the candidate for mathematical honours is examined against his competitors. At Oxford, his class determines his qualification; at Cambridge,

1853. his place determines whether he is above or below any given competitor. At Oxford his mind may, though not without certain wholesome restraint, develop itself in reading and thought dictated by its natural bent. At Cambridge the examination realises the bed of Procrustes. The Oxford system has a tendency to develop the useful differences between the varied types of human character. The Cambridge system is an unconscious effort to destroy them. I shall not be suspected of any original bias against the Cambridge system. I once thought that the race for the place in the list was a valuable part of that system, but I have slowly arrived at the full conviction that the Oxford plan is greatly superior. The system of private tutors, the drill in *writing out*, and the mode in which so many of the elementary books are *got up*, are well worthy the attention of all who are interested in the subject of this letter. They are the natural consequences of the personal competition for honours; and if ever the number of candidates in the University of London should bear any considerable proportion to that in the University of Cambridge, the same cause will produce the same effect. I hope this subject will receive some attention. Why, because political tendencies have thrown the University of London almost entirely into Cambridge hands at the outset, should all that is from Cambridge be received as of course, and without a discussion of what is to be found at Oxford?

Probably it will be objected that the medals and honours cannot be awarded without a competitive examination. To this I answer that the existence of medals and scholarships is of very small importance compared with that of the evils I have alluded to. If I am right, they had better be abolished than allowed to introduce the evils of competition into the main examinations for honours. And the natural consequence would be that they should be given, not for general proficiency, but on special grounds, to be tried some time after the elementary career has closed.

My view of the advantages of a liberal education is most assuredly not peculiar to myself. Let it be supposed that the former student has forgotten everything, that not a word of Latin is left, and not a proposition of Euclid. What remains to him? If little or nothing, then his education has not deserved its name. But if, in spite of the loss of all that acquirement which he has had no daily need to recall, he be a man of trained mind, able to apply vigorously, to think justly, to doubt discreetly, and to decide wisely, he has been well educated, and the loss of the positive knowledge which I suppose him to have lost is comparatively a small matter. I do not underrate knowledge; I would educate for it, even if it gave no powers; but I am sure that if we take care of the habits, the acquirements will take care of themselves. 1853.

Throughout the whole of the requisitions runs a tone which would give any one the notion that the study demanded is sought only for its results, and that it will be tested only by the knowledge of results shown. I look at the programme of the mathematical propositions required, and I find the implication that as long as a certain list of truths is known, it matters not how. I admit that the examiners by setting this list at defiance, by proposing questions which try the knowledge of principles, and which necessarily require them to travel out of the list, have done much to neutralise its evil tendency. But I cannot suppose the necessity for a complete alteration is thereby done away. We are informed that the principal properties of triangles, squares, and parallelograms (when did the square cease to be a parallelogram?) are to be treated geometrically. Among the principal properties of parallelograms are those of similar parallelograms; their study involves a doctrine of proportion. But only the first of the six books of Euclid are demanded. Must similar parallelograms be treated by what is called a geometrical theory of proportion? If not, how are the principal properties of parallelograms to be treated *geometrically*, as demanded? If yes, what is that geometrical theory of proportion, other than Euclid's, so well known that it may be trusted to implication? The only proportion alluded to in any part of the list is *algebraical* proportion, which, as usually understood, is the doctrine of the ratios of commensurable quantities, expressed by letters, with either every possible amount of gratuitous assumption about incommensurable quantities, or else a total refusal to consider them.

Might not what we may well hope, and what I am inclined to believe, will be the greatest University founded in the nineteenth century dare to promulgate definite views on the mode in which study should be conducted, its ends, its uses, and the proofs of its efficiency? Is it not the duty of that University to make it apparent that she receives and cherishes the sound principle so long maintained by her predecessors without pledging herself to the abuses which time and negligence have allowed

1853. to creep in? I hope the graduates will show that their colleges have trained them to ask these questions. If they be not asked, and asked to good purpose, the University may gain a parliamentary voice, but it will not gain the respect of that highly educated world to which the common sense of common people teaches them to look up for opinions on the higher education. And the old institutions which are rousing themselves into activity will have it delegated to them, a century hence, to teach the University of London what it was hoped by some the University of London would teach them.

I am, dear sir,
Yours truly,
A. De Morgan.

To Admiral Smyth.

7 Camden Street, January 5, 1854.

1854. My dear Admiral,—You probably know why your note has remained unanswered. I and Mrs. De Morgan are just beginning to recover the shock it has given us. Your sheets may come as usual if you have any to send.

I congratulate you on the news you conveyed to me, though, having mislaid your note, I cannot remember the name. You have twice had to bear a loss similar to mine, and I hope you will depart yourself in the course of nature before the distant time comes when you would have to face it a third time.

If you have anything to contribute or to suggest for the Annual Report, now is the time. Our kind regards to Mrs. Smyth and the young ladies.

Yours sincerely,
A. De Morgan.

To the Rev. Dr. Whewell.

January 24, 1854.

My dear Sir,—Your book on the 'Plurality of Worlds' reached me at a time when I could only throw it by for better days, and I believe it would have remained on one side as an anonymous attempt to prove what every one believed—without knowing anything about the matter—if I had not been told, casually, that you were the author, and that the title ought to have been 'On the Singularity of the World.' Accordingly, knowing whom to thank, I thank him; and learning that the argument is *singular*, I read the book.

I have always held that when the phrase 'there is a good deal to be said on both sides' applies, it means that we do not know much about the matter. Your book is a converse instance; that when we do not know much about the matter there is always a good deal to be said on both sides. Not that I mean to give up the poor dear lungless lunarians, or the jovial cinder-sifters, altogether, quite yet. I admit the argument from time to space; but, granting that the *human* world is only dt out of t of the whole of time existence, we may grant it to be ds out of s, of space existence; and all the stars and planets may be in their several progresses from $-\infty$ to $+\infty$, and every one at a different part of it, with at least the chance of two given ones being within m of each other, only $m = \infty - (-\infty)$. And this on the supposition that there is but one kind of progression; it being more likely, however, that this progression is infinitely varied in space, so that, instead of diminishing the immensity of creation—as usually taken—namely, for one time, the idea of one mode of existence infinitely varied in space, you have made prominent a system of triple entry, time, space, law of progression. 1854.

I find in your book the germ, or more, of a notion which I have had for twenty years—and which may have occurred to many others, and probably has. I have been laughing all that time in the sleeve at the clergy, for not seeing that the infidel geology, as they call it, is in truth the most unanswerable proof of *supernaturalism* that ever was propounded. Between an unintelligibly self-existent Creator, and an unintelligibly self-existent *order of things*—self-reproductive *natura rerum*,—my reason never saw *à priori* choice; not having the slightest idea which of two wholly inconceivable things was most conceivable. But the straightforward impossibility of human existence at some calculable time brings us to the alternative of an absolute creation—or the growth of some lizards or fishes into men—through various stages. I do not read controversies about the *pros* and *cons.* of the Book of Genesis, and this argument may for aught I know be common; but it never oozed into any conversation in my hearing, though I have frequently looked out for it when I heard the orthodox and the heterodox fighting about the matter.

Yours very truly,
A. DE MORGAN.

To Rev. Dr. Whewell.

May 21, 1854.

1854. MY DEAR SIR,—I have to thank you for the dialogue. If you deny a plurality of worlds I presume you admit a plurality of opponents—to judge by the letters.

You seem to be expressing your anti-pluralism more positively than before. Your work seemed to say—don't be so sure; your dialogue seems to maintain something more like a leaning to the other positive conclusion. I suppose you will not object to the conclusion that the stars either are inhabited, or that they are not. This is mine, with a leaning to the affirmative of this kind:—Let it be granted that each planet has upon it, or in it, or around it, some things which have a destiny of their own, for which they might be conceived to exist independently of the other planets or stars. These things I should call inhabitants of that planet, but whether conscious or unconscious, intelligent or unintelligent, &c., &c., I could have no opinion. But I cannot divest myself of the idea that they have *uses* independent of us—and these *uses* are inhabitants. I strongly suspect that, to use law phrases, these *uses* are also *trusts*, and therefore suppose *responsibilities*.

Yours very truly,
A. DE MORGAN.

To Admiral Smyth.

7 Camden Street, July 16, 1854.

MY DEAR ADMIRAL,—Here you see the balance of blue queen's heads forwarded to me on a special service. I hope a larger proportion of Napier's blues will find their way home again from the Baltic.

All is going on well as to the Government proceedings.[1] We shall not be stirred these ten years, I augur. You know the story of the birds in the nest listening to the farmer plotting how to cut the corn. Now Government is a man who cannot work for himself. He acts through people who *report*. Deep calleth unto deep—that is, one office reports to another, and the other refers back, and then they consider, and red tape becomes grey before they have settled how to proceed. And if you then give them six months' start, and set a snail after them, the

[1] Referring to a proposed removal of the Astronomical Society from their rooms at Somerset House.

snail beats them by a thousand lengths; and then there is a change of ministry and a new report to 'my lords,' and 'my lords' make a minute which means in time a year, and so on *ad infinitum*. 1854.

Kind regards all round.

Yours very truly,
A. DE MORGAN.

To the Rev. Dr. Whewell.

October 27, 1855.

MY DEAR SIR,—M. Biot presses for the meaning of Newton buying a *supersedeas*. He wants to give it in a forthcoming article. (Brewster, vol. i. p. 18.) Could you ask any one in college to see what it may have meant? I am pretty sure no such thing was for sale in college in my time, for freshmen or any other. 1855.

Excuse my troubling you again. The world has so passed by that I am not sure I know the name of any office-bearer in college. I have only an indistinct remembrance that Prof. Sedgwick is Vice-Master.

I told Biot that *China ale* was tea, and reinforced it by telling him that water was often called *Adam's ale* in England. This, he says, has amused the French philologers very much.

Pray come to the rescue of a Frenchman in a fix about a college phrase. I must send the French philologers the phrase *Henry Soph*.

Yours very truly,
A. DE MORGAN.

To Sir J. Herschel.

November 10, 1855.

MY DEAR SIR JOHN,—I am glad to see your signature, failing more, and also that you are in pretty good spirits. We shall see you come out in chemistry yet, with the discovery of a new principle, *Uncommonly-impossible-to-get-ine*, obtained by treating the singular *Takes-a-week's-cookingic Acid* with all the salts in succession of your new metal *Describable-in-six-foliopagesium*.

I shall not bother you with the proofs of your memoir.[1] I shall respect the text as if it were Horace—and there are no

[1] Memoir of Francis Baily.

1855. lections that I am aware of—and I shall add a few *editorial* n oes; one must notice the new moon-bobbery which has upset the eclipse of Agathocles and every other, and perhaps some other little matters. I made a few additions to the biography in a subsequent Annual Report, which I shall *append*, but not *incorporate*. You shall have revises—not to correct, but to protest against, *pro re nata*, and your protests shall meet with more attention than such things usually meet with.

The Sheepshanks inscription is now in Whewell's hands.

Yours very truly,
A. De Morgan.

SECTION VIII.

1856–65.

MY mother-in-law died after a long illness this year, to the great sorrow of her three sons. Though there was great difference of opinion, chiefly on doctrinal matters, between my husband and herself, there was strong mutual affection, and some resemblances of character. He shared with her the quality which he used to find troublesome when he lived in her house; namely, anxiety to a morbid degree about those she loved when they were out of her sight. If he came home an hour later in the evening than she expected, she conjured up all kinds of terrible accidents which he might have met with. One reason of this, on Augustus's account, was his want of sight on the right-hand side. He was very like her in this morbid anxiety, so that those who left the house in the evening had to be punctual in the time of their return if they wished him to be easy. From his mother he inherited his musical talent, and most probably his mathematical power, for she was the granddaughter of James Dodson, the author of the *Mathematical Canon*, a distinguished Mathematician, the friend of Demoivre, and of most other men of science of his time, and an early F.R.S. But he was Mathematical master at Christ's Hospital, and some of his descendants seem to have thought this a blot on the scutcheon, for his great-grandson has left on record the impression he had of his ancestor. When quite a boy he asked one of his aunts 'who James Dodson was;' and received for answer, 'We never cry stinking fish.' So he was afraid to ask any more questions, but settled that

1856. Death of Mr. De Morgan's mother.

1856. somehow or other James Dodson was the 'stinking fish' of his family; and he had to wait a few years to find out that his great-grandfather was the only one of his ancestors whose name would be held deserving of record. My husband also inherited his love of a city life from his mother, who declared that a night in a country house, with 'the dreary trees moaning all round,' made her sleepless.

Mrs. De Morgan's death occurred while she was living in the house of her second son, Mr. George De Morgan, the barrister and conveyancer. My husband, of course, visited her almost daily, and was struck with the reality of her conviction, constantly asserted, of the presence of Jesus Christ. He spoke to me of the frequency of this appearance, or supposed appearance, to the dying, and wished that the instances should be always carefully recorded.

Mrs. De Morgan was one of eleven children and nine daughters of Mr. John Dodson, of the Custom House. Eight of the daughters married officers of either the Military or Civil service in India. At the time of her death there was living, besides her sons Augustus and George, Campbell Greig De Morgan, who was Senior Surgeon to the Middlesex Hospital—a man much beloved and highly distinguished in his profession. He survived his brother Augustus four years, dying in 1875.

Dislocated shoulder.

A few days before our return from Eastbourne in the autumn, I was startled by receiving a long letter from my husband, written in pencil and in the middle of each page. He always wrote every day, but it was often not more than to ask after me and the children, and to tell me whom he had seen, with occasional information about the cat or the canaries. This pencil letter was a dramatic description of how he had the day before fallen off the ladder in his library and dislocated his shoulder; how the doctor had been fetched and had replaced the shoulder in the socket, which the patient said had given him no pain. His account would have amused me if it had not frightened me so much. On hurrying to London I found him reading comfortably in his arm-chair. Happily he neither suffered from pain nor fever, and the weakness in the arm caused by the accident did not last long. 1856.

Decimal Coinage. History of the movement.

As early as the year 1824 Sir John Wrottesley, father of the first Lord Wrottesley, had introduced the question of Decimal Coinage in the House of Commons.[1] His proposal was to retain the pound as the unit, dividing it by tens until it reached 1,000 farthings. The motion was not pressed to a division.

In 1832 Mr. Babbage's work *On the Economy of Manufactures* was brought out. In this the plan of a decimal system was advocated, and lesser attempts by other writers followed. In 1833 the first number of the *Penny Cyclopædia* was published, and Mr. De Morgan in the article *Abacus* gave a good summary of the advantages of the proposed change.

It [the abacus] never can be much used in this country owing to our various divisions of money, weights, and measures.

---

[1] A very early suggestion on this subject is to be found in a little book of my father's, long out of print. Speaking of the abacus, the use of which he had described, he says, 'The Chinese use this toy in the common concerns of life; and they can do it with great ease, since in their nation the decimal arithmetic is preserved in the weights, measures, and money. The French and Americans have returned to their ancient and best mode of counting; but it will be difficult to establish it in this country on two accounts: first, it would be considered an innovation, and it is almost incredible how great is the number of persons who prefer their father's *mumpsimus* to a modern *sumpsimus*. Secondly, it is a question of mere public benefit, without reference to party politics; and it must be a fortunate concurrence of circumstances to produce an individual resolute enough to bring forward a motion that would get rid of our troublesome modes of numbering, and introduce that which is the simplest, the best, and the most ancient.' (*Tangible Arithmetic*, by W. Frend, 1806.)

1856. We should need one abacus for pounds, shillings, and pence, another for avoirdupois weight, a third for troy weight, and so on. In China, however, where the whole system is decimal—that is where every measure, weight, &c., is the tenth part of the next greater one—this instrument, called in Chinese *schwanpan*, is very much used and with astonishing rapidity. It is said that while one man reads over rapidly a number of sums of money, another can add them so as to give the total as soon as the first has done reading.

First Commission on weights and measures.

General Pasley tried to bring forward the question in 1834 in a volume *On Coinage, Weights, and Measures, and on the Advantages of a Decimal System*. Four years after, Mr. Spring Rice, then Chancellor of the Exchequer, obtained the appointment of a Royal Commission on weights and measures. In the *Companion to the Almanac* for the year 1841, Mr. De Morgan showed the advantages that would arise from the adoption of a decimal coinage.[1] He insisted on the introduction of an entirely decimal system of accounts, in combination with such change in the coinage as should be best adapted to, and be the means of introducing such a system of accounts. He showed how easily our present system might be changed to a decimal one by retaining the pound sterling, and dividing it into 1,000 parts; and recommended the retention of as many of our coins as bore a relation to the pound, and the very small alteration in the value of sixpences and shillings needed to bring them into the new system. The plan of the proposed change is explained, and names of coins suggested. He strongly advised that the change be made first in the coinage, believing that the complications which would arise from carrying it into weights and measures would throw everything into confusion. He saw that the minds of the mercantile and working classes must be made familiar with the decimal

[1] 'On the Use of Small Tables of Logarithms in Commercial Calculations, and on the Practicability of a Decimal Coinage.' *Companion to the Almanac*, 1841.

reckoning in money in the first instance. 'Education,' he said, 'must promote the demand for a complete decimal system, but the application of the principle to coinage only must first promote education.' In answer to the question, 'how much of the time spent in education in Great Britain and Ireland is spent in overcoming the disadvantage of our present system of coinage?' he said, 'I believe that five per cent. is under the mark, taking in all classes; that in purely commercial schools it is a great deal more; but that in all together, from Oxford and Cambridge down to the lowest village school, more than one-twentieth of the whole time passed in every kind of learning and practising is lost, by the having two systems of Arithmetic to learn, the common decimal, and the monetary.' 1856.

Report of Commission.

At the end of the year 1841 the Report of the Commission of 1838 was made. In it the Commission strongly recommended the adoption of a decimal scale of weights and measures preparatory to a change in the money; than which, the Report says, 'no single change which it is in the power of our Government to effect would be felt as equally beneficial when the temporary inconvenience attending it had passed away.' The details of the change recommended are those set forth by Mr. De Morgan in the *Companion to the Almanac.*

In the year 1842 he gave more extensive information on the subject in the same work, and in the next year (1843) another Commission to inquire into weights and measures was appointed. It consisted of the Astronomer Royal, Lord Wrottesley, the Dean of Ely, the Speaker, Sir John Herschel, Sir J. W. Lubbock, Rev. R. Sheepshanks, and Professor Miller.

The next step was taken in 1847 by Dr. Bowring, afterwards Sir John Bowring, who brought forward the subject in the House of Commons. The florin, or one-tenth of a pound, now in circulation, was in consequence issued by Government, but no further attempt was made

1856. to decimalise the money, and Sir John Bowring soon after went to China.

There is an article in the *Companion to the Almanac* for 1848 by Mr. De Morgan.[1] He describes the state of feeling at that time on the question as compared with what it had been when it was first agitated. Referring to the debate on Sir John Bowring's motion, which resulted in the introduction of florins, he says 'the Chancellor of the Exchequer, in yielding the first step, rested his non-acquiescence in the whole extent of the motion on the want of public interest in favour of the question, and the slow growth of belief adverse to existing usages. He said, as plain as a Chancellor of the Exchequer could speak, "Force me, and here I am ready to be forced."'

Since issuing the florin Government had taken no further steps towards the complete decimalisation of the coinage, but Sir John Bowring, who was in England in 1853, was still hopeful for more, and many of the most enlightened friends to the measure, both mercantile and scientific, were anxious that the efforts already made should not be lost. In 1852 Mr. William Brown called the attention of the Liverpool Chamber of Commerce to the importance of the question as regarded currency and accounts, and a memorial was presented by the Chamber in favour of the proposal already made. About the same time the Royal Commission last appointed made its report, which confirmed all the recommendations that had been made by the Commission of 1838. The Commissioners expressed their hope that no new coins should be issued except such as should be expressible by one figure in the decimal scale, descending from the pound sterling, and that every new coin should have marked upon it its value with reference to the pound sterling.

Early in the year 1853, Mr. (afterwards Sir) William

[1] In this article the Commission of 1838 is spoken of as the *last* Commission on the subject, but it must be remembered that that of 1843 was still sitting, and did not report till 1853.

Brown, M.P., as representative of the Liverpool Chamber of Commerce, had interviews with the Chancellor of the Exchequer and the President of the Board of Trade, to suggest the appointment of a Committee of the House of Commons to consider the whole subject of decimalising the money, weights, and measures. Mr. Gladstone saw, with the most practical of those who had dealt with the question, that to include the subject of weights and measures at that time would throw needless difficulty in the way. Taking up one subject at a time, it would be more easily understood; and when the plan should be adopted and its advantages felt, the difficulty of securing a uniform system throughout would be removed. This Committee was appointed, with Mr. William Brown for Chairman.

1856. Committee of House Commons.

My husband's correspondence at this time shows how large a share he had in the uphill work in which he was at once expounder, adviser, and referee. He was applied to for information on every part of the question—on weights and measures, on foreign money, on the history of the change to a decimal coinage in other countries, and on the changes that would be required to decimalise our accounts and coinage in such a way as to cause least difficulty in our money transactions both at home and with our neighbours abroad; for references to books of authority;—in short, for every sort of information that would enable the advocates of the reform to support their cause. All this he gave freely and readily, more perhaps in answer to private inquiries even than in print; and the amount of work done by him in this way—all extraneous to his lectures and other occupations—can only be guessed at by those who were with him at the time, or who have seen his correspondence since.

Here is an instance. Sir John Herschel asks—

What book, report, or *résumé* contains what you would *refer any one to,* who wanted to get a clear view, in a short time, of the history of the change to a decimal system of currency in

1856. France, and more especially in the United States? I should like to have a reference to some authentic report as to the latter, and indeed to anything official as regards the former; but what I chiefly want to be able to point out is an historical view of the origin of the thing, the order of procedure, the course of action of the Governments, the way it was met on the part of the countries, and the steps by which it ultimately rooted itself.

I know you are as full of information as an egg is of meat.

On receiving the answer:—

I am really sorry I have plagued you about it, but I thought you would very likely have been able at once to name a work which, referred to, would do the needful. Such a work I now perceive is yet to be written, unless Dr. Bowring's now forthcoming one be that work.

He has called twice on me about it. What an ardent creature he is! He seems to me as if he lived on live birds.

International Association.

Many people who had pet schemes of their own as to the proposed coinage brought them to my husband, and several of these had influence enough to get their plans considered by statesmen. These formed impediments in the way. The various views on the change of coinage were numerous, and I shall only refer further on to that which, though well meant, formed the greatest obstacle —the *International Association for a Decimal System in Weights, Measures, and Coins.*

In the year 1854 the Parliamentary Committee reported in favour of the decimal plan which had been proposed by scientific men, and, on the issue of this report, the Decimal Association was formed. Its first meeting was held in July 1854. Sixteen members of Council were chosen, all influential in Parliament or commerce. Their number was afterwards increased by seven, one of whom was Mr. De Morgan.

The Association recommended the adoption, or retention, of the pound as the unit of account; the only new coins which would be required to complete the scale being the *cent*, a silver coin ten to the florin, and the *mil*, a copper coin ten to the cent: the decimalisation of weights and measures to be afterwards considered. 1856.

A deputation from this body (among whom were Lord Monteagle, Mr. Cobden, Mr. Bright, General Pasley, and others) waited on Mr. Gladstone, the object being to urge the coinage of a sufficient number of cents and mils to circulate with the present money, of which the coins which did not come under the system should be gradually withdrawn. Deputation to Mr. Gladstone.

Mr. Gladstone 'saw in the deputation a great deal of power, as well as of intelligence, represented,' but hesitated as to the adoption of the pound as the unit of account, and believed the nation was hardly ready for the change. Mr. William Brown, who was the commercial leader of the movement, wrote to Mr. Gladstone, showing how all the various objections had been met.

1. As to the adjustment of railway fares, by the fact that several directors of leading lines were members of the Association.

2. As to the Post Office, by Mr. Rowland Hill, who was an advocate of the measure.

3. As to the turnpike tolls, by Professor Airy.

4. As to the Customs and Excise duties, which had been supposed to be a great difficulty. Mr. Brown pointed out that nowhere would the convenience of the change be more rapidly felt, both in saving labour and securing accuracy, than in the accounts and returns of the national income and expenditure.

5. In the wages of working men, wherein the difficulty was shown to be imaginary.

But Mr. Gladstone still thought the time not ripe for the change.

Another deputation waited on the President of the Board of Trade. It consisted of men who represented every phase of the subject, each one taking his own special part in the discussion. The recommendation in which all concurred was that the sovereign should be re-

1856. tained as the unit, the florin one-tenth of the sovereign, the cent, a new silver coin, one-tenth of the florin, and the mil, or old farthing, one-tenth of the cent. There would be no difficulty as to the change of value of the farthing, for, as the copper coin is circulated at a *nominal* value far beyond its intrinsic worth, its current value might be declared each time by proclamation.

Among other things it was said that 'men with the rare facility of explanation possessed by Professor De Morgan might fix the attention of meetings of the working classes upon the value of the easy road that would be opened to a knowledge of Arithmetic by the proposed change; but by those who had to legislate for the people the fact was known, and there was no need to withhold the advantage until the masses, becoming informed of its value, should seek it for themselves; it was, on the contrary, precisely a case in which the Government, supported by those best informed, should take upon itself the responsibility of conferring a practical boon upon the people in advance of their knowledge.'

These proceedings of the Decimal Association, of which I have only made a slight mention, were printed with an introduction by Mr. De Morgan, which touches all the most important points of the subject. It concludes with these words:—

'The Chancellor of the Exchequer says that *Government holds an impartial position, and is ready to be guided by the decision of the public.* Every one knows that, in his own circle, the opinion in favour of a decimal coinage, based on the pound sterling, is that of a very large majority of all who know what it means. What, then, is left? *Nothing, but that the public should let the Chancellor of the Exchequer know that it has decided, and what the decision is.*'

The writer was over-sanguine as to the adoption of the measure. Delays, unfounded objections, and, more than all, the obstacles created by injudicious reformers of the whole, put a stop to the work, and rendered all the efforts which had been spent on it of no avail. 1856.

Renewed efforts of Parliament.

After more meetings and much correspondence it was agreed to bring the question again before Parliament. This was done on June 12, 1855, by Mr. Wm. Brown, who moved, after referring to the recommendations of the Commission of 1838 for restoring the standards, to those of the subsequent Royal Commission, and the Committee of the House of Commons, that an address be presented to her Majesty, praying her to complete the decimal scale (already existing in the pound and the florin) by authorising the issue of silver coins to the value of 1-100th of a pound, and copper coins to represent 1-1000th part of a pound, to be called respectively cents and mils, or such other names as to her Majesty should seem advisable.

The motion was seconded by Lord Stanley, now Earl Derby, who cited the authorities of Babbage, De Morgan, Pasley, and Huskisson, on the practicability and advantage of the change. After showing the defects of several plans proposed, he advocated that supported by the Decimal Association, and introduced to the House by Mr. William Brown.

Mr. J. B. Smith moved an amendment that a humble address be presented to her Majesty, praying that she would be pleased to invite a congress of all nations in some convenient place, with the view of considering the practicability of adopting a common standard of money, weights, and measures.

Mr. Lowe (then M.P. for Kidderminster) made a very amusing speech, to show, first, that the present system of coinage did very well, and that the mischief of a change would be greater than any good which could result from it. Secondly, that the method proposed of decimalising by division and not by multiplication was fallacious. He would have a low unit, and multiply by tens. He made humorous illustrations and allusions, and caused much laughter. The then Chancellor of the Exchequer,

1856. Sir G. C. Lewis, expressed his belief that the time was not come for the change, and recapitulated other plans as possibly preferable to the one before the House. He spoke of the opinion, already reported, of Sir J. Herschel, the Master the Mint, as involving great difficulties.

Finally two resolutions were carried:—

'*That in the opinion of this House the initiation of the decimal system by the issue of the florin has been eminently successful and satisfactory.*'

And—

'*That a further extension of such system will be of great public advantage.*'

But the practical resolution for an address to her Majesty, praying for the completion of the scale by the issuing of cents and mils, was withdrawn. The question was therefore left by the House of Commons much as it had been before, with the exception of the greater publicity given to the arguments on both sides by the debate. But a Commission was appointed to investigate the whole subject of weights, measures, and coinage.

In an article in the *Westminster Review*, in which a summary of the question was given, Mr. De Morgan, writing of the Chancellor of the Exchequer, 'the chief of our financial Arithmetic,' says,—

The right honourable gentleman, after saying that 'there are differences of opinion,' proceeds to give 'some of the plans' which have come under his observation. One is the tenpenny plan; others are as follows:—1st. 10 farthings or mils, one cent; 10 cents one dime; 10 dimes one prime. 2nd. 10 farthings or mils one coin unnamed; 10 of these a florin; 10 florins a Victoria. 3rd. 10 farthings or cash twopence; 100 cash a cent; 10 cents a mil. 4th. 10 farthings a lion; 10 lions a florin; 10 florins a queen. To these four 'plans' we beg permission to add two of our own invention, as distinct from the above as the above are from one another. Our first plan is 10 farthings a what's-his-name; 10 what's-his-names a how-d'ye-call-it; 10 how-d'ye-call-its a thingembob. Our second plan is 10 farthings a George; 10 Georges a Cornewall; 10 Cornewalls a Lewis.

1856. Mr. Lowe and the apple-woman argument.

All the old arguments were considered, and the answers to them repeated and enlarged. As to the witticisms of Mr. Lowe, they had been answered by Mr. De Morgan in a paper published by the Decimal Association, *Reply to the Facetiæ of the Member for Kidderminster.* 'Mr. Lowe is of opinion,' the *Westminster Review* says, 'that if a poor man owed another a penny, for which four mils is too little, and five mils too much, this mil between them would lead to a *mill* between them; and some of the conscript fathers cheered him.' In the 'Reply,' a dialogue between an orange boy and the member for Kidderminster shows, in the method of the latter, how the supposed difficulties in a money transaction with an old apple woman might be overcome.

The *Review* takes in the substance of the two Reports of Commissioners, 1841 and 1853; the Report of the Committee of the House of Commons, 1854; the debate of June 12, 1855; the publications of the Decimal Association, and the *Journal* of the Society of Arts, with a list of about one hundred publications on the subject in 1853, 1854, and 1855. At the end of this year 1855 he wrote for the *Companion to the Almanac* of 1856, *Notes on the History of the English Coinage*, giving many particulars of the history and origin of money in England and other countries, both as to coins and accounts. Speaking of the ruined and confused state which the coinage had reached at the time of the Restoration, he describes the reform projected by Montague, afterwards Lord Halifax, and carried into effect during his administration by Sir I. Newton, who only added to the pieces already in circulation the quarter-guinea, which was found too small for use. A note in the author's handwriting states, 'The next *Scientific* Master of the Mint coined a *quarter*-sovereign, which was never circulated.' Sir John Herschel gave my husband one of these pretty little gold coins, which he valued greatly. It was lost, in moving, after his death.

1856. Many sanguine persons had been withheld from joining the Decimal Association, in the hope of establishing an International Currency; 'a proposal which,' Mr. De Morgan said, 'unites the millennial and decimal systems.' They also contemplated the universal decimalisation of weights and measures. Mr. J. B. Smith's resolution for an Address to the Queen, praying her to invite a congress of all nations for this purpose, had not been carried, but shortly after, in the autumn, the Paris Exhibition took place, and the difficulty of harmonising the various weights, measures, quantities, and prices of the articles brought by contributors from twenty-two different states or countries, led to the consideration of the possibility of making a uniform system throughout the world. The advantages of such uniformity would be felt both morally and socially, in making free trade more easy, and war between nations more difficult. All these, as well as the benefits to commerce and merchandise, were fully acknowledged by all the advocates of decimal coinage. The only question was as to the first step. The mutual advances made by the French and Americans by interchange of specimens of weights and measures were followed by memorials and petitions to their respective governments in favour of a congress of delegates from the leading nations of the world. In Tract No. 12 of the Liverpool Financial Reform Association, all the evils of the present confused state of the means of effecting commercial exchanges were shown, and all the arguments for a complete decimal system throughout brought forward. Shortly after the Paris Exhibition an 'International Association' was formed, 'for obtaining a uniform decimal system of weights, measures, and coin throughout the world.' The proposals and arguments of this Association caused a great deal of extra work to the active members of the Decimal Association. Among my husband's letters are several like this from the secretary—it enclosed some new tract or report of speech:—

Pray take charge of the 'International,' &c., &c., &c., Association, and blaze away from time to time, rifle-shot being better than 60-pounders in such a cause. 1856.

The 60-pounders and small shot were cast and fired without thought of time and labour, by one who, besides his expenditure of both in daily lectures, &c., at University College, did more work than would have filled the time of an ordinary man.

Lecture at the Society of Arts.

His next public effort was the delivery of a lecture in the large room of the Society of Arts, at a meeting, to which the advocates of the various plans were invited, of the Decimal Association. The lecture was *On the Approaching Simplification of the Coinage.* We move very slowly in good and useful directions. The 'approach' of twenty-seven years ago remains in 1882 just where it was. The lecturer said :—

The various systems[1] which had been proposed had all sunk out of notice but two—the *pound-and-mil* system, and the *tenpenny* system. These terms were used sarcastically, which was no disadvantage, but then they must be correctly given. Some opponents on the tenpenny side had called themselves 'Little-endians,' and the pound-and-mil people 'Big-endians.' These had got hold of the poker by the wrong end. Samuel Gulliver, on whom all relied except the Irish bishop, who, when the voyage to Lilliput appeared, declared he did not believe half of it, stated that the Endian dispute arose out of the following dogma :—'True believers break their eggs at the convenient end.' Now, the pound-and-mil people believe that the small end was that at which the coinage ought to be broken, and a small crack of 4 per cent. in the copper served their purpose. But the real Big-endians, the tenpenny people, smashed the sovereign into tenpenny bits, making such a hole as let out all the meat in getting rid of the pound and shilling.

The lecturer showed how very small a change in the present coinage need be made to introduce the new system. He had been supposed to look at the matter from a scientific rather than a practical point of view, and many

[1] Referring only to coinage.

1856. writers had objected to his having a voice in the question. But these writers did not know or had forgotten that he was an actuary of twenty-five years' standing, besides being a teacher of monetary Arithmetic. He had also a position which made him better able to judge than he would have been either as a Mathematician or as an actuary. He had been for twelve years [1] a manager of a savings bank, and in that capacity had had, scores upon scores of times, to receive and pay out from two to three hundred pounds in a couple of hours, and in all kinds of coins, from a shilling and some halfpence upwards. When he looked at the banker's clerk, with his luxurious table and his convenient scoop, and all his other paraphernalia, he at the pay-table of the savings bank looked upon that same clerk as an aristocrat, who knew little of the difficulties of humble life.

The lecture was followed by a discussion, Mr. De Morgan in the chair, which lasted two evenings, and in which the members of the International Decimal Association, who wished to adopt the French system entire, and the 'tenpenny people' defended their respective plans. The sense of the meeting was ultimately taken, when, with one exception only, the opinions were declared to be in favour of the pound-and-mil system. At this time some articles by Mr. De Morgan *On the Approaching Simplification of the Coinage*, intended to make the subject clear to all classes, were published in the *Metropolitan*.

In the spring of 1857 Lord Overstone, one of the Commissioners for inquiring into the subject of decimal coinage, communicated sixty-five questions to the Decimal Association. These questions were answered by Mr. De Morgan, Sir J. F. W. Herschel, the Dean of Ely, the Astronomer Royal, Professor Miller, Mr. W. Miller, Mr. J. B. Franklin, and others. Those which were

[1] During our residence in Camden Street. He thought this the best way in which he could be useful to his poorer neighbours.

given by the above members in October were printed by the Decimal Association in November 1857. 1856.

Answer[1] No. 1 (Mr. De Morgan's) is preceded by some remarks by himself. He says,—

Lord Overstone's questions.

> These questions are the first attempts I have seen to bring the advocates of the pound-and-mil system to a close hand-to-hand controversy with the existing system. In such a trial of strength the weak points of both systems may appear, but the weak point of the assailant's system is sure to be discoverable from his mode of attack.

After pointing out some of the weak points (those implying statements or opinions held by him to be essentially unsound), he says :—

> It must be conceded that the questions are wholly free from some absurdities very common among opponents of the pound-and-mil scheme. They do not bring forward the usual dirge upon the fraction of a farthing which the possessor of some copper pieces must lose, for once, on the day when the change takes place. The only question asked on this point is a sensible one, fully deserving of consideration and answer. They do not enable us to amuse ourselves with the supposition that we mean[2] apple women to transact business by help of ·0041666666666 *ad infinitum*.

The 'Answer' runs through forty-one pages of small print, a portion being taken up by repetition of the questions. They are amusing and instructive even in subjects only indirectly related to coinage. The questions included references to authorities believed to be unanswerable. These were easily dealt with. Here is a specimen slightly abridged :—

9. In an old treatise on coin and coinage (Vaughan, 1675) this passage occurs :—

> Of all the numbers, twelve is the most proper for money, being the most clear from fractions and confusion of accompt, . .

[1] I give them in the order in which they are printed.

[2] Referring to Mr. Lowe's ideal apple woman.

1856. by reason that of all the other numbers it is most divisible, being divisible into units, as all numbers are; into two parts, as no odd number is; into three parts, as no even number is but six, and the numbers that consist of sixes; into fourths, into which six is not divisible; and into sixths.

In the memoir dictated by Napoleon at St. Helena is this passage:—

On avait préféré le diviseur 12 au diviseur 10 parceque 10 n'a que deux facteurs, 2 et 5, et que 12 en a quatre, savoir 2, 3, 4, et 6. Il est vrai que la numération décimale . . donne des facilités aux astronomes et aux calculateurs; mais ces avantages sont loin de compenser l'inconvénient de rendre la pensée plus difficile. Le premier caractère de toute méthode doit être d'aider la conception et l'imagination, faciliter la mémoire, et donner plus de force à la pensée.

To the quotations from Vaughan and the Emperor Napoleon this answer is given:—

When an old writer is stript of his conceits, translated into correctness, and under those changes presented as a sage of antiquity, the only way to meet his authority is to restore his true form, and to allow his whole character to be judged. Whether the divisors of 12 gain anything by Mr. Vaughan's advocacy may be ascertained by reading the whole passage. . . The punctuation may be excused, since the work was printed after the author's death by his crotchety brother, Henry Vaughan, *the Silurist*. He is speaking of the proportion of gold to silver in value, which he would have 12 to 1, because 12 has many divisors.—(Pp. 73, 74.)

'But the most, and the most judicious propositions that I have seen, both at home and in other parts, do agree upon twelve for one as the most equal proportion; and it agrees with the proportion of Spain, upon which in this subject we ought principally to have our eye fixed; and for my part I do the rather incline to this proportion, because 12 of all the numbers is the most proper for money, being most clear from all fractions and confusion of an accompt, by reason that [here the divisors enumerated as in the question]. And to the sixth this proportion seems like to square with the conceit of the alchemist, who called gold *Sol*, and silver *Luna*, whose motions do come near upon the point of 12 for 1.'

Lord Overstone had wisely left Vaughan's Sun and Moon out of his question. The answer goes on:— 1856. Lord Overstone's questions.

Vaughan was a clever attorney, who had read more out of law than he was able to digest. Sir William Petty will be allowed to have a much better judgment by all who have read in both. In his *Quantulum-cunque*, reprinted in 1856 by the Political Economy Club, it has been pointed out to me that he speaks as follows:—

'The use of farthings is but to make up payments in silver' (N.B. Copper farthings and silver pence were then in circulation), 'and to adjust accompts, to which end of adjusting accompts let me add that if your old defective farthings were cryed down to five a penny you might keep all accompts in a way of decimal arithmetic which hath been long desired for the ease and certainty of accompts.'

Decimals were not well understood in Petty's time. His system would give 1,200 farthings to the pound. But his main point evidently was that the multiplier 5, by its relation to 10, is an easier multiplier than 4.

Mr. Vaughan lived at a time when decimal fractions were not familiar to the mass of arithmeticians. It would be easy to show that, up to the year 1700 at least, the mastery over decimal fractions which is common in our day was almost confined to high mathematicians. Mr. Vaughan's statement merely amounts to this, that 12 is better than 10, because it has more divisors. The answer is that 10 is better than 12 because it is the radix of our present system of counting. Mr. Vaughan's objections would be exceedingly valuable if a new *system of numeration* were to be under contemplation.

Napoleon at St. Helena is no authority. He had never been a shopkeeper or a money calculator; and if he had been, his position at St. Helena was not favourable to sense or candour. He was grumbling at all creation; nobody knew his own business, not even the General who commanded against him at Waterloo. He is very unfortunate in his expression. He adopts the erroneous supposition that the decimal system is only useful to scientific calculators, and styles them *astronomes et calculateurs*. Now the Astronomer is the only scientific calculator to whom, as such, decimalisation is impracticable. He is in such continued connection with the records of his science that he cannot afford to decimalise angular measure. He is a

1856. merchant who wants all his ledger for 2,000 years past. The French tried it and failed; it is the only part of the metrical system which is not in use at this day, except indeed the *tenth* day of rest, which, being an attack on religion, disappeared with the return of Christianity. Scientific calculators in general use decimals, and they want the world at large to have the advantage which they feel every day of their lives.—(P. 7.)

These appeals to the authority of Vaughan and the French Emperor were answered by some of the other members of the Association. The Astronomer Royal's reply is much to the point:[1]—

Napoleon's infantry, cavalry, guns, rations for troops, requisitions in francs for their support, were all expressed on a decimal scale; and he never failed in 'distinctness of conception,' 'facility of recollection,' 'readiness and ease in mental calculations,' or in 'making fractional parts,' in dividing an army, from these circumstances. So far as I see, the same applies to smaller numbers.

Sir J. Herschel's answer to Lord Overstone's questions filled thirty-seven pages. Those of the Astronomer Royal, the Dean of Ely, Professor Miller, Mr. Miller (Cashier of the Bank of England), and Mr. J. B. Franklin, being much shorter than the other two, were collected in one pamphlet of fifty-five pages. The preliminary Report of the Commission of 1855 was published about the same time (1857). It contains, besides the answers given by advocates of various schemes to Lord Overstone's questions, a great deal of evidence obtained from well-informed witnesses, of the operation of the decimal system in other countries. It is a very large blue-book. The Government was now in full possession of all the information that could be gained upon the question.

Much discussion was naturally excited at this time,

[1] It will be apparent that the work of other friends to the decimal cause is not made so prominent in this memoir as that of Mr. De Morgan. Space would not allow of my mentioning it at any length, even if my husband's share were not that with which I am concerned. But I believe that his share was the largest.

and the clear-headed advocates of the measure, and those who 'darkened counsel,' all contributed their share. In a long article in the *Literarium* of October 7, 1857, Mr. De Morgan sums up the history. He says, 'the Decimal Coinage discussion is now in its third phase.' After narrating what had been done up to the year 1855, he adverts to the *penny* scheme, the plan of counting by tens and hundreds of pence, leaving the shillings and sovereigns in which the pence were to be paid, to take care of themselves. After the debate of 1855 had been followed by the Royal Commission, the penny scheme came fairly into discussion, 'for such, in a manner, is evidence with cross-examination. The preliminary Report of the Royal Commissioners shows the result. One of the chief advocates for the penny fairly bolted out of the course, and declared himself for remaining as we are; another did nearly the same thing. This was the end of the second phase. Lord Overstone completed the downfall of the penny scheme by proposing sixty-five questions, mostly in advocacy of the plan of remaining as we are, and led the way to the third phase of the discussion, thereby doing the best possible service to the cause.' At this time the dispute turned upon the question whether it would be best to adopt the pound-and-mil system, or to remain as we are.

1857. Article in *Literarium*.

Mr. De Morgan again set forth the nature of the change in the currency which would be necessary on the adoption of the pound-and-mil system.

It would be embarrassing to have two sets of coins not fairly interchangeable. No such thing is proposed. The *penny*, even if still called a penny, would be lowered four per cent., and would become 4 mils; the halfpenny would become two mils. The cent, $2\frac{4}{10}$ of old and extinct money, would be five halfpence, and ten farthings, if the mils were called farthings. The poor would know it as $2\frac{1}{2}d.$ from the outset, for they would never know the difference between a penny and 4 mils.

The paper concludes, speaking of those who had

1856. believed our present system to be better than a decimal one:—

If this school be a logical one, it ought to be prepared to maintain that a country with a decimal system already established ought to abandon its coinage, and to introduce the succession of 4, 12, 20. This is a conclusion at which all parties would have laughed three years ago, and at which those who are to come after us will well laugh when the objections to this salutary reform are written down in histories after the happy, and we hope speedy completion of the change.

It is now 1882, and our reckonings and payments are still made in pounds, shillings, pence, and farthings.

Let me add one argument, which I have never seen used, in favour of decimals. It touches the morality of the question. Small shopkeepers, especially haberdashers, find the full benefit of the present system by the use they make of farthings, and the difficulty of bringing them fairly into the calculation. At this time articles in small retail shops are often priced $\frac{1}{4}d.$ less than an even sum of pence. The result is that to persons unaccustomed to reckoning they appear cheap, while in reality the impossibility of halving and quartering farthings in accounts where farthings are of frequent occurrence gives a gain over the professed prices of some halfpence on every bill. The shopman cannot be expected to do a difficult sum in small fractions for every customer, and each customer loses very little, but where these customers in each day count by hundreds, as in many of the large retail ready-money shops, the gains in this way must be considerable, as those who understand business well know.

The Commission sat through the years 1857, 1858, and 1859. From resolutions passed on March 1, 1859, it appears that nothing had been ascertained which rendered the change desirable; that while the weights and measures remain as at present the coin could not be touched; and as the weights and measures could not be interfered with, the coin must be left alone. These, in few words, were the substance of the decisions of Parliament. 1856.

The energetic promoters of what would have been both morally and socially a most useful measure were naturally disappointed. Their efforts did not entirely cease at the time, and Mr. William Brown, Mr. De Morgan, and others, still did their utmost to keep the public interest alive on the whole question. Mr. Brown, whose philanthropy showed itself in many ways, died in 1864. Of him my husband says,—

'The agitation for a decimal coinage was put to rest by the illness and retirement of Sir Wm. Brown, whose recent death has revived the memory of his splendid benefaction (a fine library) to the town of Liverpool. A parliamentary leader of weight and energy is absolutely necessary to the success of any public measure; and as soon as the man shall be found who combines with William Brown's great energy his interest in the subject, the agitation will be revived. All the work that has been done is good material for a new attempt, and a new beginning will be made under great advantages. All the discussion about the metrical system works to the same end. We may be well assured that our system of calculation will not always be cramped by counting in one way and measuring in another. And our firm belief is that the way to work the change will be by beginning with the coinage, in which decimals are most wanted and most easily obtained.'[1]

We have now to help us in this work the aid of the Board schools. There is no reason why the strong efforts of the Decimal Association should be lost; and in the hope that the revival to which my husband looked forward may not be far off, I have given this history. All his writings on the subject are instructive, and the series of articles in the *Penny Cyclopædia* on Weights, Measures,

[1] *Athenæum*, review of 'Battle of the Standards,' April 9, 1864.

1856. and Coins, would form a large volume, containing information on every part of the question in all its relations.[1]

Sir Isaac Newton.

The memoir of Newton by Mr. De Morgan, which appeared in 1846 (Knight's *British Worthies*), was, after Baily's *Life of Flamsteed*, the first English work in which the weak side of Newton's character was made known. Justice to Leibnitz, to Flamsteed, even to Whiston, called for this exposure; and the belief that it was necessary did not lower the biographer's estimate of Newton's scientific greatness, and of the simplicity and purity of his moral character. Francis Baily's discovery of the correspondence between the Rev. John Flamsteed, the first Astronomer Royal, and Abraham Sharp, as well as between Newton, Halley, and Flamsteed, on the publication of Flamsteed's catalogue of stars, had thrown a new light on the character of Newton. It appeared that the practical astronomer had been treated ungenerously by Newton, who failed to observe the conditions of publication agreed to by all parties; and afterwards when remonstrated with omitted the name of Flamsteed in places where it had formerly stood in the earlier editions of the *Principia*.

My husband entered into the inquiry with keen interest, and with a power of research possible only to one who was fully master of the history of Mathematical discovery.[2] With reference to Newton's character he says:—'We

[1] I am told that a movement in this direction is now thought of. Whatever documents on this subject were left by my husband are still in existence.

[2] Several of the works in which the questions relating to Newton's Mathematical discoveries were treated are as follows:—

1. 'Memoir on Newton,' by J. B. Biot. *Biographe Universelle*, 1794.

2. 'Life of Newton,' by Sir D. Brewster. *Family Library*, No. 24, 1831.

3. *Life of Flamsteed*, by F. Baily, 1835. Article in *Quarterly Review* by Barrow, No. 109. Remarks on the same by Dr. Whewell. Article in *Edinburgh Review* by Mr. Galloway.

4. 'Life of Newton,' by Lord Brougham. *Library of Useful Knowledge*.

must differ in some degree from our guide,[1] as well as from all those (no small number) whose well-founded veneration for the greatest of philosophical inquirers has led them to regard him as an exhibition of goodness all but perfect, and judgment unimpeachable. That we can follow them a long way will sufficiently appear in the course of this sketch.' Later on he says, 'The great fault, or rather misfortune, of Newton's life was one of temperament;[2] a morbid fear of opposition from others ruled his whole life. When, as a young man, proposing new views in opposition to the justly honoured authority of Descartes and lesser names, he had reasons to look for opposition, we find him disgusted by the want of an immediate and universal assent, and representing, as he afterwards said, that Philosophy was so litigious a lady, that a man might as well be engaged in lawsuits as have to do with her.

1857. Newton's moral peculiarity.

5. (Art.) 'Newton,' by A. De Morgan. *Penny Cyclopædia*, 1840.

6. 'Life of Newton,' by A. De Morgan. *Knight's Cabinet Portrait Gallery (British Worthies)*, 1846.

7. Tract upon Keil and Leibnitz, by A. De Morgan. *Cambr. Phil. Trans.*, 1846.

8. 'A Short Account of Recent Discoveries in England and Germany relating to the Controversy on the Invention of Fluxions,' by A. De Morgan. *Comp. to the Almanac*, 1852.

9. 'On the Authorship of the "Account of the Commercium Epistolicum,"' by A. De Morgan. *Phil. Magazine*, June 1852.

10. 'On the Early History of Infinitesimals in England,' by A. De Morgan. *Phil. Magazine*, November 1852.

11. *Life of Newton*, by Sir D. Brewster, 1855.

12. Review of Brewster's *Life of Newton*, by A. De Morgan. *North British Review*, 1855.

13. Articles in *Notes and Queries* by A. De Morgan, 1853 and 1856.

14. *Correspondence of Sir Isaac Newton and Prof. Cotes*, by J. Edleston, Fellow of Trin. Coll., Camb. 8vo., London, 1850.

15. *Historical Essay on the First Publication of Sir Isaac Newton's 'Principia,'* by Prof. S. P. Rigaud. 8vo., Oxford, 1838.

16. *Gentleman's Magazine*, lxxxiv., p. 3.

17. Weld's *History of the Royal Society*. 2 vols. 8vo., London.

[1] Sir D. Brewster, from whose 'Life of Newton' in the *Family Library* the facts are taken.

[2] My husband always used this word for what I should call original character or inborn disposition.

1857. How could it be otherwise? What is scientific investigation except filing a bill of discovery against nature, with liberty to any one to be made a party to the suit? Newton did not feel this, and, not content with the ready acceptance of his views by the Royal Society, a little opposition made him declare his intention of retiring from the field. He had the choice of leaving his opponents unanswered, and pursuing his researches, committing it to time to show the soundness of his views. That this plan did not suit his temper shows that it was not the necessity of answering, but the fact of being opposed, which destroyed his peace. And he steadily adhered, after his first attempt, to his resolution of never again willingly appearing before the world. His several works were extorted from him; and, as far as we can judge, his great views on universal gravitation would have remained his own secret if Halley and the Royal Society had not used the utmost force they could command. A discovery of Newton was of a twofold character—he made it, and then others had to find out that he had made it. To say that he had a right to do this is allowable; that is, in the same sense in which we and our readers have a right to refuse him any portion of that praise which his biographers claim for him. In the higher and better sense of the word he had *no right* to claim the option of keeping from the world what it was essential to its progress that the world should know, any more than we should have a right to declare ourselves under no obligation to his memory for the services he rendered. To excuse him, and at the same time to blame those who will not excuse him, is to try the first question in one court and the second in another. A man who could write the *Principia*, and who owed his bread to a foundation instituted for the promotion of knowledge, was as much bound to write it as we are to thank him for it when written.'

The principle here expressed governed the writer's own life. What he knew belonged to the world, if by the knowledge the world could benefit; and no sooner had it become his own than he felt as much bound to give it to whomsoever could receive it, as he did to repay to the uttermost farthing, and with interest, a debt that he believed himself to have incurred. 1857.

Newton as an undergraduate.

Newton went to Cambridge as a sizar—'a student whose poverty compels him to seek to maintain himself in whole or in part by the performance of some duties which were originally of a menial character. By this means a youth could live by the work of his hands while he pursued his studies. In our days there is little distinction between the sizars and those above them; except in college charges, none at all. Those who look upon Universities as institutions for *gentlemen* only—that is, for persons who can pay their way according to a certain conventional standard—praise the liberality with which poorer gentlemen than others have gradually been emancipated from what seems to them a mere badge of poverty. But those who know the old constitution of the Universities see nothing in it except the loss to the labouring man and the destitute man of his inheritance in those splendid foundations. If sizarships with paid personal services had not existed, Newton could not have gone to Cambridge, and the *Principia* might never have been written. Let it be remembered, then, that so far as we owe this immortal work and *its* immortal work to the University of Cambridge, we owe it to the institution which no longer exists, by which education and advancement were as much open to honest poverty seeking a maintenance by labour as to wealth and rank. *Let the juries, who find on their oaths that scores of pounds' worth of cigars are reasonable necessaries for young students, think of this, if they can think.*'[1]

Proofs of all the writer's assertions on the jealousy and even vanity of the man whose intellectual work he

[1] Note to *Memoir of Newton*. The italics are mine.

1857. prized so highly had been brought forward in the memoir with a distinctness that left no room for doubt, though the question was not entirely set at rest at that time.

Another question, that of Newton's religious belief, is discussed in the *Life*; 'not,' the writer says, 'from any particular interest in it, for there are too many great minds on both sides of the controversy to make one more or less a matter of any consequence to either; but because we have a curious matter of evidence, and an instructive view of party methods of discussion.' Notwithstanding this disclaimer I believe my husband felt more interest in the question, from its own nature, than he was himself aware of. Whether I am mistaken in this may be surmised by those who have read his own letter to his mother in this volume. He says, 'Whatever Newton's opinions were, they were the result of a love of truth, and of a cautious and deliberate search after it.'

That Newton was a firm believer in Christianity as a revelation from God is very certain, but whether he held the opinions of the majority of Christians on the points which distinguish Trinitarians from Arians, Socinians, and Humanitarians, is the question of controversy.

The generic name *Unitarian*, with the specific names *Arian*, *Socinian*, and *Humanitarian*, are 'bandied about in interpretative discussion until they are so misused that the chances are many readers will need explanation of them. An *Arian* believes in the finite pre-existence of Jesus Christ before His appearance on earth; a *Socinian* believes him to be a man who did not exist before His appearance on earth, but who was a proper object of prayer; a *Humanitarian*, with all others who come under the general name *Unitarian* (the *personal* unity of the Deity being a common tenet of all), believes him to be a man, and not an object of prayer.' Having given the arguments on all sides, and taken into consideration Newton's great fear of discussion and of opposition, the biographer shows that the weight of evidence goes to prove that Newton, the devout Christian believer, was an Arian at the least. Besides this evidence, he was the friend of Locke, Clarke, and Whiston, all distinctly Antitrinitarians. 1857.

Newton's religious belief.

It is to be remembered that during the whole of Newton's life the denial of the doctrine of the Trinity was illegal, the statute of King William III. (which relaxed the existing law) making the offender ineligible for any position of trust for the first offence, and liable to three years' imprisonment with other penalties for the second. In these days, when few men of science do more than tolerate the idea of the existence of God, we may dwell with satisfaction on the fact that the world, if not wiser, is better. For whereas in 1696 a man was hanged for denying the Trinity, in 1881 men deny their Father in Heaven, and their fellow-men, instead of hanging them, are content to leave the opinion of each to find its own place in philosophy. If our intellectual conclusions are chaotic, our moral sense, in this respect at least, is clearer than it was 200 years ago.

The 'Life of Newton' concludes with these words in reference to the failings which truth compelled the writer to disclose:—'Surely it is enough that Newton is the greatest of philosophers and one of the best of men; that all his errors are to be traced to a disposition which seems to have been born with him; that, admitting them in their fullest extent, he remains an object of unqualified wonder, and all but unqualified respect.'

Brewster's biography.

In the year 1855, three years after the appearance of Mr. De Morgan's tracts on the Fluxional Controversy, Sir David Brewster's *Life of Newton* was published. Some of the statements of the biographer in the *British Worthies* were controverted, though not strongly; some were softened, and some ignored. Sir David, in his great veneration for his subject, had fallen into hero-worship, and my husband's critique of his *Life of Newton* in the *North British Review*, No. 46, 1855, shows this very clearly.

1857. In this review[1] Brewster's objections and omissions were refuted and supplied. The faults of Newton in the matter of Flamsteed and Leibnitz are proved, and in relation to the order of the discovery of Fluxions the character of Leibnitz received its due testimony:—

'We shall not stop to investigate the various forms in which Sir D. Brewster tries to make him out tricking and paltry. We have gone through all the stages which a reader of English works can go through. We were taught, even in boyhood, that the Royal Society had made it clear that Leibnitz stole his method from Newton. By our own unassisted research into original documents we have arrived at the conclusion that he was honest, candid, unsuspecting, and benevolent. His life was passed in law, diplomacy, and public business; his leisure was occupied mostly by psychology, and in a less degree by mathematics. Into this last science he made some incursions, produced one of the greatest of its inventions almost simultaneously with one of its greatest names, and made himself what Sir D. Brewster calls the "great rival" of Newton in Newton's most remarkable mathematical achievement.'

The reviewer speaks of the pleasure he derived as a boy from Brewster's invention, the kaleidoscope :—

'The two deans of optical science in Britain and in France, Sir David Brewster and M. Biot, are both biographers of Newton, and take rather different sides on disputed points. Sir D. Brewster was the first writer on optics in whose works we took an interest; but we do not mean printed works. We, plural as we are, remember well the afternoon, we should say the half-holiday, when

[1] Speaking of the titles of all the parties concerned, Mr. De Morgan says, 'Sir David never neglects the knighthood of Newton. . . . Should we survive Sir David, we shall *Brewster* him. We hold that those who are gone, when of a certain note, are entitled to the compliment of the simplest nomenclature.' In the tracts on the *Commercium Epistolicum* he reverses the usual phrase, saying, "Inasmuch as knighthood was not honoured with Newton until," &c.'

the kaleidoscope, which our *ludi-magister* (most aptly named for that turn) had just received from London, was confided to our care. We remember the committee of conservation, and the regulation that each boy should, at the first round, have the uninterrupted enjoyment of the treasure for three minutes ; and we remember, further, that we never could have believed it took so very short a time to boil an egg. A fig for Jupiter and his satellites, and their inhabitants too, if any! What should we have thought of Galileo when placed by the side of the inventor of this wonder of wonders, who had not only made his own telescope, but his own starry firmament?' . . . .

1857. Article in *North British Review.*

'Since his own scientific sensibilities are keen, . . . we hope they will make him fully feel that he has linked his own name to that of his first object of human reverence for as long as our century shall retain a place in literary history. This will be conceded by all, how much soever they may differ from the author in opinions or conclusions; and though we shall proceed to attack several of Sir D. Brewster's positions, and though we have no hesitation in affirming that he is too much of a biographer and too little of an historian, we admire his earnest enthusiasm, and feel as strongly as any one of his assentients the service he has rendered to our literature.'

The two biographers had differed in their estimate of Newton's religious belief, Sir David Brewster, in the first instance, maintaining his orthodoxy, 'by which,' Mr. De Morgan says in his *Review*, 'we mean a belief of as much as the Churches of England and Scotland hold in common.' He himself believed Newton to be an Arian, and the MS. creed of Newton found in the Portsmouth papers showed that he was right. M. Biot, who had been a worshipper of Newton early in the century, wrote to Mr. De Morgan at the time, expressing his satisfaction and concurrence in the statements of the *North British Review*. He received from my husband a copy of the memoir, with which he was greatly pleased. The author and M. Biot had met

1857. in Paris twenty-five years before, but this M. Biot, now in his eighty-second year, had forgotten.

In the course of the inquiry a question had arisen, which my husband felt to be an important one, nearly concerning Newton's moral rectitude. Catherine Barton, the niece of the philosopher, was known to have kept the house of his friend Lord Halifax. It was, however, never publicly known whether they were or were not privately married. From all that Mr. De Morgan could gather on this subject, which amounted to a great deal of presumptive if not quite conclusive evidence, he was convinced that, except as a wife, Newton would not have countenanced his niece's connection with the Prime Minister. This conviction was strengthened by the production of a MS. letter of Newton's bought by our friend M. Libri at a sale, in October 1856. But the opinions of scientific and literary men were not unanimous.

All the facts and arguments connected with this question were carefully embodied in an article for the *Companion to the Almanac* for 1858. This article was objected to by the publisher, Mr. C. Knight, on the ground of its not dealing with a subject of general interest. It was suggested to Mr. De Morgan to alter or curtail his writing, or to furnish another article, and he refused to do either. This was the cause of his discontinuing his contributions to the *Companion to the Almanac*. His reasons will be found in the correspondence.

1858. My husband's time was too thoroughly filled to allow of his taking an active part in many public movements, but he was always glad to give what help he could to benevolent schemes. He has mentioned, *à propos* of decimal coinage, his work at the Savings Bank. Another design which he much wished to see carried into execution was the opening of playgrounds for poor children. Our friend the Rev. David Laing, well remembered in his parish of Trinity, St. Pancras, N.W., for incessant efforts

1858. The play-ground movement.

to improve and assist his parishioners 'in mind, body, and estate,' devised a plan for opening playgrounds wherever land could be obtained in London, in which poor children might play harmlessly and happily, uncontaminated by street influences. Mr. Laing asked me to join his committee, and my husband fully shared the interest felt in the scheme. We had a dinner at the Freemasons' Tavern, at which Mr. Charles Dickens presided, and spoke as warmly as he was known to feel for the little vagrants, who, like the dweller in *Tom All-alone's*, were always being 'chivied' away. A meeting, too, was held for the same object, when Lord Shaftesbury was in the chair, supported by Lord Ebury and Mr. De Morgan, who both spoke warmly in our favour.[1] Our object was not attained, partly from the difficulty of exciting general interest, partly from the want of workers on the committee, for Mr. Laing's health gave way, and the society ceased to exist. The want is now in some measure but not entirely supplied by the playgrounds of the Board schools.

It was a few years before this time that Mr. Dickens and Mr. De Morgan had met at the house of Mr. Charles Knight at Broadstairs. I heard that the meeting gave pleasure to both, but I was not myself of the party. It was in the autumn, on one day of which the forty drowned bodies of cattle were heaped on the little pier, as described in the volume of letters recently published.[2] I well re-

[1] I wrote in *Household Words* 'A Plea for Playgrounds,' and a longer article in *Good Words* some time after for Miss Octavia Hill's playground.

[2] Some little time ago I came upon a letter from Mr. Dickens to us both, dated 1840. A difference of opinion had arisen between my husband and myself on the meaning of one of the illustrations in *Nicholas Nickleby*, that in which Mrs. Kenwigs, her four daughters, with Miss Petowker the fireman's daughter, and the reciter of the Blood-drinker's Burial, appear. Mr. De Morgan believed that the stout lady was the fireman's daughter, and the thin lady the mother of the little girls who were 'too beautiful to live.' The dispute ran so high that it could only be settled by an appeal to head-quarters. Accordingly Mr. De Morgan sent a letter to the author from 'a lady and

1858. member Mr. Dickens's active interest in the scene, and I heard the proposal to make the dead cattle into sausage meat, which probably was carried into effect 'unbeknown,' for several of the carcasses were bought by butchers. I was strictly enjoined not to let 'sassingers' come into the house.

1859. Our three youngest daughters had been born in Camden Street, and there were many associations with the house which were, with the sad exception of our eldest child's death, pleasant to my husband and myself. But we wanted a more roomy house, and it was thought that my severe illnesses might be averted by a better air.

After we were settled at No. 41 Chalcot Villas, Adelaide Road (at that time nearly surrounded by fields, and fully two miles from the College), he left the house always before eight o'clock in the morning, and met the omnibus in the Hampstead Road, which took him to Grafton Street a short time before the lecture began. He returned to dinner at five o'clock; and as he only gave himself about half an hour's rest after dinner before going to his library, where he wrote or read for four or five hours,

gentleman who, being husband and wife, seldom agreed about anything, though they were in one mind in admiration of the novel,' entreating the author to adjudicate the question. We received the following reply :—

'1 Devonshire Terrace, York Gate, April 12, 1840.

'Mr. Charles Dickens sends his compliments both to the gentleman and the lady who do him the honour to differ upon an illustrated point in *Nicholas Nickleby*, and begs to inform them that the lady sitting down is intended for Mrs. Kenwigs, and the lady standing up for the designing Miss Petowker. But Mr. Dickens begs the gentleman and lady unknown to take especial notice that neither of their portraitures is quite correct, Mrs. Kenwigs being constitutionally slim and delicate and of a slight figure (quite unimpaired by her frequent confinements), and Miss Petowker a young female of some personal attractions, set off by various stage effects and professional captivations.'

This was according to my husband's impression, so he was triumphant and I crestfallen.

1859.

he seldom gave up an evening to friends without feeling that his work for the next day had accumulated. When he had nowhere to go to between the lectures he spent an hour or two in the Professors' room with some of his colleagues in the middle of the day, and this was his chief rest and recreation. I cannot record any of the conversation held there, but I know that for many years it was very pleasant and sociable, and many a good anecdote and riddle (generally traceable to Dr. Sharpey) have come to me from that little conclave.

One day my husband told me that he had only been in the Professors' room for a few minutes for two days, for that a poor man driving a cab had been thrown from his seat just as he was himself crossing the road. 'They picked him up,' he said, 'quite insensible, but he recovered when we got him home. I saw him comfortably in bed, and that he did not want anything. Yesterday he was much better.' I found that the injured man's home was more than a mile from the College.

Essay on Probability.

Mr. De Morgan's friend Mr. Drinkwater Bethune held a high Government appointment in India. Besides being a statesman and a distinguished scholar, he was a philanthropist, and was active in promoting the education of Hindoos and Mahomedans. The lives of Galileo and Kepler in the *Library of Useful Knowledge* were by him, as was also an essay *On Probability*, which he wrote in conjunction with Sir J. Lubbock. On the back of this little book the binder had by mistake printed Mr. De Morgan's name, and the attribution to himself of a work to which he had no claim troubled the reputed writer, who, of course, was uneasy till he had thoroughly disowned it. On the cover of his own copy he substituted for his name those of the real authors, adding after the printed words 'by Augustus De Morgan' the comment, 'the last named is *homo trium literarum*.'[1]

[1] F. U. R.

1859. Ramchundra.

In 1850, shortly before Mr. Drinkwater Bethune's death, he had sent to England some copies of a work by Ramchundra, a native teacher, and head master of Science in the school at Delhi. A copy was presented to Mr. De Morgan, who saw in this treatise on problems of *Maxima and Minima* 'not only merit worthy of encouragement, but merit of a very peculiar kind, the encouragement of which . . . was likely to promote native effort towards the restoration of the native mind in India.' With his lively interest in all that belonged to India, the land of his birth, and with his still wider interest in every effort of original thought, especially where, as bearing the impress of national character, it had place in the history of mental progress, he called the attention of the Court of Directors to the work of the Hindoo teacher, suggesting that Ramchundra should receive a reward for his work, and that the work itself should be brought under the notice of Mathematicians in Europe. After some correspondence with the authorities and with Ramchundra himself, the Court expressing entire concurrence in his views, his offer to superintend the reprint of Ramchundra's work was accepted. It was a work of some labour, because it was done so thoroughly. His preface, consisting of twenty-three closely printed pages, gives a short but scientific history of the rise and progress of Mathematical science in Greece and India, and an analysis of the mental character of the two nations in their respective leanings to geometrical and algebraic thought, with the causes of the entire extinction of Mathematical speculation in India, while it remained to some extent active in Greece as long as that country existed as a nation. The extinction of active speculation, and its replacement by a taste for routine, 'to which,' he says, 'inaccurate thinkers give the name of *practical*,' he shows to be coeval with the death of Science in a nation. This was the fate of Hindoo philosophy, and how to revive active thought in the land of his birth was to him a question of deep interest. His counsel may still be valuable. 'Some friends of education,' he says, 'have advised that the Hindoos should be fully instructed in English ideas and methods, and made the media through which the mass of their countrymen should receive ideas in their own language.' This plan has not succeeded, and a deeper knowledge of the psychological aspects of national character might have predicted the result. 'My conviction,' my husband says, 'is that the Hindoo mind must work out its own problem, and that all we can do is to *set it to work*—that is, to promote independent speculation on all subjects by previous encouragement and subsequent reward.' Ramchundra had a stronger leaning to geometry than could have been expected by a person whose sole knowledge of the native mind was derived from the Vija Ganita, but he had not the power in geometry which he had in algebra. 'Should this preface,' the writer says, 'fall into the hands of some young Hindoos who are systematic students of Mathematics, I beg of them to consider well my assertion that their weak point must be strengthened by the cultivation of pure geometry. Euclid must be to them what Bhascara or some other algebraist has been to Europe.' It may be that the prevalence of algebraic thought among the Hindoos naturally accompanies their power of computation, and their love of symbolism (without beauty of external form) shown in the mythology and astronomy; while the mathematical reasoning of the Greeks fell, as might naturally be expected from their love of symmetry and proportion in form, into a geometrical method.

1859. **Ramchundra.**

At this time, Lord Brougham being Lord Rector of the University of Edinburgh, the degree of LL.D. was offered to my husband. He appreciated the honour, but declined it with thanks, saying to me that it did not suit him: he 'did not feel like an LL.D.'

Offer of Edinburgh degree.

The International Statistical Congress was held in London in 1860, and he joined a Committee for the

1859.

Inspection, I think, of Scientific Instruments, but he did not give much work to this.

1861. Royal Astronomical Society.

An uncompromising character like my husband's must, while the world is what it is, sometimes make its owner appear combative. And indeed it always needs some kindness and geniality in a dissentient to produce the conviction that he is not actuated by love of opposition in the part he takes.

Mr. De Morgan's fellow-workers in any cause soon knew him well enough to feel sure what part he would take in any occasion of difficulty involving self-sacrifice for principle. An occasion of this sort occurred in 1861, when he left the Council of the Royal Astronomical Society.

The following list will show the offices he had held since he joined it in 1828 :—

Feb. 1830. He was elected a member of the Council.
1831. Secretary with Hon. Mr. Wrottesley.
1832–38. Secretary with change of colleague.
1839–40. One of the Vice-Presidents.
1841–42. Member of Council.
1843–44. One of the Vice-Presidents.
1845–47. Member of Council.
1848–54. Secretary with Admiral Manners.
1855–56. One of the Vice-Presidents.
1857. Member of Council.
1858. Vice-President.
1859. Member of Council.
1860. Vice-President.
1861. Elected a Vice-President; declined to act.

From this list, taken from the *Monthly Notices*, it appears that he held the place of Secretary for fifteen years. By his own statement in the letter given farther on, he filled it for eighteen years.

In 1860 he retired from the Club.[1] After our removal

[1] This consisted of the most influential members of the Society,

in 1859, from Camden Street to Adelaide Road, the distance made it difficult for him to be in London late at night, on account of the necessity before mentioned of his leaving home early in the morning. His resignation was received with regret. The meetings of this friendly Club had been a great pleasure to him, but he had latterly been unable to join them. His friend Mr. De la Rue, then treasurer and secretary, wrote:—'Regret was universally expressed at the announcement of your retiring, and also because it recalled to mind how little you had been with us of late. The members could not reconcile themselves to the withdrawal of a name so intimately connected with the Club for the last thirty years, and you were immediately proposed as an honorary life member. I have to announce to you that you were elected by acclamation, and that the Club hope that you will dine with them whenever your leisure and inclination permit of your doing so; and this wish I endorse on my own account most heartily.' I do not think he ever found leisure, however much he might have had inclination, to dine with his old friends again. 1861.

Election of Dr. Lee as President.

During this and the following year occurrences took place which affected his happy relations with the Society, though not with the friends who continued to belong to it. In the year 1861, six members of the Council determined to place Dr. Lee, of Hartwell House, in the President's chair. Dr. Lee was a respectable and estimable man, who, by the maintenance of a private observatory, had shown great interest in Astronomy; but he was himself more of an antiquarian than a scientific man, and, but for his wealth, would not have been eligible as President. It was the manner of proposing the candidate at a packed meeting, and the canvass for his election by his supporters, in place of the open election by which,

who assembled after the meeting. Their entertainment consisted of coffee, cigars, sometimes a bowl of punch, and always much friendly talk.

1861. in compliance with the recommendation of the Council, a President had hitherto been chosen, that constituted a departure from the course hitherto followed. It was no longer 'our little honest Society,' as my husband had called it in writing to Admiral Smyth; and all its oldest friends—those who had made it what it was—declared against the innovation. The other party prevailed, and on the election of Vice-Presidents after the President's election, Mr. De Morgan was informed that his name was on the list. His friends, when his intention to leave the Council was made known, entreated him to remain, several of them urging that his was the name which could be least spared. But he believed that his resignation would cause the smallest possible shock to the Society, while it would be a protest against the unconstitutional tendencies which he wished to arrest. In answer to the official announcement of his election as one of the Vice-Presidents, he sent the following letter:—

41 Chalcot Villas, Adelaide Road, March 1, 1861.

Letter on quitting Council.

GENTLEMEN,—I have received from Mr. Williams a notification that I was elected a Vice-President of the Society at the General Meeting of the 8th ult. This election, legally valid, is morally defective in one essential particular. The appointment of a voluntary officer must be a result of concert between the choosers and the chosen. It is a matter of prudence, when a new system is established by a contested election, that the first should inquire of the second whether he will be willing to take the office on the terms which the altered circumstances of the case expressly or implicitly lay down. Failing such inquiry, any election, however good in law, is but an offer; and to the offer I reply, in all good humour, first by thanks, secondly by non-acceptance.

Here I might close this communication; but the regard I feel for the Society, in whose affairs I have taken part from early manhood up to a time when old age is within signal distance, impels me to explain myself farther. In placing before you the ground of my retirement, I am fully satisfied that the retirement itself, accompanied by reasons, will do the cause more good than any services of mine have ever done yet, and vastly more than my continuance on the Council could do. Were I not so well assured of this, that the course I take becomes imperative, I should content myself with silent acquiescence, and should defer my secession until the actual arrival of the state of things which I fear is on its way.

1861.

Letter on quitting Council.

Before I enter on the subject I will premise—first, that there is nothing in my refusal to serve which has reference to the new President, for whom I entertain, as always, high esteem and regard. Had Mr. Airy been preferred to Dr. Lee in the manner in which Dr. Lee has been preferred to Mr. Airy, my course and my reasons would still have been what they are. Secondly, I fully admit, should any one suppose I question it, that the Society has not exceeded its rights a bit more than I shall have exceeded my own rights in sending this letter. Thirdly, I have not acted in concert with any former colleague, and have not even given a hint of my resolution to any Fellow of the Society.

The Astronomical Society has gained a high position by sheer hard work. It is the *plainest* of all the scientific associations, and the one which least glitters by the show of rank and wealth. Its sole thought has been the promotion of Astronomy. And undivided attention to its real business has been rendered easy by such harmony as is very rarely found in public bodies. Nevertheless, during the last two or three years there has not been that entire unity between the Council and the Fellows which had always existed in time past. I believe that such interruption of the usual concord as has taken place was the consequence of adverse feeling in a very small number of the Fellows; how generated I do not know. Perhaps that bias towards initiation of political action by which Englishmen spoil so many of their extra-political associations may have taken hold of some minds. I thought I saw symptoms of the Council being a corrupt aristocracy, who made pocket boroughs of the planets, and deprived the moon of her due share of the franchise. I will make no further allusions to the manifestations, which satisfied me, independently of all I knew besides, that the side they came from was the wrong side.

The feelings I have mentioned soon took the form of a desire to facilitate combined opposition to the list of officers which each retiring Council had always recommended to be their successors. The existing Council met the expression of this desire by the proposal of those by-laws on the subject which now govern the

1861. Society. That I was not against the change may be inferred from my being the acting Secretary of the Committee of Council which drew up the new by-laws, and the mover of their adoption at the general meeting which passed them. My assigned reason was that nothing could be too democratic for a scientific society. I think so still. But I saw in silence, though with great satisfaction, that the changes would speedily bring to issue and to settlement questions, on the right and speedy settlement of which the prosperity of the Society depends. And this is a much better way than could be opened by the results of a period of smouldering discontent. I saw that the good sense of the whole body would soon be turned to reflection upon the question whether the Society could exist in honour and in usefulness upon any other basis than that of peaceful government by *a* Council: not *this* Council, nor *that* Council, but *a* Council. That is to say, government by *one* Council, under the statutory changes of its details, until that Council ceases to be in harmony with its constituents, and then government by *another* Council, composed, in strong part at least, of different men.

I did not suspect that the Society would be so fortunate as that reflection on this point should first be promoted by so slight a pair of matters as the substitution of one President for another, and the retirement of so dispensable a member of the Council as myself. It is just as I could have wished. My own secession will not—whatever it might have done twenty years ago—impede the action of the Society, while the retirement of a person who has known it so thoroughly as I have done for thirty years, during eighteen of which he has officiated as Secretary, will certainly lead to that reflection which I desire to promote.

The circumstances under which I retire are, in my view, as follows. Half-a-dozen of the Fellows, desiring to pay a compliment to a highly respected member of the Council, proposed him as President, in opposition to another Fellow proposed by the Council. That this, and no more than this, was their first intention, I feel well assured. But, as they grew warm in the business, they availed themselves of the usual resources of opposition—the personal canvass, the newspaper article, and the invention of a principle to justify a course which, in the first instance, had no reason except the innocent one which I have stated. They lay it down that A B having been President three times, and C D not at all, it is now, as it were, C D's turn. That is to say, they propound a law of rotation, independently of any reason which the Council might have had for their recommendation. Should I have misrepresented, or rather under-represented, the Fellows to whom I allude, the misrepresentation comes of their own fault. They had the power of stating the principles on which they acted to the general meeting; it pleased them to prefer the partial and private canvass for their only mode of action, the paragraph and its reason for their only statement of view.

1861.

Letter on quitting Council.

They succeeded, and I am convinced that one element of their success was that modicum of adverse feeling towards the Council, of the existence of which I had seen proofs. Their success transfers the responsibility of their course of action to the Society as a whole; and thus, and thus only, does that course of action become a legitimate object of my criticism. I maintain and uphold the *individual Fellow* in his claim to propose a President on any criterion of superior fitness which he pleases. I will support him, acting for himself, in the assertion of his right to use the canvass, the newspaper, or anything which a civilised man can have recourse to, in preference to placing his views before the assembled Society. But when the Society *adopts* a proposal, the result then becomes my affair, with its reasons, if any be assigned; or, failing such assignment, with the reasons deduced by myself from circumstances.

The power of organising opposition, recently and most properly conceded by the Society to individuals, is one, the corporate assent to any use of which should both be governed and defended by reason. It was intended, as the Society itself was intended, for the encouragement and promotion of Astronomy. When employed and privately argued against a deliberate recommendation of the Council, it should not be sanctioned by the collective Society except upon avowed grounds. I speak of concerted plans, not of votes of individual Fellows, each acting on his own judgment. If the existing system of administration be not in harmony with the corporate feeling to an extent which requires united action, it is expedient that the Council should know the how and the why. This is reasonable, because it will discourage and *retard* Astronomy, so far as the Society can do it—and it can do something—if the Council and the Society should take to working against each other in the dark. Should the system continue, it is my fixed opinion that the harmonious and useful body to which I have so long been attached will go through a series of faction fights, compared to which the recent matter is hardly worthy to be called a contest;

1861. and will emerge, if indeed it emerge at all, with crippled utility, diminished honour, and wasted resources. This is my opinion. I desire no one else to assent unless his own experience of collective action should make him see the danger as I do, after that reflection which I know my proceeding will excite. I do not affirm that the substitution of one President for another is in itself a dangerous act. I look at the whole history of the Society for some years past, and on that whole I am irresistibly impelled to form a very strong opinion.

I will add something which may tend to prevent a greater calamity than my own refusal to act. It ought to have been evident to the promoters of the recent division that the new by-laws have the effect of making what is called *the* balloting list into two or more, whenever two sets of nominators are therein exhibited. Usually when parties contest such a question each gives the whole of its own list, and any two lists are not the less two because there may happen to be many names common to both. It is most expedient that those who originate a second list should take care to ascertain that those whose names they take from the other list are willing to serve in either event. For two lists which differ by one name only, especially when the name is that of the proposed President, may symbolise two very different principles. It might easily happen that *many* of the common names might decline to give such assent to the principle of one of the lists, as would be inferred from their accepting office at the hands of a majority. It cannot reasonably be expected that any nominee should rise at the meeting, and declare his intention of not serving except on one contingency. I duly considered the propriety of such course, and rejected it for three reasons. First, it would have had such an appearance of disrespect to an old friend as it would have been impossible to neutralise, save by such explanations as would have brought on a discussion which it was not for me to originate. Secondly, because such a declaration—a dictation, as it would have been called—would have been a firebrand thrown into the meeting by way of commencing the discussion. Thirdly, because the course announced would have caused difference of opinion among those who thought as I did concerning the policy of the opposition to the Council.

The true way of securing a working Council would be for those who differ to take care, each side for itself, to present a list of those whom they have reason to know to be willing to

1861. Letter on quitting Council.

serve upon the views which the nominating party has put forward. In the present case it has been proposed to me, by the mere fact of election, and in no other way, to act as Vice-President on the principle of filling offices only to pay compliments, or else upon a principle of rotation, the compliment and the system of rotation being settled by private canvass. Or if not one of these, then upon the principle that the deliberate recommendation of the Council may be set aside on no reason, either assigned or discoverable. I disapprove equally of the compliment reason, the rotation reason, and the nullity reason. I consider them all as fraught with danger when applied without deliberation. I am not prepared to say that, on due consideration of all circumstances, the case might not arise in which any one of these reasons might be sufficient. But I feel compelled to decline action on either reason on no better support than the *pro ratione voluntas* of the balloting-box. If, indeed, the promoters of the change had come forward and had justified their course in public meeting, I might possibly, though not concurring in their reasons, have been able to accept the *deliberate conclusion* of the Society, in lieu of the *deliberate conclusion* of the Council. I have much respect for the result of argument, even when I do not feel convinced by the argument itself. But I will not act in the affairs of a Society which rejects the recommendation of its best advisers on grounds which those who promote the rejection do not submit to discussion; and, were I not satisfied that the successful majority do not comprehend the character of their own proceeding, I should look upon them as almost wanting in courtesy for not taking the pains to ascertain whether I could meet their views, or whether they would have to substitute for my name that of one of the gentlemen whose advice they were disposed to prefer to mine. As it is, however, I can thank them without reservation for the honour which I decline.

I cannot help saying that it will give me much satisfaction to hear that no one but myself, however much he may be convinced that a new and perilous period in the history of the Society has commenced, judges it necessary to carry matters so far as I have done. And with this, coupled with the expression of the deep regret with which I separate myself from those with whom I have so long acted in the most friendly concert, I remain, gentlemen,

Yours sincerely,

A. De Morgan.

1861. P.S.—It will be seen that I write as if I were addressing the Fellows at large. This means that I expect that my letter will be entered on the minutes of the Council, which are by express by-laws accessible to the Fellows at their meetings.

These suggestions on government might apply to larger bodies than the one addressed. They give an idea of his political principles, which demanded the utmost possible freedom for the individual, subject to a conservative respect for law. I have been told that the Astronomical Society suffered at this time from the causes which led him to leave it. But these things occurred twenty years ago, and their effects have, no doubt, long passed away.

1862. His correspondence with M. Biot had been chiefly on the *Life of Newton*. In this year M. Biot died, and Mr. De Morgan received from his son-in-law, M. Lefort, some particulars of his early life as material for a biographical notice.

Introductory lecture at University College.

An introductory lecture, the last he ever gave at University College, on the opening of this session, was never printed. How greatly this was regretted by many of his hearers I am unable to say. Seldom was an address listened to within those walls with a more lively interest, or received with such mirth and hearty acclamations. The subject was a branch of his favourite one—'Education,' and that branch was the method of examining at Cambridge. The attacks on the system were made with so much humour, and so much good humour, that the attacked could hardly have resented what they must feel was so well deserved. Great amusement was caused by the description of the self-satisfaction of an examiner after he had set a question well fitted to show off his own ingenuity and cleverness, but unfitted to elicit the thought or power of the student. The illustrations which half filled the lecture were taken from common sayings, old ballads, and nursery rhymes; and if the grave body in the centre of the theatre, a few of whom could appreciate the force and humour of the quotations, felt their dignity at stake, the indisputable truth of every sentence justified the utterance. Many of the students who felt the importance of all that had been brought forward in this playful guise signed a request that he would print the lecture. A similar request was also sent privately from Cambridge. He meant to rearrange and add to it before publication, but he never had time for this, and the slightest notes only are left. 1862.

I ought not to leave out his work in Life Assurance, but of this I can say very little. I have heard it said that his attention was drawn to this subject by my father, who, from his own pursuits, was supposed to be interested in it. This was not so, however. The two had always mutual subjects of more interest to discuss, and, as far as I recollect, it was scarcely mentioned between them till Mr. De Morgan, who had been consulted on some Company's business, referred to my father for information as to their way of doing it, a subject on which he had been consulted before. My husband frequently gave opinions on insurance questions. He was a contributor to the *Insurance Record*, and gave many valuable papers to it and to the magazine of the Institute of Actuaries. One of the longest articles in this was a severe criticism and exposure of Mr. T. R. Edmonds, who had given to the world as his own a discovery which was made by Mr. Benjamin Gompertz. The latter was distantly related by marriage to Mr. De Morgan's family.

Failure of Albert Life Office.

The last paper in the *Insurance Record* to which his name is appended relates to the Albert Life Assurance Company. In 1861 he had made the valuation according to the data furnished to him, and in 1862 gave his opinion that the Society was in a condition to give a bonus. When, eight years after, the Society was declared bankrupt, under circumstances which were far from creditable to the management, his name and opinion were brought forward as a justification of their proceedings. This was

1862. Failure of Albert Life Office.

first known to him by paragraphs extracted in our newspapers from the *Bengal Hurkaru* and other Indian papers, and afterwards by the *Overland Summary*, by which it appeared that severe censure had been passed upon him for the part he had taken in the monetary affairs of the Albert. The question was not unimportant, because much ruin had been caused by the failure of the office; and though this could not be traceable to his advice, his name had been used as a screen by those whose mismanagement, if nothing worse, had caused the calamity. As was said, he had but to fight shadows, but this might be worth while when the shadows rest upon a good name. When all this came to his knowledge in 1870, he was too weak and ill to care much about erroneous statements respecting himself, but his friends prevailed on him to write an explanation of the case, which was simple enough. He had, he said, given an opinion upon the data laid before him. He had not been required to investigate the affairs of the office. Had this been asked, he would have perceived that the managers were counting as realised capital large sums which they believed would be paid to them from various quarters, an error against which he had strongly cautioned them, and into which he afterwards suspected they had fallen. His letter, the last he wrote upon public business, contains a little touch of his old humour.

> When a scientific opinion is given, it is intended that 'it'—the whole opinion, remember—may be used in any way the receiver pleases. Let him give 'it' as it was given, without alteration or suppression, and he may speak of it as he pleases, may call it what he pleases, and may infer from it what he pleases. He may call my life office valuation a receipt for mince pies—which I certainly never *intended* it to be—but he must not mix up with it anything out of Mrs. Rundell or Mrs. Glasse. Let him give it but fairly, and I am content, if he will do the same, to take all the consequences of his change of description.

Mathematical Society.

The last occurrence connected with Science which gave him pleasure was the formation of the Mathematical

1863.

Society. Our second son, George Campbell, had gained the highest prizes for Mathematics and Natural Philosophy in University College, though his father's scruples were strong as to adjudging the first prize in his own class to his own son. My husband told me that the papers, which he knew to be George's, were much ahead of the other competitors. 'But,' he said, 'I don't see how I can give him my prize.' I reminded him that the sons of other Professors had frequently taken their fathers' prizes, and that justice to George required it. He said that was true, and he would show the papers to another Professor, who was enough of a Mathematician to judge. The arbitrator, who did not know the writing, adjudged the prize to our son, as his father had done, without the slightest hesitation. George afterwards took his degree in the University of London, and obtained the gold medal for Mathematics and Natural Philosophy. He likewise took the Andrews Scholarship in University College, but the work required, especially that for his degree, was too much for a delicate frame, and a severe cold caught about this time was the forerunner of his last illness. The last three years of his life were alternations of illness and partial recovery, though a winter at the sea-side and a subsequent voyage to the West Indies with his brother Edward gave him strength for a time, and we hoped he might have outgrown his delicacy of constitution; but this was not to be.

Origin of the Society

It was in the year 1864 that Mr. Arthur Cowper Ranyard and George were discussing mathematical problems during a walk in the streets, when it struck them that[1] 'it would be very nice to have a Society to which all discoveries in Mathematics could be brought, and where things could be discussed, like the Astronomical.' It was agreed between the young men that this should be proposed, and that George should ask his father to take the chair at the first meeting. I have a list in his

[1] The words in which he told me of the occurrence.

1864. handwriting of those Mathematicians who were invited to join, with the names marked of the gentlemen who accepted the invitations. Among these were many of the first Mathematicians in England, and their number rapidly increased.

One of the first documents relating to this Society—it may have a value in its future history—is the following, lithographed from George's writing:—

University College, Gower Street, Oct. 10, 1864.

SIR,—We beg leave to request the honour of your attendance at the first meeting of the 'University College Mathematical Society,' which will be held at the College in the Botanical Theatre on the evening of the 7th of November, at eight o'clock precisely.

Professor De Morgan has promised to take the chair, and will give an introductory address, and the general objects and plans of the Society may then be discussed.

It is proposed that the ordinary meetings of the Society should take place once a month, and that the papers then read should be lithographed and circulated among the members.

The annual subscription will not exceed half a guinea.

We have the honour, &c.,

G. C. DE MORGAN,
ARTHUR C. RANYARD,
Hon. Secs. *pro tem.*

1865. The first meeting was held January 16, 1865. My husband was the first President, and his inaugural speech contains so many of his own leading thoughts, that I may give a few sentences. The first conveys his old opinion, formed early in life, upon the constitution of public bodies, and the inexpediency of crippling their future action by legislation at the outset.

There is much discussion about what our Society should be. But this cannot be settled and marked out; it must be determined by the disposition of its members. All scientific societies are in danger of getting into a groove, and settling into a routine which possesses small interest to the great body of their members. . . . On the other side, there is always the danger of guiding the Society off the rail, as it were, and getting it out of the way in which the momentum of its members can be applied to move it. . . .

1865. Inaugural address.

Our great aim is the cultivation of pure Mathematics, and their most immediate applications. If we look at what takes place around us, we shall find that we have no Mathematical Society to look to as our guide. The Royal Society, it is true, *receives* mathematical papers, but it cannot be called a Mathematical Society. The Cambridge Philosophical Society seems to fulfil more nearly the functions of a Mathematical Society, but it is in an exceptional position. It is the Society of the place which may be regarded as the centre of the Mathematical world; it is a Society in which almost all the members are able to relish its highest discussions. But in London we have no Mathematical Society at all.

He had a few words for his old object of attack—the Cambridge examinations:—

The Cambridge examination is nothing but a hard trial of what we must call problems—since they call them so—of the Senior Wrangler that is to be of this present January, and the Senior Wranglers of some three or four years ago. The whole object seems to be to produce problems, or, as I should prefer to call them, hard ten-minute conundrums. These problems, as they are called, are necessarily obliged to be things of ten minutes or a quarter of an hour. It is impossible in such an examination to propose a matter that would take a competent Mathematician two or three hours to solve, and for the consideration of which it would be necessary for him to draw his materials from different quarters, and see how he can put together his previous knowledge so as to bring it to bear most effectually on this particular subject. It is, I say, impossible that such a problem as this should be set in these examinations.

It must be one of our objects to introduce into our discussions something more like problems properly so called, and, if possible, to keep ourselves from entertaining an undue number of the questions just described. In some quarters the Mathematics are looked at, I may say, almost entirely with reference to their applications. These applications are not only physical applications or commercial applications, which may be termed *external*, but there are also what may be termed *internal* applications. Those very questions to which I have alluded already,

1865. respecting curves of the second order, are applications of the principles of pure Mathematics. They form, in fact, a particular branch of the application of first principles. . . .

Inaugural address.

We must not mistake and misapprehend these internal applications; we must not regard them as constituting entirely what we are to turn our attention to. We have several things before us besides these, which are very little attended to. One of these is what may be called Logical Mathematics. We want a great deal of study of the connection of Logic and Mathematics. Where is any consideration of this question to be found? If I may be allowed to say more on a subject to which I have devoted a good deal of time and thought, I would make a few observations on this very important and yet very much neglected one.

There is exact Science in two branches: the Analysis of the necessary *Laws* of Thought, and the Analysis of the necessary *Matter* of Thought. The necessary Matter of Thought, that without which we cannot think, consists of Space and Time. These exist everywhere, and we can imagine no thought without them. Space and Time are the *only* necessary Matter of Thought. These form the subject-matter of the Mathematics. The consideration of the necessary *Laws* of Thought, on the other hand, constitutes Logic. These latter have been little studied hitherto, even apart from the study of the necessary matter of thought.

We mathematicians may very easily improve our reasoning from the very beginning. For, though the Logic that Euclid used is very accurate, there has been no inquiry made with regard to it; and the consequence is that for two thousand years we have been proving, as we go through the Elements of Geometry, that a thing is itself. That is to say, we have been proving, in the Elements of Geometry, by help of a syllogism, a thing which must be admitted before syllogism itself can be allowed to be valid. Thus, does Euclid not prove that, when there is but one A and but one B, if the A be the B, then the B is the A? He would not take such a thing as that without appearance of proof. 'A thing is itself;' that is the assertion, that is what Euclid would not take without proof!

To take an example. Let us suppose that there is a village which contains but one grocer and but one Post-Office. Then, if the grocer's be the Post-Office, the Post-Office is the grocer's. For, if it be possible, let the Post-Office be somewhere else, say at the chandler's. Then, because the Post-Office is the chandler's and the grocer's is the Post-Office, it follows that the grocer's is the chandler's, another place, by hypothesis; which is absurd; and so with every other place except the grocer's. Therefore the Post-Office is at the grocer's. 1865. Inaugural address.

Is not this mode of proof in the third book of Euclid, being the way in which proposition 19 is deduced from proposition 18? Yet anybody who should use it out of geometry would be laughed at, though Euclid used it, and all those who have studied his Elements have been proving things in this manner for two thousand years.

There is no doubt about the matter, I say—and it will appear more distinctly on further thought—that you are proving by help of a syllogism what must be admitted before syllogism itself is valid.

As to the chances of the Society finding for itself lines of original work, he says:—

The higher Mathematics may be carried on with much greater effect by elementary students if they will but study points of Logic, History, Language, and perception of propositions by simple common sense. Mathematics is becoming too much of a machinery, and this is especially the case with reference to the elementary students. They put the data of the problems into a mill, and expect them to come out ground at the other end—an operation which bears a close resemblance to putting in hemp-seed at one end of a machine, and taking out ruffled shirts ready for use at the other end. This mode is, no doubt, exceedingly effective in producing results, but it is certainly not so in teaching the mind and in exercising thought. If it should chance that we find a disposition among the members of this Society to leave the beaten track and cut out fresh paths, or mend the old ones, we make this Society exceedingly useful. But if not, if it be our fate only to become problem makers and problem solvers, there is no harm done; we shall but add one more association to the list of journals, colleges, &c., devoted to this object. The only objection is that this branch of the subject is sufficiently well appreciated and more than sufficiently well practised already.

Original papers by both its first President and Secretaries appear in the first pages of its earliest reports, and

1865. some brilliant mathematical discoveries by Professor Sylvester were communicated to it soon after its foundation.

In 1866 George was teacher of Mathematics in University College School, and in January 1867, the last year of his life, the Tract *On the Proof of any Function, and on Neutral Series*, read before the Cambridge Philosophical Society the year before, has this note appended to it: 'My son Mr. G. C. De Morgan recently showed me this case of failure of development.' The algebraic operation is given.

His father had a high opinion of the power of George's mind, which in some ways resembled his own. Our friend M. Libri called him Daniel Bernouilli, in reference to the two Bernouillis, father and son. It gave his father pleasure to think that although he died so young, his son's name should have been associated with his own.

SECTION IX.

CORRESPONDENCE, 1856–66.

To Sir John Herschel.

Camden Street, June 8, 1856.

MY DEAR SIR JOHN,—I have long had the idea of a piano- 1856.
forte in which each set of strings belonging to one note is to communicate with a pipe for resonance; and sometimes I have thought that a spring at the mouth of a pipe struck by a hammer would make a good instrument. In this case we might have various pedals opening and closing the upper end of the pipe. But I never imagined anything so grand as the introduction of a vast force by means of electro-magnetism. I should propose to call your instrument the *electro-magnetic* whack-row-de-dow.

What is the reason why thirds and sixths, major or minor, are more pleasant to the ear than fourths and fifths, which are consonances of simpler ratio of vibration? Fifths, by themselves, have a certain something which the ear does not like much of, and consecutive fifths we all know are forbidden. But thirds and sixths are very pleasant. If Dr. Smith's theory of beats be *true*, I almost suspect I spy a way to explain this. But I must get hold of an organ tuner, and learn whether they are actually effective.

Yours very truly,
A. DE MORGAN.

To Sir John Herschel.

7 Camden Street, August 15, 1856.

MY DEAR SIR JOHN,—What can you say for yourself? Your last note was written in a good strong hand. This is hot weather. When I was a boy I read how Cato the Censor used to allow himself a little vinegar in his water when the heat was

1856. great, and when a young man I used to imitate him. A few drops of vinegar, half or a whole teaspoonful to the tumbler, I found the most refreshing addition possible to a tumbler of water, and I have revived the habit. Try it. Raspberry vinegar is a great deterioration of the principle. Cato would have *censored* it prodigiously.

I have got a clincher about Cath. Barton and Lord Halifax. Last Sunday Libri showed me a letter of Newton which he had bought. The handwriting is indisputable. It came out of some Newton papers which Rodd picked up in 1847. It is written four days after the death of Halifax to a Sir John of Lincolnshire (probably Sir J. *Newton* of Westby, Newton's distant cousin). It excuses him (J. N.) from paying a visit for these reasons:—

'The concern I am in for the loss of my Lord Halifax, and the *circumstances in which I stand related to his family*, will not suffer me to go abroad till his funeral is over.'

Not a scrap of evidence exists that Newton was ever acquainted with the other Montagues, though of course it is very likely he knew them; but relation of any kind, whether *rapport* or *parente*, is utterly unknown, much as Newton has been poked into. Newton was not an executor of Halifax. *Quære* whether Halifax's *family* means family in the usual primary sense of wife and children? Did Halifax leave a widow? Was that widow Newton's niece? If so, a very natural reason for keeping the house occurs. Macaulay, who used to battle the point, and fought for the Platonics, now says he does not *entirely reject* my hypothesis. Brewster has never written to me since I reviewed his book, so I cannot send it to him. Lord Brougham is brought up by it; says it is very curious, and he must think about it. I believe this letter will be Cath. Barton's marriage certificate.

Here is another letter which I picked up in sorting my letters to-day:—

'SIR,—Please give me information on the following points:—

'1. A course of mathematical study by which an accurate and comprehensive knowledge of the abstract principles of the science shall be gained, and at the same time such a course as will prove an efficient instrument in the study of physical science.

'2. The best works—Continental, classical, and English—on the several branches of mathematics, and where I can get complete lists of books.

'Yours . . . .'

I never heard of the man in my life. I made the following note on the letter, and did not send it:— 1856.

'Ēst mŏdŭs īn rēbūs, sūnt cērtī dēnĭquĕ fīnēs.
Hōw coŭld Ī ānswĕr thĭs lēttĕr ĭn lēss thăn fīve shēēts fŭll ŏf līnēs?'

Yours truly,
A. De Morgan.

To Sir John Herschel.

7 Camden Street, Sept. 10, 1856.

My dear Sir John,—This letter is not written to be framed and glazed, but because, *quoad* the form, I must have a line no longer than the sway of the wrist, because the arm is in a sling; because on Friday last I put out the shoulder, because I and my book-ladder slid down together, because the angle at which said ladder may be trusted on a beaten and tightly nailed carpet is very different from the same when the carpet is dusty and what seamen call loose in stays, because the coefficient of friction is vastly altered. However, I am thriving apace, and my wife, who would not believe my report of exceeding good health written with a pencil an hour after the replacement of the shoulder, but ran up to see how things were, went back to the children satisfied that I was a convalescent. . . .

Now mark, dislocations are among the minor evils which step in to shut the door in the face of greater ones. If the bone had been invincible, and all the wrench withstood by the muscles, I should have had a long fainting fit or fever, a sprain of six months, and it would have been a question whether I should have used the arm again; whereas, after a flash of fire and a bump, *cogito ergo sum* began to act, and I got up as much as ever alive to all things, and especially to the necessity of sending for a doctor. However, all this is merely as to the form of the writing.

Do you remember Sir H. Davy's habits? Was he in the habit of rubbing his hands together in any peculiar way and frequently? I want to know, because verification of a story too long to write now depends on it.

I shall hope to hear a good account of yourself. With kind regards to all around you, believe me,

Yours very truly,
A. De Morgan.

To Rev. Dr. Whewell.

Sept. 16, 1856.

1856. MY DEAR SIR,—I have been prevented from answering your note by an effect of gravity which brought me and my book-ladder to the ground together, and dislocated my right shoulder. However, it is getting right now, and the whole thing is not so bad as it is called. As to Galileo, I have inquired of Libri, who is up to that case above all men, and he says that though various statements have been made, he never could find the least ground for supposing that any Pope had done anything in the matter. And this was my impression also. I feel confident that all the rumour is a mere sham.

The following (not to be used) may confirm you. The narrator is Biot, to Libri, long ago.

A little before 1830 Biot was at Rome, conversing with the chief Inquisitor, who said, 'You men of science think that the Inquisition is opposed to scientific statements, which is quite untrue.' 'Then,' said Biot, 'I suppose the Professor at the Sapienza College may teach the motion of the earth?' The Inquisitor shook his head and said, 'No, that could not be allowed.' Depend on it, if anything had been done, it would have been widely promulgated.

Yours very truly, however illegibly,

A. DE MORGAN.

To Rev. Dr. Whewell.

7 Camden Street, Nov. 7, 1856.

MY DEAR SIR,—For your thin volume of additions just received many thanks. These little supplements are like giblet pie—a collection of all that is most racy: only, mind, I don't mean to say they are goose giblets; on the contrary, they shall be owl giblets if the owl is still to be the bird of Minerva.

Now for notes and remarks. Who can *unlatinise* in our day? Who answers for Nicholas *Cusa* or Adam Marshman?[1] He might be Adam Marsh, or Fen, for aught we know. Have not the French made *Viète* of M. de Viette, who would have been horrified at his prefix of gentility being abolished? We happen to have plenty of evidence to *De Viette.*

[1] Adam de Marisco de-latinised.

Smyth is *rather* slapdash sometimes. He killed Mezzofanti in the R.A.S. Annual Report long before that polyglot *bipes implumis* set off to leave his card on all the builders of Babel to see if he could manage to pick up a stray dialect or two more than he carried with him. I feel sure that if there had been any revocation of the decree—above all in 1818—Libri must have known it. But I will get something yet. The Catholics are evasive on the point. Smyth heard it, no doubt, so have I; and there is a disposition to have it believed among the R. C. 1856.

Page 51. In p. 19 of my notes on the Antegalileans I have given a better account of Digges, and especially of the edition of 1594, which you seem not to mention. I have it. It has an actual defence of Copernicanism (physical).

P. 33. I have never seen the perspective of Bacon separately, and so say nothing. But I am not without a silent suspicion that the work published separately is by John Peccam, afterwards Archbishop of Canterbury, a pupil probably of R. Bacon.

P. 47. The acceptance of the motion of light is not *pro re natâ*; the motion of light is first *proved* by Jupiter's satellites, which establish geometrically a motion of the *effect* called light; and then, with the velocity inferred of the *effect*, the aberration is explained in quantity and quality both. There is not even the assumption that light is material. Measured motion is geometry, not physics. . . .

Yours very truly,
A. De Morgan.

To Sir John Herschel.

7 Camden Street, Nov. 13, 1856.

My dear Sir John,—I am glad to hear you are getting on. I think you should not do too much in the way of being a free body in the morning. The evening phenomena are fatigue.

I doubt your prognostic about the coinage. I think you *may* be eatable by the time it comes. It is getting into country schools and colleges, and I think people are learning it. The House of Commons will probably be tried again next session.

As to how you might cook next century, you remind me of an experiment I have often thought of—mummy soup. The muscular fibre which remains must be partially soluble in water, I should think. I should like to catch some of the Fee-Jee Islanders—if that be the way to spell it—and feed them on the

1856. strongest mummy soup made in a Papin's digester. They are cannibals, and would enjoy the idea. Fancy the souls of the poor Egyptians, who preserved their bodies with such care for a future resuscitation, seeing their remains devoured by ferocious savages, and incorporated literally into the bodies of the same!

I have been looking at a 10-inch bar of aluminium which Graham has lent me. Queer stuff. Costs at *present* its bulk of silver, being $\frac{1}{4}$ of the weight. Smells a little, and rings like Scylla and Charybdis—I mean the Sirens. It makes a very pretty noise. They are making it at Paris in earnest. They put 40 lbs. of sodium into some preparation at one go, I am told, during the manufacture of a lot.

Our kind regards to all the ladies.

Yours very truly,
A. DE MORGAN.

To Sir John Herschel.

7 Camden Street, Nov. 8, 1856.

MY DEAR SIR JOHN,—Are you doing well and getting on? I am going on with my shoulder famously, chalking on the board;[1] but I can't do everything yet. I have learnt that whoso putteth out his shoulder, him shall his shoulder put out.

I think that if you were to give me an answer before Friday, I might talk of you at the Astronomical, where people ask after you. Next, how are all your party?—no small one.

I saw Warburton a week ago at his own house, working away at $\Delta^n 0^m$, in utter falsification of the maxim that out of nothing can come nothing.

I have nothing in the world to say about mathematics except, musing idly, I found that Euclid has *not* demonstrated the way to bisect angles.

I send you a lecture on decimal coinage. It is a slow-moving subject, but it must be carried sooner or later, in spite of ministers. They might have mentioned it at Paris, while they were talking about everything. I mean the Plenipotentiaries.

Yours very truly,
A. DE MORGAN.

[1] In his lecture-room.—ED.

To Sir John Herschel.

7 Camden Street, Oct. 9, 1857.

MY DEAR SIR JOHN,—. . . I am very glad to hear your son is safe. You say nothing about yourself, so I hope there is nothing to say. Subdivide yourself into any number of *egos* or *ichs*, but take care you don't get the double identity, which sometimes happens, in which a person is one person for a time and then another. Did you ever read a novel called *The Devil's Elixir*? If not, try for it at the circulating library. This *Elixir* has the effect that if two persons drink thereof their identities get mixed up in a very odd way. Each one becomes the other to a considerable extent. 1857.

By the way, you and I may be cousins all this time without knowing it. One of my mother's ancestors—her mother's mother, I think—was a Pitt, belonging to a family which considered itself an elder branch of the Chatham family; and I think I remember some expressions of hers being quoted which seemed to savour of thinking it very presumptuous in the younger line to come out as prime ministers, &c.

Multa renascentur quæ jam accidere. Among them is a fact which I discovered a few days ago—that I, A. De M., have sailed under our friend Beaufort's orders. He commanded the convoy in which my father and mother brought me home to England in 1806, he being in the Woolwich frigate. I was then four months old. So you see I was at the Cape long before you.

Yours very truly,

A. DE MORGAN.

To Sir John Herschel.

7 Camden Street, Feb. 7, 1858.

MY DEAR SIR JOHN,—*King Cole* (2nd edition) is nearly perfect. *Carmina cum fumo* is not idiomatic, and that is all.[1] 1858.

When did 1858 begin? on what meridian? An insidious question, demanded in the interests of equinoctial time. Why, 1858 had as many beginnings as there are meridians. When did Wednesday begin?

[1] Speaking of a Latin translation of *King Cole*, which Sir John had sent him.—ED.

1858. If a year begin with a day there must be as many beginnings of years as there are beginnings of days. Now I ask another question. From 0° to 360° of terrestrial longitude, how many meridians are there? The political world will be content with defining the year by the place. The astronomer, when he names time, always names a *place*. If you like to begin the year with the centre of the mean sun, in the mean equinox, it can be done, but the poor almanac makers must not be puzzled, and I protest against any more $185\frac{8}{9}$, or the like.

I returned to the Royal Society the other day a book which was given to them in 1728, and had probably wandered the world for more than a century. That book and others satisfy me that about the years 1734–40 the R. S. library was expurgated—purged of all anti-Newtonian and infinitesimal books. This is curious, but they were curious people in curious days.

There is a very marked absence of all materials for studying the Newton and Leibnitz controversy in the R. S. library.

Yours truly,
A. De Morgan.

To Sir John Herschel.

7 Camden Street, Feb. 11, 1858.

My dear Sir John,—I see your drift. But that *year* question set me off on equinoctial time. Why, the question is an ethnological, not an astronomical one. When the Portuguese and Spaniards met in the Philippines—*viâ* India the Portuguese, *viâ* S. America the Spaniards—they differed a day in their reckoning, kept their Sundays on different days, and cursed each other as only real Christians can curse. I never could learn how the Pope settled it.

Taking Christendom as a point of departure from whence all have gone whom we are concerned with, we shall find the Americans beginning their Sunday after us, and the Anglo-Indians before us; the New Zealanders after us, the Australians before us, owing to the way they go. But when New Zealand goes to Australia there is a change of day for them, and *vice versâ*.

Rule.—Do at Rome as they do at Rome. And what they do at Rome depends on the direction of travel by which they got to Rome. If there were a constant meeting at the meridians opposite Christendom by people of different modes of coming there,

there must be an arrangement made somehow. By a wise arrangement of things it will be long before the opposite part of the world is fighting the question. I don't see how there could be three days current at once, unless some chaps had gone twice round the world and never made a correction. 1858.

Did I ever explain to you how it is that the opposite hemisphere to the one which has London for pole is nearly all water? You might go ahead in science many a day before you would find the true reason. Yours very truly,

A. DE MORGAN.

To Sir John Herschel.

7 Camden Street, May 25, 1858.

MY DEAR SIR JOHN,—When your excessively rare animalculæ had been duly studied, it struck me that all creation is full of life; and though I have neither pond, tank, nor aquarium, yet I have access to divers atramentaria which in common life are called inkstands. Out of these I soon fished some specimens,[1] which I send you greatly magnified. I begin to have a suspicion that the style of writing depends somewhat upon the monsters which live in the ink, and that people would do well to examine the fluid before committing articles. Most of the specimens are difficult to make head or tail of—which is very frequently the character of other products of the inkstands. Care, however, must be taken how such things are published, for the world is very incredulous. And Fabricius, in his *Philosophical Entomology*, says, 'Damnanda vero memoria Johannes Hill et Ludovici Renard qui insecta ficta proposuere.'

As to the algebra, you are proving that you *won't* look at symbols. What!!! When nk is the number of vibrations in *one* second, and ma the time of each vibration, you pretend to tell me that you don't see—

$$ma \times nk = 1;$$

or do you dispute—

$$ma \times nk = mnka?$$

You will not easily make me believe that you were doing anything but laying a trap for me to make a pun that you

[1] Figures made by scribbling, and then folding the paper in half, by which both sides are made alike and resemble strange insects.

1858. might be down upon me if *I* missed it. But I see through you—you were pretending to labour under a *mank d'intelligence.*

As to the little dees—De Mogorgon—it is not the first time. My old friend Farish (the professor's son) could not call me anything else; it went against his conscience up to the day of his death. 'But why is the gentleman not called De Mogorgon?' I am constantly tempted to make a mistake in one Greek name, because in the second-hand book lists it always comes after mine. Look into any book list of a miscellaneous character, and you will see the succession following:—

De Moivre
De Morgan
De Mosthenes.

Yours very truly,
A. De Morgan.

To the Rev. Dr. Whewell.

Oct. 10, 1858.

My dear Sir,—Many thanks for the Bacon which you found in the Barrow. It all amounts to wondrous little, if, as you say, Bacon was known to the Cambridge men generally. How could Bacon be so little quoted? The conceits of which that age was fond were taken out of puerility by him, and made into wit and covered with taste. And yet they knew nothing of him to speak of. Newton's silence is emphatic. When I have time and opportunity I intend to work out the thesis, 'That Newton was more indebted to the Schoolmen than to Bacon, and probably better acquainted with them.'

The question whether I wrote the *two* articles in the *Athenæum* is entirely the question whether personal identity lasts through time.

Cowley I had forgotten. I have looked him up again, and see that he merits Harvey's satire. Gassendi I knew of. He is a Baconian *prononcé.* I dare say you have received Mansel's vol. of *Bampton Lectures.* I tell him by this post that it is the best argument I have seen against subscription at matriculation. Can you detect that the printer has punctuated Bampton's will into full Priestleian heterodoxy? . . .

Yours very truly,
A. De Morgan.

To Sir John Herschel.

Oct. 15, 1858.

MY DEAR SIR JOHN,—It's all very well for people to be clever, and go to the British Association, and talk philosophy and chemistry and—confound the hard words!—transcendity, which transcends all entity whatever, and is next of kin to the two German equations— 1858.

Everything = God.
God = 0.

This, I say, is all well as far as it goes. But can your philosophy answer me this?—Suppose the Northern Hemisphere to be all land, the Southern Hemisphere all water: is the Northern Hemisphere an island, or the Southern Hemisphere a lake? Crack that.[1]

Yours very truly,
A. DE MORGAN.

To Sir John Herschel.

Camden Street, Nov. 15, 1858.

MY DEAR SIR JOHN,—I have as much chance of meaning to try for the Lowndean as I should have of getting it if I tried (=0). Knew you not that I am a heretic who is B.A. of thirty-one years' standing by reason of subscriptions being unsubscribable? Moreover, I have other fish to fry. Cayley is a capital man for it.

I hope you have been asked to do a memoir of Peacock for the R.S. anniversary. He is lost at the time when he is most wanted.

I heard of your frisking about the country like a young

[1] The geographical question was answered with another by Sir John Herschel.

'Suppose all was water except a patch of land of an insular form round the North Pole, N° in radius (N = 3), would that be an island? I should say yes, because it *is land*. Next, let N = 4, same question. Next, N = 5, N = 6, 20, 30, 90°.

'At what value of N does it cease to be an island?

'Then N = 95, 100 . . . 179° 59′ 59″. At what value does the sea cease to be an ocean, a lake, a pool, a pond, or a puddle? I pause for a reply.'—*Collingwood*, Oct. 18.

1858. gentleman, and was very glad to hear of it. You are now 'discharged cured,' and are, I hope, meditating some proof of violent health. A new edition of the *Differences* would be a very pretty step in the proof.

Did I send you this riddle ?—the answer is in itself a riddle. If a comet were to take a much more elongated orbit, and the King of Naples were to prohibit the importation of malt liquor into his capital, in what particulars would two empty heads differ ? Answer overleaf.

Kind regards to all your circle.

Yours very truly,
A. DE MORGAN.

Answer : The comet *would* have double eccentricity, and the King of Naples would *not.*

What can it mean ?

To Sir John Herschel.

7 Camden Street, N.W., Jan. 1, 1859.

1859. MY DEAR SIR JOHN,—Many thanks for your dates. I want one more thing—Who was Peacock's father ? I have before me a book by a clergyman of the name, whom I suspect to be the one—Rev. *Thos.* Peacock, author of *The Practical Measurer* and of *Walkingame's Arithmetic Modernised.*

I am very completely set up by your dates. The Prolocutor of Convocation is, in fact, the speaker of the *Lower* House ; for the clergy have their higher House, made of bishops, and their lower House, made of dignitaries and *proctors*, so called, elected by the lower clergy ; and they all have a hankering to be what they once were, when they persecuted books as heretical, and excommunicated the writers, and kept the pot boiling to the wonderment and amusement of men and angels. And this was called *synodical action,* but at last the State voted it *tom-nodical,* and put it down. In our own day it has been revived to the extent of allowing a day or two of talk, and appointment of committees to organise talk for next time ; but no measures have been allowed to pass. And Peacock, as prolocutor, was, I understand, very useful as a king of order and a stifler of pranks, he being himself favourable to the revival of synodical action on the principle of all the clergy being as discreet as himself, a theory which beats out of the field Homœopathy, Mesmerism, Table-turning, Parliamentary Reform, and Perpetual Motion.

What they want is that Parliament shall not legislate for the Church without consent of Convocation, and this Parliament will never agree to. In the meanwhile they are allowed to use *logarithms*, not Napier's, but of the sort which Sophocles mentions, λόγων ἀριθμος, which is translated a *set of words*. 1859.

Now you may guess what the Prolocutor of Convocation is. Until very lately, and from George I. or thereabouts, they did nothing but walk in procession, at the meeting in Parliament, to St. Paul's or the Abbey, where they heard a sermon, and were prorogued. A happy New Year to all.

Yours very truly,
A. De Morgan.

To Sir John Herschel.

Camden Street, May 2, 1859.

My dear Sir John,—Did your Southern dealings ever bring you in contact with the history of the formation of the south polar constellations? Do you know any work which treats that subject especially? Did you ever come in contact with the doings of Frederic Houtmann, the real framer of these constellations, as far as giving their star materials is concerned? Next, answer the question *Come sta*, curiously Englished by *How goes it*, with reference to yourself and selves. I suppose the Italian phrase to be Ptolemaic, and the English to be Copernican.

What have I to do with Houtmann? He adjoins himself *more slantendiculari* to the question of a manuscript sold in Libri's sale, as written by Galileo on the doctrine of the sphere. Some question has arisen about the evidence, external and internal, and I have been looking up the points. The internal evidence is to me very satisfactory; and as to the handwriting, there is a hitch about the letter *r*, which all the Galileos we have to compare with make *r*, and the MS. makes *i*. In spite of this I cannot help believing that the MS. is Galileo's. It would be much better worth its money if it were *not*.

As to the state of things in general I have nothing to say. I don't know whether the Austrians have forced a *tête du pont* at Buffaloroary, or some such place, or not. I hope some bright nebula or other is of a white heat, and is set apart for all who make or instigate wars of ambition. I wish Lord Rosse could find it out, and could show III. that I.[1] has his spirit herme-

[1] Meaning the first and third Napoleon.—Ed.

1859. tically sealed in a bottle, and heated up to a pressure of 100,000 atmospheres. It might make him behave himself, but perhaps not. With kind regards to all,

Yours very truly,
A. DE MORGAN.

To Sir John Herschel.

Camden Street, May 18, 1859.

MY DEAR SIR JOHN,—Thanks for your pamphlet. I have not had time to do more than glance at it, but will say what I think when I have got through a heavy job of calculation—a job of life and death, as one may say, for it is all about premiums and claims and assurances, &c., &c.

Maurice de Biran, who died in 1824, aged about sixty, was a *philosophe*, who speculated and died, even as a silkworm spins and dies. He will be a gaudy moth, I dare say. His cocoon was published by Victor Cousin in 1841 in four volumes. He was very much against Napoleon in 1814, which means, I suppose, that he had been his parasite theretofore. He was a public man of some kind. Probably his will was an impulse to better his condition, or butter his condition. He passes for an acute thinker in France, but I have never seen a line of his writing.

I believe that so much of cause as is not mere notion of precedent and consequent is derived from our own consciousness of power exercised at will. If we had been rational posts, incapable of motion, chewing the cud on what passed before our eyes, and if with a will incapable of action, I do not see how we should have had any real notion of cause. What the will is I have not the least idea, or whether it ought to be called the *shall* or not. Query, if it be really correct to call it the *will*, how is a person whose will is undecided said to be *shilly-shally*? Ought it not to be *willy-wally*? Kind regards to the circle.

Yours very truly,
A. DE MORGAN.

To Miss Sheepshanks.

41 Chalcot Villas, Adelaide Road, N.W.,
August 24, 1859.

MY DEAR MISS SHEEPSHANKS,—I do not know what you have done with my dear friend's books. There are one or two which I should recommend to be given to Trinity College library, if they can be detected. They contain handwriting of Horrocks, the famous predecessor of Newton in the Lunar Theory. They will no doubt have the book-plate of either Christopher or Richard Towneley with these arms. . . .[1] 1859.

Such a book-plate in a book with handwriting in it is very likely to be Horrocks's, if astronomical. I found one of these books at a sale, and gave it to Trinity College, and I remember your brother having two at least. But the thing does not press. A life of Horrocks just published reminds me of it.

August 27.—We are all pretty well, and I have got my books into something which is *not dis*order. But two negatives do *not* make one affirmative. . . .

Yours very truly,
A. DE MORGAN.

To Sir John Herschel.

41 Chalcot Villas, Sept. 9, 1859.

MY DEAR SIR JOHN,—You puzzle me. I always took it that you and our friend Francis Baily foregathered at the Astronomical Society, and then and there concocted a private friendship, as many good men have done, and more by that same token will do. But if Miss Baily was a friend of early boyhood, you can hardly miss to have the same to say of Francis. She must have been, I should say, ten years older than you: if *Men of the Time* be correct in time, eight years. Explain this point, I pray you. Where did you first make Francis B.'s acquaintance? As you have let out the name of the person who sold you cakes, there is nothing that you can have any excuse for being secret upon.

I am not clear in my memory about the names of any of *my* purveyors till about fourteen years of age. I think I took their names to be immaterial, and their sweetmeats the real thing. But at the age I named I was introduced to Mother Fudge—her *real* name—who could carry in her head the debts of any number of boys, and no mistake. How she managed to remember the several little accounts from $1\frac{1}{2}d.$ up to half-a-crown I never knew, nor she either; but those who really wanted to *do* her, and those who pretended to want it, found her utterly uncheatable. She would run over their tradings for a week past with a confidence

[1] The arms drawn appear to be—on a field argent, a bar sable surmounted by three stars sable.—ED.

1859. which was sure to get a verdict from the little mob round her. She could neither read nor write.

'Mother Fudge,' said a boy to her, 'can you make poetry?' 'I, sir?' said she; 'if I could make English-Poeters, do you think I'd be here selling apples and pears?' 'Now, Mrs. Fudge,' said I, who had been reading Johnson's Lives, 'I could name you three poets who did not get as much among them as you get by this one school.' 'Then I wouldn't give nothing apiece for them,' said she; 'they must have been regular bad ones.' I think the copyright of 'Paradise Lost' sold for about a year of her profits from that school.

I have had a slight touch of gout—nothing to hinder my walking, with a little pain, but just a straw to show which way the wind is blowing. If you could have known of your own consciousness how regularly the homœopathic medicines alleviated it when I stuck to them, and how it got back again when I forgot them, as I often did in the moving, &c., you would be satisfied that the *post hoc* of little globules was *propter hoc.*

I have now got rid of my books; that is, have shelved them. And behold! for the first time these two months I have walked my fill without feeling lame.

Yours very truly,
A. De Morgan.

To Rev. Dr. Whewell.

41 Chalcot Villas, Adelaide Road, March 3, 1860.

1860. My dear Sir,—I have received the copy of the *History of Discovery*, for which many thanks; also one from the *Athenæum* for review, which will go back uncut. I fully expect some one will some day prove by instances that the *Athenæum* reviews books without reading them, and the instances will be copies which reviewers have not cut, preferring to take that trouble upon their own copies.

I see you have at last admitted that Induction=Induction, means Induction< >Induction. Whether you have gone as far with *logic* I have not yet found out.

And I see, with great satisfaction, that the name of *Friar Bacon* begins to appear. Brother Roger has always been a great favourite of mine since as long ago as B in the *Penny Cyclopædia*, and he ought to be allowed his share of the name of Bacon.

Bacon is a queer name; but there seems to be a providence which watches over names, and makes the great men have bearable sounds. None of the hypotheses that, be the names what they may, the exploits will sanctify them in time! If Bacon and Newton had been Wiggins and Figgins, would any time have taken off the ridicule of the rhyme? Could anybody with a grave face have argued the question whether the immortal Figgins was or was not indebted to the great Wiggins? 1860.

. . . .

I see a little theological philosophy. A rationalist Nonconformist a few weeks ago opened one of his paragraphs thus:—'Now, my brethren, let us proceed to make a logical incision into the psychology of God.' This was making a subject of the Deity. *Subject* is the sense in which the word was used in reply to an argument of mine. I maintained that a good teacher must have his heart in his subject, and his subject in his head. Not if he teach anatomy, says Dr. Sharpey, for he could neither have his heart in his subject, nor his subject in his head.

To Professor Kelland.[1]

41 Chalcot Villas, Adelaide Road, April 2, 1860.

MY DEAR SIR,—I have received your note announcing that the Senate of the University of Edinburgh intend me the honour of a degree of Doctor of Laws on the occasion of Lord Brougham's installation as Chancellor.

I hope I shall give no offence by very respectfully declining the honour. I mean the diploma. The honour lies in the good opinion of the Senate, and that your communication gives me a right to say I have already earned.

My reason for declining the degree is my own peculiar dislike of conventional titles, which are not what they seem to be. If I had studied civil law I should be very glad to be styled fit to teach it by any competent body; but as I never have studied, I object to call myself a teacher, and should object to others calling me so, and I would not consent to accept a degree in law from any University in the world. This is for myself, without impeachment of the conduct of others for adopting any conven-

[1] In reply to an offer to confer on Mr. De Morgan the honorary degree of LL.D. of the University of Edinburgh.

1860. tional misnomer in which they may see good, or for which they may find reason.

I am, dear sir,
Yours very truly,
A. DE MORGAN.

To a Friend.

41 Chalcot Villas, Jan. 19, 1861.

1861. MY DEAR ——,—Two days ago I heard from Miss Sheepshanks of the heavy blow which has fallen upon you and ——, a dreadful shock, and, I must suppose, wholly unlooked for. With my strong and increasing disposition to congratulate those who leave this world, I feel an increasing amount of sympathy for those who are left behind; and you and your wife's share of this mournful event will command the sympathies of thousands you know nothing of in addition to those of your friends.

I trust you both bear up, and try to balance what is left against what is gone. That this is not easy I know. A few days more than seven years have elapsed since it was my turn, and I could not then *feel* that six left made any set-off against one gone. I could only *understand* it. But time will do for you what it has done for me.

My wife unites her sympathy and kind regards with mine.

Yours sincerely,
A. DE MORGAN.

To Rev. Dr. Whewell.[1]

41 Chalcot Villas, Jan. 20, 1861.

MY DEAR SIR,—This very (Sun)day—how I do not know, but I suppose the Parcels Company holds it a work of necessity and mercy to distribute the parcels which have accumulated during the frost; most likely they have been obliged to suspend some work by the state of the streets—I have received the volume of Barrow, for which I have to return thanks either to you or the College; *c'est égal, l'état c'est moi!* I say nothing as yet, except that it is exceedingly handy and time-saving to have these books

[1] Wafered into a volume of tracts left by Dr. Whewell to Trin. Coll. The succeeding letter refers to the memoir entitled 'On the Syllogism, No. IV., and on the Logic of Relations,' in the *Transactions of the Cambridge Philosophical Society*, vol. x., part ii., 1860.

in one volume. For B. is the man of all others, according to my experience, who is referred to by citation of one work for what is in another. How could aught else happen to a cove who called one of his writings *Lectiones Mathematicæ*, and another *Lectiones Geometricæ*, and then treated what is considered as exclusively *Geometrical* (as Euclid V. wrongly is) in his *Mathematical* lectures, pp. 8, 9 (*i.e.* one word shared between them)? For *metaphysical* read psychological. I don't object to the word a few lines higher up. 1861.

Do you know the use of the word *metaphysical*, which is growing up among the writers of the examination books which have taken the place of all others?—I mean at Cambridge. It means *requiring thought*, and proceeding without symbolic calculation. When a proof of two pages of symbol drumming is avoided by an act of reasoning, it is said to be 'too metaphysical.' This is one of the consequences of the death and burial of psychological thought in Cambridge. There seems to be a complete acquiescence in the maxim that Oxford shall settle what the world shall think, and Cambridge shall settle who is to be Senior Wrangler. It is getting worse and worse from day to day. Are any of the younger men alive to the facts? With best remembrances to Lady Affleck,

I am yours very truly,

A. De Morgan.

To Rev. Dr. Whewell.

41 Chalcot Villas, Adelaide Road, Jan. 20, 1861.

My dear Sir,—There are some mistakes which are too small to be of any consequence, and some which are too large. Extremes meet; $-\infty$ is curiously a comrade of $+\infty$. . . .[1]

The reason I call $x^3-2x-5=0$ a celebrated equation is because it was the one on which Wallis chanced to exhibit Newton's method when he first published it; in consequence of which *every* numerical solver has felt bound in duty to make it one of his examples. Invent a numerical method, neglect to show how it works on this equation, and you are a pilgrim who does not come in at the little wicket (*vide* J. Bunyan).

Newton was anything but illiterate. He *knew* Bacon. His silence is most marked. How could he avoid every possible

[1] Referring, I think, to some error in a figure in a former letter.—S. E. De M.

1861. amount of mention of Bacon on every possible subject? I never said he did not *know* Bacon; I only said he could not be *proved* to have known of his existence. Nor can he. I think he has taken such pains not to be known to know him as cannot be attributed to accident.

I am glad to hear there are logicians at St. John's. It is a college at which more pains are taken to make the men write ⊙ for 'circle' in their *writing out* than to prevent their reasoning in a circle. There is no attention given to *writing in*. Nevertheless, St. John's has preserved the shadow of a teacher of logic. When I published my syllabus last year, I sent a copy to every college in Cambridge, directed 'to the Tutor in Logic,' just to make them stare. I got an answer from St. John's from Mr. Mayor, who acknowledged the title.

It is not examination that is wanted, but good teaching and example. A paper of logic conundrums would be just as useful as one of those fearful mathematical papers, to prepare for which private tutors drill men in passing examinations. Thank Heaven that I was at Cambridge in the interval between two systems, when thought about both was the order of the day even among undergraduates. There are pairs of men alive who did each other more good by discussing x versus dx, and Newton versus Laplace, than all the private tutors ever do. With kind remembrances to Lady Affleck,

I am yours very truly,

A. De Morgan.

From Professor Alexander Bain.[1]

University, Aberdeen, Feb. 7, 1861.

Dear Mr. De Morgan,—As two copies of your paper on the 'Logic of Relations' have reached me, I beg to return you one of them, and to thank you for the other. I am very much interested with this new subject which you have entered upon, being convinced that the greatest omission both in logic and psychology is the not seeing how far the principle of relativity goes. So far

[1] This letter was given to me by Professor Alexander Bain. I have departed from the general rule of not giving letters *to* my husband in the correspondence, because in this instance the value of his own to general readers is greatly enhanced by being accompanied by that to which it is a reply. I wish it had always been possible to give both sides of the correspondence, but this would have rendered it too voluminous.—S. E. De M.

as I am able to judge, relation goes into everything; no quality 186
existing except as related to some other, which we sometimes call its negative, at other times its contrast, and again its correlative. The *straight* line has no meaning without its contrast, the *bent* line; the occurrence of the two kinds is necessary to our recognising either property. Every quality, every cognition of the mind, implies an antithesis or couple. Hot—cold, up—down, &c. If I say *red*, I mean to exclude all other members of my 'universe' (to use your own well-chosen designation); and if that be 'colour,' I exclude all other colours. The important inferences deducible from the essential doubleness of all cognition are, I am sure, very numerous, and I have no doubt you will convince us of this if you continue the subject.

Yours faithfully,

ALEX. BAIN.

To Professor Alexander Bain.

41 Chalcot Villas, Adelaide Road, N.W.,
Feb. 9, 1861.

MY DEAR SIR,—I am sorry to have given you the trouble of returning my second copy. I see I must have entered you as of two Universities.

I quite coincide in your view as to a quality being unthinkable except in company with its *non*. I forget where I said, long ago, every name designates every object of thought as either *in* the class or *out*; but I did say it, and the equipollence of X and non-X is the foundation of completeness even in common syllogism. I hardly like to claim the word *universe* as mine, though I have brought it down from its modern sense (the το παν) to the old sense. Those who have derived the word from a mixture of *unum* and *diversum* (strange etymologists!) certainly very much favour my plan of making it any aggregate of X and non-X which is in hand. But the old *universal* was any name which had plurality of things signified: of *two* only, the name turned the two into *one*, *in unum versa*. I have made some people stare by telling them that universality begins at *two*.

The combinations of relation are the ambiguities of language. Looking on a little into compound relation, I come to such a sentence as the following:—

'He is the father of a friend of every one of my children.'

Do I mean that one of his children is the friend of every one

1861. of mine, or that every child of mine has a friend among his children?

Here is L(MN)′ as distinguished from (LM)N′. This door is a very little way open as yet.

Yours very truly,
A. DE MORGAN.

To Admiral Smyth.

41 Chalcot Villas, Adelaide Road, Feb. 19, 1862.

1862. MY DEAR SMYTH,—I am not very especially busy just now.

. . . .

The obscure men are, as we know, precisely the men that future necrologists *will* look out for. I find that the *Biographiæ Obscurorum Virorum* are very useful. I have an old Italian *Glorie degli Incogniti*, which I find very useful for information about men who are merely there to be shown up for non-notoriety.

. . . .

Yours sincerely,
A. DE MORGAN.

To Sir J. Herschel.

April 19, 1862.

MY DEAR SIR JOHN,

. . . .

No news here; not even a riddle. Here is a poem which was given me:—

Ἄειδε εἰδύλλιον δέα,
Felis adest cum cithara,
Vacca lunam transivit,
Hoc jocoso motus visu
Rumpitur canellus risu
Cum cochleari lanx abit.

Which do you believe in, metal plates or guns? I have just received a pamphlet on the subject from *Michael Scott.* If this be the great wizard, then we know how it was that—

When in Salamanca's cave
Him listed his magic wand to wave,
The bells would ring in Notre Dame.

The wand was a long match, and his range was so good

that he would hit the belfry at Notre Dame from as far off as Salamanca. The end of it I have long foretold to be that the different capitals of Europe will shell one another without the trouble of sending out soldiers. 1862.

With kind regards to Lady Herschel and the juniors all,

I am, yours very truly,
A. De Morgan.

To Sir J. Herschel.

91[1] Adelaide Road, April 29, 1862.

My dear Sir John,—Many thanks for the hexameters. They are as good as they can be, but all the logic in the world does not make me feel them to be English metre, and they give satisfaction only by reminding me of the Greek. Just as, mark you, a flute-player—which I have been these forty-five years—only plays Haydn and Mozart because he has the association of the orchestral accompaniment, which arises in his head with the melody.

The idea of the Scott ballad metre is not recent. When I was at school, forty-two years ago, our *ludi-magister* read out about 100 lines of Homer, which he said were versified by Scott himself as a specimen. They were decidedly Scott, and I thought not a little Homer.

The hexameter, it is clear, does not fix itself in the popular mind. If it has done so proofs can be given, but I have not met with them; the popular mind knows neither quantity nor accent, but that which is to last bites its own way in, without any effort. Is the hexameter making any way? Do people quote any hexameters?

It seems to me that the problem of a metre for translation of the hexameter is not yet solved. The English hexameter is not a better reminder of Homer than the usual metres of our language.

I have discharged my conscience. *Richard's visual organ.*[2]

Yours very truly,
A. De Morgan.

[1] The number of the house had been changed.
[2] Dixi!—S. E. De M.

To Sir J. Herschel.

91 Adelaide Road, May 3, 1862.

1862. MY DEAR SIR JOHN,—A great many years ago you stood up after dinner at our club, and gave strong hints that your time was nearly up. But you brisked up, took the Mint, overworked yourself, got an illness worth prophesying about, got over it, and committed mathematical papers. Now here you are again, talking about softening of the brain, and a knacker's yard, and all kinds of incommensurables. If this mean that you are going to be Chancellor of the Exchequer, why, take off the income tax. If it be really melancholy foreboding, take on a little quinine or brandy and water, and give up the hexameter for six months. It is a mournful metre.

As to your catalogue of Greek ships and of *nebulæ*, take care you do not mix them accidentally, 'A catalogue of ships which sailed against Troy, reduced to the year 1862,' by Sir J. H., &c. People will stare to see how 2,500 years of precession turn a trireme into a steamship. All our progress may be only precession of the equinoxes, motion backwards of the zero of reckoning.

As to the hexameters, it is only now and then remembered that verses among Greeks and Romans were not for *recitation*, but *recitative*. An hexameter is a natural measure for a chant. I dare say the rhapsodist in the streets of Athens gave it out something like as a Puseyite parson gives out the Litany, only with more taste. A famous hexameter might be made out of the opening line of the hymn to the Virgin in 'Masaniello,' but our most natural measure is a foot too long, and the last spondee is doubled.

Yours very truly,
A. DE MORGAN.

To Sir J. Herschel.

91 Adelaide Road, N.W., May 30, 1862.

MY DEAR SIR JOHN,—I should not wonder if Sylvester and you were at one without any intercommunication of your particles. *I* have had the same idea a long time. I have even hinted at it through a glass darkly. In my third logic paper there is the following passage: 'It is easy to frame hypotheses which no one can of knowledge deny, under which attributes in the brain should be as real as the sun in the heavens, or the rocks on the earth, and this without a denying either the existence of matter or the separate existence of mind.' If the things of the universe be affections of the immovable primary particles of space, the impresses on the brain may be veritable copies, as real as the things themselves. A very pretty system of pre-established harmony might be established. If all the matter-universe be in motion of translation through the space-universe or in *transference*, and if an individual in a certain part of a certain nebula be to have a headache at a certain date, he may at that date find the space particles, which are to keep up his head, ready supplied with the adjunct affections—confound them, whatever they are!—which are essential to an ache of predestined intensity. 'How charming is divine philosophy!' 1862.

Of course all this means that I have received your letter and book. I will look at the latter, and let you have it back soon. I never heard of the dialogue between Hermogenes and Hermione. The puzzle about ∞ arises much, I think, from a want of distinction between the subjective and objective infinity. But before I fairly tackle the subject I have to superintend and, *en bloc*, to calculate a valuation of about 30,000 life policies; but not 30,000 calculations—Heaven in its mercy forbid! But I must leave off. With kind regards,

Yours very truly,
A. DE MORGAN.

To Sir J. Herschel.

91 Adelaide Road, Aug. 9, 1862.

MY DEAR SIR JOHN,—I return you with thanks your MSS. on algebra. There are little bits here and there that I wish had been published. Did it chance to you that the first thing you

1862. wrote never was published? It did so to me. The first thing ever proposed to me was a treatise on mechanics for the U. K. S. I wrote a few chapters, and, chancing to become a candidate for what I now hold, I sent my MSS. in as a testimonial, and I believe it greatly helped me. At any rate, I was picked out of fifty candidates, being known to be only twenty-one last birthday. I think Brougham and Warburton were the people who dared a thing so bold, considering the danger of making any ventures in an institution beginning under so many evil eyes as the University of London. Olinthus Gregory was against it; S——, who always had a wonderful faculty of getting something against somebody, though he did not know me, and had never seen me, either concocted or retailed to Stratford a story which I never heard from elsewhere, namely, that my appointment was the doing of Mr. Frend, then an acquaintance of mine of a few months' standing, not on the Council, and at Cheltenham for health all the time, and who learnt my candidateship and appointment from the newspapers at one and the same moment.

Who shall escape? Mark the following. In some journal in 1851, M. Bertrand, in a paper on the convergence of series, is charged with *suppressing* what I had done on the same subject. It is hinted that he had used what I had done. The facts are—

1. M. Bertrand invented a set of rules before he had seen mine, so he says, and I believe; his method has all the marks of independent thought. After he had observed the identity of his rules and mine, in effect and each to each, it struck him to try a hint of a M. Raube, and he thereupon constructed a third system. 2. He announced my rules in half a quarter-page of translation from me, with inverted commas to every line, and mentioned my name *eleven* times in his descriptions and comparisons. 3. He gave my book the date 1839 instead of 1842, 1839 being the date of the *number* in which the rules of convergence appeared. 4. He sent me a copy of his paper as soon as it appeared.

What could he have done more? Nevertheless, he is unblushingly charged with *unfair suppression* by a man who knew nothing of my book but what he himself had told him, for he (the critic) gives the wrong date of 1839.

As to infinity, I hold $\frac{1}{0}$ to be the infinite of infinites.

For 0 marks the change from $+$ to $-$, which ∞ does not.

As we generally use ∞, we admit ∞^1, which is not negative,

and ∞^3, &c. But quantity which changes sign through infinity passes though $\frac{1}{0}$. This will become a very important distinction. 1862.
The $\frac{1}{0}$ of common algebra is high up above the $\frac{1}{dx}$ of the differential calculus.

I am rid of all fear about ∞^2. I believe in ∞, ∞^2, ∞^3, &c., &c.; and I intend to write a paper against the skim-milky, fast-and-loosish mealy-mouthedness of the English mathematical world on this point. My assertion is that the infinitely great and small have *subjective reality*. They have objective impossibility if you please; or not, just as you please.

I have first to remove an ambiguity, which has played a large part in causing confusion. To *imagine* is originally to *form an image* in the mind. But it has been transformed into a synonym of to *conceive*, to form a *concept*. The distance from here to the sun is a *concept*. I have no image of it. But of six feet I have both image and concept when I shut my eyes. Now many persons, when they cannot *image*, speak as if they could not *conceive*, and use the ambiguous word *imagine*. We cannot, they say, *imagine* infinite space. I grant they can't *image* it, but I am sure by their modes of denial that they have a *conception* of it. Locke and others affirm that we arrive at the notion of infinity by finding out that when, say, we add number to number, we *find* the succession incapable of termination, and so fashion interminability in our minds. I say the process is precisely the reverse. If it were not for our *conception* of infinity we should not know the interminability.

Who ever tried up to 10,000,000,000,000,000? It is certainly not experience. If any one were to affirm that 10^{16} is only a symbol, and that any one who should try would find himself brought up by the nature of things, Locke has no answer, unless, as would probably be the case, he should ask permission to bring on the conception of infinity.

I therefore affirm the concept *infinite* as a subjective reality of my consciousness of space and time, as real as my consciousness of either, because inseparable from my consciousness of either. When, therefore, I think of a finite space—say a cubic foot—if I compare it with the totality of space, I say *infinitely small*; if I compare the *totum* with *it* I say *infinitely great*.

Now comes a postulate on which there may be a fight. Let A and B be two magnitudes, any whatsoever, and C a third, also any whatsoever. Let these magnitudes be concepts, imagin-

1862. able or unimaginable. I take a right to affirm the conception of D such that

$$A : B :: C : D$$

The notion of ratio is a fundamental thing, not dependent on, though only definitely expressible by number. A person who cannot count, and who does not even know that language can turn *multitude* into *number*, has the idea of ratio, relative magnitude. He sees, feels, and knows that if A be the house, B is too small for the chimney, and C too large.[1] I claim the existence of D, so that A : B as C : D is a concept.

This being premised, then I have the infinitely small part of any magnitude, and of that again, &c. For instance, if $dx : x ::$ a pint pot : all space, dx is an infinitely small part of x.

All this I mean to develop and fight for. So with kind regards to Lady Herschel and the next generation,

Yours sincerely,

A. DE MORGAN.

To Sir J. Herschel.

91 Adelaide Road, Aug. 15, 1862.

MY DEAR SIR JOHN,—Many thanks for the dialogue.[2] From the parties to the dialogue I deduce the equation—

$$\frac{\text{mogenes}+\text{mione}}{2}=\text{schel}.$$

What a quantity of arguable propositions! I cannot see how you deduce your account of Descartes. As to at'ems—I spell the name thus, considering it as a challenge to attack them, and make your boast of it—I suspect that if you look back into the world a thousand years hence you will find the remote posterity, as we call it, fiddling away at the creatures, and knowing about what we do.

I have often thought of the minimum of extension endowed with attraction, &c., and adjusting his accounts with $\{(10)^{1,000,000,000}\}\ (10)^{1,000,000,000}$ brethren instantaneously. It is a wonderfully fine hypothesis for expressing what we see and

[1] A rough sketch of a house, with a too small chimney on one side and a too large one on the other.

[2] *Dialogue on Atoms, by Hermogenes and Hermione.* Private circulation.

know; but for an actual objective truth, oh my! And we call them *blind* atoms! Why, the fellows see faster and farther than we do, by the above to 1, at least. 1862.

If a malevolent being could create one single atom more than is in the plan, he would of course bring the whole thing to smash at last. Query, in what time?

I hope we shall know more about it next world. We can't know much less than we do now, that's one comfort.

Yours very truly,
A. DE MORGAN.

To the Rev. Dr. Whewell.

91 Adelaide Road, April 1, 1863.

MY DEAR SIR,—I am not going to take the privilege of the day, but it reminds me, though it is not the occasion of my writing, that this very day ten years I made a sort of speculation which I thought many would attribute to the influence of the day. While I was chuckling at the idea of having quite succeeded in a new metaphysical insanity, and before the pen was out of my hand, there actually came in from a bookseller Heywood's *Analysis of Kant*, 1844, and there I found the very same notion. It occurred in a description of the 'paralogisms of reason,' as they occurred in the *first* edition. You can tell me whether there is any allusion to the subject in the later editions, and this is my question. 1863.

I was considering a syllogism in which a term is a *class* of which the individuals are the subject at different *moments* of its existence. For instance,—

No black ball is ever a billiard ball; this ball has always been black; this ball has never been a billiard ball.

The individuals of the class are the *balls* which we call *one ball at different times.* Thereupon it struck me to think, how is it that we call this ball *the same ball all the time?* Whereas, if we had a number of fac-simile balls in different places, we should not say it is the same ball *all the space.* I suppose we borrow a notion from our personal identity, in which we feel sameness. Consequently, if our presence had *multipresence*, if the *ego* knew himself for himself in all the different parts of a space without being able to say, I am one person here and another there—any more than he can say, I am one person *now*

1863. and was another *then*—he would be very much inclined to doubt the difference of balls, when the only difference is that of place.

This supposes a faculty altogether beyond our comprehension—if, indeed, anything be within our comprehension on the question of what is what—which ties spaces together, just as memory ties times together.

All this I found in Kant (Heywood, *ut supra*, p. 109). He uses it to prove, as he thinks, that *ego* 'I think' might be identical, though the thinking subject is variable. He will not admit the space-string to constitute the *nos* of different places one *ego*. I cannot find any of this in Heywood's or Tissot's translations, and I think it possible that he may have learnt better in the interval of the editions. But you may be able to refer me to some notice of it.

With this metaphysical reduction of omnipresence to depend upon an incomprehensible something which has at least an analogue in our own consciousness, I have looked for ten years at various ontological writings about 'the unconditioned,' and various religious works about 'the Almighty,' and I think I see a very great tendency to confuse omnipresent personality with infinite extent. At least there is a want of power to put the distinction into language.

Are you aware of any Roman Catholic speculation on the subject? They must give multipresence to the saints whom they invoke, and by whom they expect to be heard. And I should suppose that some of their writers have touched on this gift.

I am now writing on the subject of *Infinity*, trying to burn the candle at both ends. I have found out for some years that I am a full believer in the infinitely great and small, both. I mean in the *subjective reality* of both notions.

I cleared off much obscurity by a distinction which I find very faintly shadowed by the psychologists—that of a concept which has *image*, and a concept which has none. I can *image* a horse: I can't image the right to a horse, but I can *conceive* it. I cannot *image* infinity, but I can conceive it—that is, I recognise a notion *with predicates*. So that when a metaphysical writer says, as some have said, that we cannot conceive space to be finite, and are equally unable to conceive it as infinite, I say they ought to have said that we cannot conceive space to be finite, nor image it as infinite. But neither can we *image* a million of cubic miles, though we can conceive it, as proved by our knowing truth and falsehood about it. 1863.

I am, yours very truly,
A. DE MORGAN.

To Rev. Dr. Whewell.

April 3, 1863.

MY DEAR SIR,—Did *I* provoke you to an ontological discussion? Did I chalk my hat and say, 'Now I'd like to see the man who says that this is not silver lace'? That's what I call provoking a discussion. I asked you, who are Kantescient, whether you knew of a certain speculation in the later editions of Kant; and you say No. I am pretty sure you would have remembered it at once if it had been there.

I am quite sure we shall never solve the problem which my analogy went to suggest. But for all that, if we only envisage a quality acting through space as memory acts through time, we put multipresence upon a definite basis of unintelligibility—there, I have managed to spell the word, and that is something gained.

I value the analogies of space and time—the two indismissible extensions; and I have before now made much profit of the very remark you quote.

For aught I know, a body may act *where* it is not; it may leave consequences behind it. An annihilated star, which is seen by light emitted during its existence, may be said, for aught we can tell, to *act where* it is not, in as true a sense as matter, in attracting distant matter, can be said to act where it is not.

But *presence* is a very ill-used notion. If a particle really do attract all others, it is present throughout the universe. It is present in *one* quality—in others, for aught we know. The presence of matter is the presence of all its qualities—the only things we know. Now who is to say that the spheres of the qualities have the same diameters or even the same centres?

Grant one centre to qualities of a particle, and there may be millions of centres, all effective in spheres of different radii. The sphere of attraction may be the biggest, or it may not.

Mansel, I detect partly by private, partly by public evidence, is in the state of the old logicians about infinity. He cannot separate the mathematical notion from the old mixture of infinite in quality and in everything. Leibnitz had it to a considerable extent, in spite of his power over the mathematical notion.

1863. I bring the chief difficulties of quantitative infinity to something like this:—

When of A and B one and one only must be, when A is visibly self-contradictory and B only incomprehensible, I vote for B.

I shall be very glad to see anything of Ellis's. The thoughts of his long illness would be valuable. He gained an enormous power of thinking about mathematics without pen and paper. I repeat my wish that his preface to Bacon could be separately published. With kind regards to Lady Affleck,

I am, yours very truly,
A. De Morgan.

To Rev. Dr. Whewell.

April 7, 1863.

My dear Sir,—Now I *will* provoke an opinion, not an answer to a matter of fact; and if anybody in College has ever thought about the subject, I wish he would think about it again.

Aristotle has a chapter in the 'Metaphysics' about the ἄπειρον, translated 'the infinite.' The chapter opens with a description of the meaning of the word, which has come down without any strong objection that I can find. I give the sentence itself, with a literal translation from MacMahon. I hardly ever had to look closely at a sentence of Aristotle without finding what reason might take either for a gross corruption or an obvious interpolation. This chapter I suppose to have had much sway in determining the logician's obstinate confusion between the infinite, unlimited in qualities, powers, &c., and the simple infinite of magnitude. Now from 'Metaphysics,' lib. x. or xi., cap. 10:—

Τὸ δ' ἄπειρον ἢ τὸ ἀδύνατον διελθεῖν τῷ μὴ πεφυκέναι διιέναι, καθάπερ ἡ φωνὴ ἀόρατος, ἢ τὸ διέξοδον ἔχον ἀτελεύτητον, ἢ ὃ μόλις, ἢ ὃ πεφυκὸς ἔχειν μὴ ἔχει διέξοδον ἢ πέρας· ἔτι προσθέσει ἢ ἀφαιρέσει, ἢ ἄμφω.

'But the infinite is either that which it is impossible to pass through in respect of its not being adapted by nature to be permeated, in the same way as the voice is invisible; or it is that which possesses a passage without an end, or that which is scarcely so, or that which by nature is adapted to have, but has not, a passage or termination. Further, a thing is infinite from subsisting by addition or subtraction, or both.'

Here is a nice kettle of fish! Now I try to put a sense upon it. 1863.

1. I take ἄπειρον to mean rather without *boundary* than without *end*; indefinite. But the quantitative notion of *having no end* does intrude and confuse that of having no *determinate*.[1]

. . . .

To Rev. Dr. Whewell.

April 11, 1863.

MY DEAR SIR,—Many thanks for your letter. I feel helped by the word διιέναι, because it is a very thoroughfaresome word. As the lexicon says, it is used for going *through* a country, or for running a man *through* the body, which is a process very definitely suggestive of in at one side and out at the other. It points very distinctly to the idea of bounded on all sides, being that which ἄπειρον denies. And this, combined with an ἄπειρον gained by *subtraction*, confirm me in the notion that Aristotle is treating of the indefinite—not necessarily, though possibly, infinite in magnitude.

I agree with you that the adjective infinite without a substantive is like all other adjectives similarly situated.

On infinity—*i.e.*, infinite quantity—a concept necessarily connected in our minds as an attribute or predicate, with space and time, I have come to the conclusion that we must treat it as a concept without image. Throwing away the word *imagine*, as spoiled by becoming a synonym of *conceive*, I distinguish the concepts which we can *image* from those which we cannot. We can put before the *mind's eye*, or the mind's mode of remembering sensible things, an image or likeness; a man, for example, as he appears when alarmed. But *alarm* is a concept without image; it has predicates, it is the subject of true and false propositions, but not as a thing having an image.

Now of quantity of space, and even of time—for succession of things is among our sensible relations—we have images; but not when too small or too great. The infinitely small and the infinitely great are below and above our imagining power, but they are concepts with attributes. Those who reject both or either because they cannot *form an idea*—by which they mean an *image*—ought equally to reject those *entia rationis*, the $\frac{1}{10^{100}}$

[1] I much regret that I have lost the rest of this letter. But I insert it as it stands, as the same subject is spoken of in the next.—S. E. DE M.

1863. of an inch and the length of 10^{100} miles. But these are subjects of which we can predicate; and so, I hold, is infinity.

I have made all manner of efforts to repudiate infinity of quantity for forty years, in obedience to the dicta of people who assured me I did not possess any idea of it, and I have failed. And I have observed that no people seem so clearly to have the idea as those who argue against it, while engaged in their task. And I begin to lean towards the notion that the difficulties of infinity of quantity arise from our having more knowledge of it than of things, for which we depend on attributes—as mind or matter.

The absolute, as you say, really has no predicates; and it is a very circular idea. Is not the being unconditioned, if *per se* and necessarily, a *condition*? *Cannot* is a word of limitation and condition. Can the Creator commit suicide? if not, he is, to our thoughts, *conditioned*. I should like to know what Hamilton would have said to this. Seeing that the Germans shine as smokers and also as metaphysicians, and also that in the former capacity they *blow* a cloud—which was the word for taking a cigar in my day—it is worth while to think about transferring the phrase, in a transcendental sense, to their other pursuit.

Yours truly,

A. De Morgan.

To Sir J. Herschel.

91 Adelaide Road, May 10, 1863.

My dear Baronet,—There's change for your 'Professor.' Everybody attaches some ideas to a word derived from early associations. The first 'learned Professor' I read of under that name was Olearius Schinderhausen, of Leyden, who disparted with his cast-off suit biennially. I did not think I should live to match him; but as I never go out, and always work at home in a dressing-gown, I also have but one coat in two years.

Seventy-one, eh? Go on to eighty, and then apply to me for further directions, if I should be in a condition to give them. Addition of the same to a ratio of greater inequality diminishes it. So says Jemmy Wood; and the life of man confirms it. When you were preparing $\sin^{-1}x$, I was learning numeration from my father on a *zahlenbreitstein*—a pebble, of diameter and flatness, picked up in the road. And I remember that when it was lost I refused all arithmetic till another was found; which, considering that no one had told me the etymology of *calculation*, showed a kind of natural philological acumen. 1863.

Yours very truly,
A. DE MORGAN.

To the Master of Trinity.

91 Adelaide Road, October 21, 1863.

MY DEAR SIR,— . . . . If you read *Notes and Queries*, look out for a few notes I have given on Robert Robinson. I shall send down a cutting to paste into his 'History of Baptism,' or any work the library of Trinity may chance to have.

I have been in communication with Mr. Wright about a book he used out of that library. He was the most remarkable Cambridge (town) man of the last century, I suppose; at least he comes next after Maps and Jemmy Gordon.

I do not know whether you know Crabb Robinson (no relation), to whom I am indebted for anecdotes. He is eighty-six, and pours out anecdotes about everything and everybody, especially his especial friends Goethe, Wordsworth, Southey, Coleridge, Charles Lamb, *et id genus omne.* He tells me that Wordsworth agreed with Samuel Parr that Dyer's 'Life of Robert Robinson' is one of the best biographies in the language.

Yours very truly,
A. DE MORGAN.

To Sir J. Herschel.

91 Adelaide Road, November 19, 1863.

MY DEAR SIR JOHN,—Thanks for the paper on the standard. Hurrah for anything which preserves our great units! On that condition I will accept the earth's axis or anything else. But I stipulate for the foot, as for common usage. A yard is too long to start from; a foot I hold too long; I should like better $\frac{1}{10}$ of 2 yards.

Many years ago I demanded of my bootmakers the lengths of foot of 100 adult men, taken as they came in his books. The result was 10·26 inches as the average foot of man, English, measured in Bedford Street, Bedford Place. This is rather surprising, seeing that a bootmaker *gives* a little additional rather than otherwise.

I hope the metrical people will continue to agitate. I do not

1863. fear *their* metre, which will never go down; but they will do good to the decimal principle. I think we shall get the decimal coinage up again by their help.

I am dry of information of every kind.

Yours very truly,
A. DE MORGAN.

To the Rev. Dr. Husenbeth.

91 Adelaide Road, December 31, 1863.

MY DEAR SIR,—I am very much obliged for the excerpt. But how do you manage with rubric letters on a dark morning or by candlelight? The first opening I made of the creed showed me, by candlelight, '*œ*que confondentes personas,' which, said I, must be a misprint.

I am glad to be set right about the *filioque*. I was once—but it dropt—puzzled to know how the Greeks could *reject* two words of the Athanasian when they rejected the whole creed. But like most (Western) others, I had but a cloudy notion of the Greek Church. My Latin Prayer-book is certainly the old Latin. A new Latin translation, the veritable original being Latin, strikes me as would a Greek Homer translated from Pope. . . .

I shall certainly attack *reliable*. One of the tale-writers in *All the Year Round* has introduced it into a document purporting to be of the early reign of George III. This is adding insult to injury.

When I say a journal cannot refuse advertisements, I mean that it cannot do so without danger to its prosperity. It is found that any check to influx of business is bad policy.

If the *Athenæum* of thirty years ago be examined, it will be found that the reader's portion is increased relatively more than the advertisement portion. It is astonishing how many persons delight in running over columns of advertisements. . . .

Yours very truly,
A. DE MORGAN.

To the Rev. Dr. Husenbeth.

January 12, 1864.

1864. MY DEAR SIR,—Many thanks for the drama, which, not knowing the original, I cannot divide between you and the

author of the second. I read it through, which is a point on 1864.
which most reviewers of books are apt to be evasive, and, as I did so at one sitting, I need not say I liked it. As there are only two Christians in it, or at most three if we count Agellius, there is no scope for a fault which I noticed in a book of the Cardinal's, of which I forget the name, all about St. Pancras (saint, not parish). That fault is that though the author can draw characters, yet the moment he puts on the religion, it is a domino of the same form and colour for all, which makes them look just alike. I thought all the while of the great magician who could make the Calvinism of David Deans and Jeanie as distinct as their characters, sexes, and ages, without anything that would bring either under the censure of the Presbytery. Either Dr. Newman or yourself, or both, have managed three tolerably different phases of religious character.

You may easily get far enough into my syllabus to see the meaning of the symbols –)))·(&c., which is all the book was intended for—I mean the copy sent to you. In)) (·), for instance, I see—

1. Premises.
2. Proof of validity.
3. Conclusion.
4. Quantities of all the terms.

Dr. Watts's book, which I call the English Port-Royal logic, deals little in purely logical exercise. As to your reading it in spite of what it says against the Pope, &c., I should have read it all the more, for I enjoy being assaulted and batteried. But the book is a good one. Had he been of the Parliamentary form of religion, it would have been a great work, as great as any of Paley, in the two Universities. . . . I am, dear sir,

Yours truly,

A. DE MORGAN.

To Sir J. F. W. Herschel.

91 Adelaide Road, August 18, 1864.

MY DEAR SIR JOHN,—If you happen to recollect you were Master of the Mint—I am sure you had reason enough to recollect it—you cannot have forgotten that Newton was there before you. *His concessis*, I find that N. and H. added each one coin to the list; N. the gold quarter-guinea, which was in circulation until towards the end of the century; H. the gold quarter-

1864. sovereign, which was never circulated. Tell me, I pray, when you quarter-sovereigned did you know that Newton had quarter-guinead, or was it an accidental coincidence?

Here I am, as usual in August, with one son in the house and all the rest at Aldborough. My wife is mending, but has been seriously ill. I hope I shall have a good account of your section of the human race.

You know Tristram Shandy—who does not? But you may not know his views of proportion. He says his father was sometimes a gainer by a misfortune; for if the pleasure of haranguing about it was as *ten*, and the misfortune itself only as *five*, he gained 'half in half,' and was as well off again as if the misfortune had never happened. Cipher this out. I call it a splendid *bévue*; as good as the two last lines of the song about the young man who poisoned his sweetheart in sheep's-head broth, and was frightened to death with—

> Where's that young maid
> What you did poison with my head?

at his bedside.

> Now all young men, both high and low,
> Take warning by this dismal go!
> For if he'd never done nobody no wrong,
> He might have been here to have heard this song.

Babbage's Act has passed, and he *is* a public benefactor. A grinder went away from before my house at the first word. The New Zealander shall sit on the remains of his doorstep to sketch the fragments of a broken barrel-organ. 'O si sic omnia!'

Yours very truly,
A. DE MORGAN.

To Mrs. Smyth.

91 Adelaide Road, September 1, 1864.

MY DEAR MRS. SMYTH,—I see by the title-page that work is on hand—when was it not? It is thirty-seven years since I became personally aware of the fact—that is, I have known the Admiral nearly half his life, and yet there was a lot of it when I first knew him; he had retired from a life's work.

I wanted to see how you swallowed the Pyramid.[1] I do not

[1] Prof. Piazzi Smyth's work on the Pyramids, published just before this time. S. E. De M.

see any occasion to feel depressed about it. The world is not what it was when the Admiral and I were young. There are strange break-downs of opinion and build-ups of system. I do not know that it is more extraordinary than that I, of all persons, should declare and publish my actual experience of phenomena which so many will have to be actually the work of disembodied spirits, which I can neither deny nor affirm. The Christian world is actually tending towards the belief that a great many mythologies and idolatries are really diseased revelations. Wait a few years until it begins to be generally apprehended that Juggernaut and Cham Chi Thaungee—if you happen to know him—were originally divine, though a little altered by time and priest, and it will then seem a very natural thing that the measures in the Pyramid should have come from the same quarter. I wish the reasoning had been a little more sound and the mind less influenced by bias of system, but this is not peculiar to primeval inspiration advocates. It is the beauty of these extreme vagaries that they show off and illustrate the methods which are most in vogue among savans who are quite in the groove. But the moral courage which ventures upon a trip off the line is not so common.

1864.

The work itself is part of an impulse which is doing strange marvels, which will make Bishop Colenso die a heretic, and which has made Robert Owen die a Christian; nay, which has made a Christian of Dr. John Elliotson, the strongest materialist almost that I ever heard of. If we go on in this way twenty years longer, the name of God will be heard at a meeting of the Royal Society, from which Dr. Cumming will take occasion to declare that the millennium has commenced.

There are educated persons by thousands not in the little knot, who will look on Piazzi's book without much surprise as to his mere conclusions. It is astonishing how little the world of science knows about opinion outside.

Kind regards to the Admiral, who, I suppose, has his work well-nigh done. What a beautiful feeling the proof of the title-page gives!

Yours very sincerely,

A. De Morgan.

To Sir J. F. W. Herschel.

91 Adelaide Road, September 20, 1864.

1864. MY DEAR SIR JOHN,—So honey-bees have stings as well as wapses. So much the better. I do confess to a baker's dozen of readings of your letter.[1] Saving your presence, your usual mildness and diffidence, and all those things of which I know no more than my grandmother's tom cat knew of the differential calculus, bring out the points with such strength that there is no leaving off. It will do an abundant deal of good, and will make the declaration a blessing. It is in good contrast with Bowring, whose letter is capital in its way.

Truly you did express yourself with tolerable precision and to much purpose in the old *discourse*. I have been pasting a copy into that discourse, and it notes time well to find that the copy wherein I paste it is one which poor Stratford gave my wife long before she was even my sweetheart, and my eldest son is twenty-five years old.

The inpasted copy is cut out of a reprint in the *Bath Chronicle*, done for the special benefit of the Br. Ass.

What is the collection of names they have got? I suppose they have handed out their best fifteen. . . . I think I had as good a right to be asked as . . . . ;[2] but possibly they took me for too great a heretic. I dare not think they respect me more than they do you; but the temptation is great. How I should strut!

They are now pledged to publish this declaration and the names they have got. They might have dropt the thing after Daubeny's letter.

I hope your bronchial state is better. To my last letter I add that the little dust does not confine itself to your books; it gets into your throat.

Yours very truly,
A. DE MORGAN.

[1] Many scientific men had been requested to sign a declaration of their belief that science did not contradict revelation, &c. By the 'honey-bee's sting' is meant Sir J. Herschel's answer, which contained a trenchant reply to the arguments contained therein.

[2] I think he *was* asked to sign perhaps rather later.

To the Rev. Wm. Heald.

September 27, 1864.

MY DEAR HEALD,—I suppose you and family are pretty well, 1864.
as you say so little about it.

In re Colenso, you can easily demolish him by assuming the points. He has never told us what residue of the Xtian religion he still believes; why, whether the historical truth of the O. T. *is* any part of the N. T. is just the point. If I remember right, he at one time *could not* use the ordination service; but he can now, I believe, after Lushington's decision in Williams's case.

The case about C. D. and A. B. stands thus. All the reviews, &c., declare that A. B. is self, and C. D. wife. Neither of us has contradicted it, which leads me to suspect that we cannot. The style of A. B. is, I am assured by good judges, unmistakably my own, and I certainly do see a strong likeness. All these things put together cannot be got over. Were I you I should assume the report to be true. I have not heard anything about a second edition.

Yours sincerely,
A. DE MORGAN.

To J. S. Mill, Esq.

91 Adelaide Road, October 10, 1864.

MY DEAR SIR,—Ten years ago I asked you whether certain abridged dialogues of Plato were yours, to which you answered in the affirmative. Have they been reprinted? I very much wish they were, if they are not. The presentation of Plato is now frequent; but there is nothing I know of in which a picture of Plato is given, and remarks kept distinct.

I am reminded of this by a translation of the *Gorgias* which has just appeared, by Mr. Cope, Fellow of Trinity College, Cambridge. A good translation, I dare say; but, to tell the truth, I and many others like to have the pith of Plato extracted, and find both Greek and full translations rather wearisome. *Nostra culpa*, no doubt, but you must have thought such sinners not quite below a missionary, or you would not have published your abridgments. I suppose it is not given to man to relish both Aristotle and Plato.

1864. I often cut from a review and paste haphazard on to the fly-leaf of a book. In your Plato articles I find, in this way, a curious accidental paraphrase of the Trinity that may amuse you. The Druse system is described as historically identified with the Caliph Hakem, the Persian Hamzé, and the Turk Davagi—Hakem the 'political *founder*,' Hamzé the 'intellectual (λογος) framer,' and Davagi the 'expositor and propagator.'—(*Athenæum*, August 27, 1853.) I am, dear sir,

Yours truly,
A. DE MORGAN.

To J. S. Mill, Esq.

91 Adelaide Road, February 5, 1865.

1865. MY DEAR SIR,—The Algebra, which I am much pleased to find you approve of, is to be divided between several. I have no doubt I may claim to have first presented it complete, and the connection of A^{∞}[1] with the rest is my own. But Warren, Peacock, and others of those mentioned in my list at the beginning of the work are real predecessors. You are perhaps aware that Peacock published two works on algebra. The first, in one volume, is that which treats the subject most generally. He is in full possession of all except what relates to the exponent, and here he is obliged to have recourse to interpretation, that is, discovery of meaning from results.

With regard to the acceptance of the system, the time is not yet come. The algebraists almost all make algebra obey their preconceived notions. They have laws which algebra must obey. Peacock had very nearly attained the idea of algebra as a *formal science*, in which every result of the form is to have meaning. His *permanence of equivalent forms* would have developed itself into formal algebra capable of any number of material applications, if he had been a logician—I mean a student of logic. So long as an algebraist has preconceptions which his science must obey, so long is he incapable of true generalisation. Macaulay said of Southey that what he called his opinions were his tastes, and this is true of many persons, and of a great many algebraists. Algebra must, *à priori*, be subject to this or that limitation, upon what is really an acquired taste of the legislator. Pure logic is in the same predicament.

[1] I am not sure about this exponent.—ED.

For myself, it is my taste, if you please, that *I will have* a formal algebra, in which every form, every law of transformation is *universal*. I admit that we have not yet arrived at it, but I have entire faith in the future. I was in the spirit on the day of Wellington's funeral, when, wife, children, and servants being away to see the remains of the glorious old man carried upon what was so justly called a cross between a locomotive and a fire-engine, I was sole master of my house and myself. So I sat down to eviscerate the following difficulty, and I believe I did it. If the forms of algebra be universal, then $2x=x$ gives $\frac{2x}{x}=\frac{x}{x}$ or $2=1$. I should not have been ashamed of myself if resolved on a formal algebra, I had *invented* a generalisation of $=$ to meet this case. But I had no occasion for any such thing. I found that, by only taking permission to lay down as a canon what mathematicians never scruple to do when they want it, I was master of the field. 1865.

I have a paper now at Cambridge which explains my views, so you see I have taken twelve years to think about it. In brief as follows. Admitting in theory as full and free a use of infinites, finites, and infinitesimals, as is made in practice, I say that—

1. The sign $=$ is that of *undistinguishability*, say *indistinction*. $A=B$ means that A and B are not distinct. Equality is but a case, though the most common one.

2. Distinction implies the use of a standard or metre. Quantities infinitely great with respect to the standard, or infinitely small, or unmeasurable by it, are undistinguishable. And this is the origin of Leibnitz's equation $dx=dx+dx^{-2}$, the metre having finite ratio to dx.

3. Whenever we divide or multiply both sides of an equation by anything above or below the standard, the new equation takes a different standard. When we multiply by an infinite we must take a standard which was infinite, &c. Now $2x$ and x can only be undistinguishable when x is infinitely small or infinitely great. Let x be infinitely small. In passing from $2x=x$ to $2=1$ we change our standard into an infinitely small quantity, and $2=1$ is *true*, that is 2 and 1 cannot be distinguished. The same when x is infinitely great.

The difficulties which will suggest themselves are many and obvious; but I think that they are superable, and also that, looking at actual algebra, they are not so great as the difficulties which actually occur. For all these things I refer to the paper

1865. when printed, to which so brief an account is more like a preliminary objection than anything else.

Yours very truly,
A. De Morgan.

To J. S. Mill, Esq.

March 26, 1865.

My dear Sir,—I thank you for the article on Comte received a few days ago. It gives me a much more definite idea of Comte than I ever had patience to get from himself. His writing always had to me a smack of that unequalled prolixity which he showed in his algebraic geometry, where he *discourses* in page after page on the equation of a straight line, without a symbol. I settled that he *was not* a psychologist. I make out from you that he *would not be*, prepensely. I thought he was impatient of the subject, but did not fathom all his guilt.

I am confirmed in my view that his philosophy is, so far as it is distinctively his, *negativism*. For his positivism I find in all thinkers, or nine out of ten; his rejections, hardly anywhere. Positive, because no more than positive. When understood thus, he is a bearable companion, for one has a right to be as anti-positive with his philosophy as he is positive with mind and matter; *i.e.*, as he has taken part for the whole, I take his whole for part.

I shall soon send you a paper in which I find I am a sort of Positivist. There are those who reject all but phenomena; there are those who reject phenomena because they cannot have more. Comte is the assailant of those who accept more because they think they can get more. In the mathematical treatment of infinity, small and great, most mathematicians reject the absolute treatment because they cannot image, or treat as phenomena, all the attributes of the notion. My notion is that ∞^7 and $\frac{1}{\infty^7}$ have a subjective reality, of which various phenomena are proper subjects of direct reasoning. The mathematicians have virtually denied that *A is B* is a component of reasoning when we know it to be true even though *A* and *B* should be porcupines with difficulties for quills. In my paper in two parts, ∞ is a porcupine, and $=$ is a hedgehog.

I see you are in England again by your complimentary letter to the Westminster electors. You pay them a higher compliment than they pay you. I am always in doubt about the origin

of the word *compliment.* It looks like a formation from *comply,* 1865.
but I doubt it. I suspect that complément is the original, though the present spelling and usage is as old as the Academy's Dictionary. I suspect that old forms of civility were at last described as *complements,* fillings up; and that *complim*[ts], at the end of a letter, meant that all usual forms are to be understood. My theory receives a little support from *comply* not being a French verb.

Yours very truly,
A. DE MORGAN.

I have as yet read your article for Comte only, for I really wanted to know what he had been at. I must read it again for criticism.

To J. S. Mill, Esq.

91 Adelaide Road, April 27, 1865.

MY DEAR SIR,—I received the *Examination* to-day, and write my thanks that I may not forget it, as I shall have no consecutive reading for a week or two.

I hold, from observation, that a question is never fairly put before a public meeting until it has been moved, seconded, and opposed. A great many persons are really only half informed by the mover and seconder what it is all about; but the first person who rises on the other side puts some light into it. It is just the old law knowledge: the points at issue come out of the pleadings on both sides. In like manner with controversies. Hamilton has moved, Mansel has seconded, and now you rise to take objections. And I also observe that the first opponent very often puts his view of things into a much more attainable-in-a-given-time form than if he had been the mover of a counter measure. A dip into several unconnected pages makes me think that may be the case here.

One of my dips is into 'All oxen ruminate.' I have shown in my fifth paper on logic that Aristotle and all his real followers never collected *all the oxen.* Their phrase was, 'Every ox ruminates,' or, any ox ruminates. That the predicate is what I have called in my third paper *metaphysical* I am satisfied. And I have maintained that the common predicate of the world at large is so. I say, 'man is born and educated a mathematician as to the subject of his proposition, and a metaphysician as to the predicate.'—*Ex. gr.* 'Every man is biped,' *i.e.*, is of biped quality or

1865. attribute. I contend in that paper that the logician's form of *extensive* reading stands or falls with the numerical syllogism, which is the true genus.

Yours truly,
A. De Morgan.

To Chief Baron Pollock.

My dear C. Baron,—First as to *subjective* and *objective*. The thinking mind is the *subject, id quod subjicitur*. The external thing, relation, &c., is the *object, id quod objicitur*. Or the acts of the minds may themselves be objects; *i.e.*, one mind may be the object of another mind.

I see an apparition: is it *subjective* or *objective*? If it be a thing of my mind its existence is subjective.

This use and others are not very sound. But, generally, objective has that relation to the thinking mind which subjective has to the exciter of its thoughts.

When Kant makes space and time pure concepts of mind to put things into, the things themselves being in some unintelligible sense *external*, he is said to make space and time *purely subjective*.

As to *your* being old, you are the youngest I have lately heard of. On Saturday, at University College distribution of prizes was Lord Brougham, eighty-seven, much broken, but still himself, and able to deliver himself as fluently as ever, and with that powerful delivery of *the one word* which makes his sentences so effective. And there was Crabb Robinson, ninety in May, and quite alive to everything. And he will last for ever if he will only take advice I heard given to him, *i.e.* not to talk more than two hours at a time.

With your note came an acknowledgment from General Perronet Thompson, B.A. of 1802, and Fellow of Queen's before he was an ensign. And he works at acoustics as hard as ever he did at the Corn Laws. I say nothing of boys of seventy who are scattered about. Our kind regards to Lady Pollock and the family.

Yours very truly,
A. De Morgan.

To Sir J. F. W. Herschel.

September 13, 1865.

My dear Sir John,—The world is changing round us very fast. I have lost two very old friends on two successive Saturdays—W. Rowan Hamilton and Smyth. Smyth went off very quietly from mere exhaustion, without any long illness. He had lost a great deal of strength in the preceding fortnight, but had rallied a little, and was driven out on Friday, and went calmly off on Saturday morning with effusion on the lungs. Poor Mrs. Smyth is very calm; it is the break of a tie of fifty years. 1865.

W. R. Hamilton was an intimate friend whom I spoke to once in my life—at Babbage's, about 1830; but for thirty years we have corresponded. I *saw* him a second time at the dinner you got at the Freemason's when you came from the Cape, but I could not get near enough to speak.

They will take care that a full life shall be published. Of forty members of the Nautical Almanac Committee of 1830, there remain now, besides our two selves, Airy, Lord Shaftesbury, Babbage, Lee, Maclear, Robinson, South, Wrottesley—ten in all; this is fair vitality for thirty-five years.

I am, yours sincerely,
A. De Morgan.

To Sir J. F. W. Herschel.

91 Adelaide Road, September 21, 1865.

My dear Sir John,—You must have been much in the way of hearing all the rumours about George III. and his malady. Do you remember hearing a story which I have heard from boyhood, which I find numbers have heard, and which I believe has been in print, but I do not know where? It is stated that one of the earliest indications of his complaint was a formal announcement to those about him that he intended to begin his speech from the throne with 'My lords and *peacocks*.' To all remonstrance he did nothing but persist, and they say that accidental noises—as tumbling down books, &c.—were prepared to drown the words. If the story be really true, I have no doubt I have arrived at the meaning of his phrase—that is, at its origin. But first, query, did it happen? If it did, there was method in his madness. With kind regards all round,

Yours very truly,
A. De Morgan.

To Sir J. F. W. Herschel.

91 Adelaide Road, September 27, 1865.

1865. MY DEAR SIR JOHN,—Then the story *is* a story, and has a place in history. Now for its explanation. Old George III. knew Shakespeare pretty well—much better than any other literature. In 'Hamlet' there are several places in which Hamlet seems on the very point either of disclosing his step-father's villany or giving him some reproach, but breaks off and substitutes something. In one case where ass is clearly coming, he makes it *peacock.*

> For thou dost know, O Damon dear,
>   This realm dismantled was
> Of Jove himself; and now reigns here
>   A very, very (ass) peacock.

Now George III. had old score recollections of the House of Commons. I suspect that when his mind was in his wanderings he determined to be revenged and to say, 'My Lords and Asses,' and he remembered and imitated Hamlet's substitute.

I am, yours sincerely,
A. DE MORGAN.

To Lord Chief Baron Pollock.

91 Adelaide Road, December 22, 1865.

MY DEAR L. CHIEF BARON,—I don't believe in December 21 as the shortest day. Put on the twilight at both ends, and calculate the shortest day. What care I for geometry? It is day as long as I can see to read, and even then the type must be stated. There is Large Pica day, Small Pica day, Bourgeois (pron. Burgice) day, etc. Diamond day is considerably shorter than the geometrical day, in winter.

But now to your questions.[1] 'Αμος is an obsolete word = τις. Πως, the common word. 'Αμωσγεπως is a word of Plato and Aristophanes (two people as opposite as L. C. B. Pollock and L. C. B. Nicholson), meaning *somehow or another.* So it is, as you say, 'some unknown cause capable of producing.' We cannot coin in English, I suppose, because it is the Queen's

[1] As to the meaning of *amosgepotically* in the preface to *From Matter to Spirit.*

English, and 'uttering' one's own coin is a felony. We cannot 1865.
say 'somehow or otherically.'

Yours sincerely,
A. De Morgan.

To Sir J. F. W. Herschel.

91 Adelaide Road, September 27, 1866.

My dear Sir John,—I write without much purpose, unless a 1866.
funny fact be one. Ulugh Bey is the end of the chain. An Arabic dictionary maker came to talk to me about getting his names of the stars in Arabic; thence came various Turks, thence Turkish pupils *he* had had—whom I had had also; thence Ameen Bey, whose father was Hassein Effendi, who trisected the angle, and who was taught by Ingliz Selim Effendi, who translated 'Bonnycastle.' This Englishman who translated B. must have been Richard Baily, Francis B.'s brother. He certainly ran away and turned Turk, and both F. B. and Miss B. told me that he translated Bonnycastle. Do you happen to remember anything about him? I remember that Miss B. had his picture.

I hope you have fought the weather with needle-gun success. By the way, where is it set forth that a leech is a famous barometer, by indications to be seen in his behaviour in the water?

I have been looking at the writing of South about Babbage and Davy. There is an exceeding patness of recollection about him on this occasion as on former ones. Are we really to believe that when he called on Sir H. Davy, and wrote a note which he left in Davy's study, he took a copy of that note, so as to be able to give it in complete form? Or has he done what Thucydides must be held to have done about the speeches of his generals—made them as they might have been and ought to have been?

What is the real origin of the severance between South and Babbage? In all Babbage's 'Autobiography' I cannot find a hint of South's existence!

Do you possess Hyde's edition of Ulugh Bey's 'Catalogue' (in Arabic and Latin)? If so, I wish you would lend it to me. I should not wonder if it were in your patrimony, though I should hardly think you would have wanted it yourself.

With kind regards to Lady Herschel and all around you,

Yours sincerely,
A. De Morgan.

SECTION X.

1866–1871.

1866. I COME now to the last important event of my husband's life—the cessation of his connection with University College. In recording this I wish to dwell as little as possible on the fact, undoubted by all who were near him at the time, that his last illness resulted from mental trouble consequent upon it, in at least as great a degree as from the losses which befell us later. But however painful it is to write it, and however painful it may be to read for the survivors among those who were indirectly responsible for it, I have no choice but to state what was the belief of all who had the means of forming a true judgment.

He had joined University College in his early youth, in opposition to the advice of some of his nearest friends, who believed that his interests would not thereby be promoted, and to the satisfaction only of those in whose minds the upholding of a high principle was a more weighty consideration than worldly success or affluence. He was fully aware how much less lucrative a Professorship in a new institution was likely to be than many appointments which he might have obtained elsewhere. The associations, too, inseparable from a perfectly new institution were less congenial than those in which he would have found himself at either of the two Universities, where he would have worked under and with men whose habits of thought (in some ways) would have been more in harmony with his own.

He had worked in the new institution with untiring energy for six-and-thirty years, *because* he trusted the assertion of its founders and subsequent Governors that the essence of its being was absolute and complete religious equality in every portion of its organisation. His anticipation that the pledge, so often and so emphatically given, would be fulfilled, was destined to a complete and final disappointment. 1866.

Candidature of Rev. Jas. Martineau.

The occasion arose on the appointment of a new Professor of Mental Philosophy and Logic, in June 1866, in place of Dr. Hoppus, an Independent minister, who had held the chair from the beginning. In accordance with the laws of the College the testimonials of all candidates were submitted to the Senate of Professors, who examined and reported on them to the Council, in whose hands rested the final appointment. From the first foundation the Unitarians had been among the most powerful supporters of the College, which could never have risen to its then condition without their assistance in money and effort. When it was first known that the Rev. James Martineau, a Unitarian minister and a distinguished scholar, was a candidate for the chair of Mental Philosophy and Logic, a gossiping rumour came to the ears of my husband and myself that the Unitarians on the Council were working to bring in their own candidate. This was merely foolish talk among a few persons, but I mention it to show what my husband's feelings were on the subject of the appointment. When he heard the report he declared his disbelief in it, but said he would make inquiries, as there must be no suspicion of the preponderance of any one party in religion in that place. He inquired about the rumour, and, as he expected, found it false. No member of the Council at that time knew anything of the relative merits of the candidates. It was evident, even if any one who knew him well could have supposed it possible, that friendship for Mr. Martineau, for whom he had a sincere respect, did not influence his subsequent conduct.[1]

[1] I note this, as one of the newspapers spoke of my husband's

1866. The report of the Senate, after enumerating Mr. Martineau's qualifications for the appointment, as shown by his writings, by his examination papers, and by the testimony of his pupils, concludes with the words, '*All these considerations evidently lead to the conclusion that Mr. Martineau is the most eligible candidate. He appears to be at least equal to the other candidates in ability and learning, while he is superior to them both in reputation, and in experience and success as a teacher.*' The question was, however, raised at this early stage, whether Mr. Martineau's position as a Unitarian minister would be injurious to the class; and of this doubt the Council, some of the influential members of which were bent on appointing a Professor far lower in the scale of orthodox belief than Mr. Martineau, availed themselves. They postponed the appointment for a time, and the Senate was called upon to make a second report in consequence of new candidates having come into the field, and some of the old candidates having sent in additional testimonials. Their report of the second of the candidates was given in these words :—'Upon the strength of this singularly strong testimony we have no hesitation in concluding that Mr. Croom Robertson is exceedingly well qualified to fill the vacant chair; and that of the candidates whose claims we have examined up to this point, he is the ablest, and, as far as we can judge, the most learned, and the most likely to rise to eminence, and to raise the reputation of the College. But there yet remains upon the list the name of Mr. James Martineau. As the Senate has already recommended the appointment of Mr. Martineau, and the Council has declined to appoint him, the Senate does not think it necessary to present a second report concerning him.'

In the hope and belief that the position of affairs was not yet past remedy, fourteen Fellows of the College, including some of its most distinguished alumni, sent a

'chivalrous advocacy of his friend's cause.' This conveyed an inexact idea of the facts.

requisition to the Council asking that a special Court of Proprietors might be summoned to consider the course taken by the Council in deferring the appointment of the best qualified candidate for special reasons to the vacant chair. An objection to this on legal grounds was raised by the Council; they referred the question to the law officers of the Crown, and before the opinion of these gentlemen (which was in favour of the requisitionists) was announced, settled it their own way by the appointment of Mr. Croom Robertson, who was a pupil of Prof. Alexander Bain, and an adherent of the school of thought upheld by that gentleman, and approved by the leading members of the Council.

1866.

Appointment of Mr. Croom Robertson.

When Mr. De Morgan heard that the Council intended to reject Mr. Martineau for reasons connected with religious belief, he openly declared that should the College make such a departure from the principle on which it was founded, he should feel that his connection with it was at an end. He waited with anxiety for their decision, and when the news came that the acknowledged best candidate was set aside on the ground of his Unitarianism, and one below him appointed, he said that the College had committed a suicidal act, and would never hold its old place again. He did not hesitate as to his own course, but at once sent in his resignation.

Resignation of Professorship.

His letter to the Council, which follows, I know to have been written without any intention of publication at the time, or rather with a distinct intention of non-publication during his lifetime.

To the Chairman of the Council of University College.

91 Adelaide Road, November 10, 1866.

Sir,—I feel much sorrow in notifying to the Council that my connection with the College must close at the end of the current session.

For some years the returns of my chair have been so

1866. Letter of resignation.

small that, taking into account the time I give, my stay has been an imprudence. I had nevertheless calculated that I might, without too great an injustice to my family, draw upon my capital, if I may use so grand a word, for the means of retaining my post during this and the next session, in the hope of the dawn of better days.[1]

The recent vote of the Council in the case of Mr. Martineau renders it unnecessary for me to settle when I shall leave the College; it proves that the College has left me. I am, as heretofore, strong in the determination not to be overlooked, and not to be controlled in any matter of religious thought, speech, or teaching. The Council has decided that a certain amount of notoriety for advocacy of an unpopular theology is a disqualification. Whether a distinction was intended between the case of a candidate and of an installed Professor I neither know nor care. I assume that such a body as the Council would never entertain this distinction. I concede that A is not B, but I maintain that those who surrender to expediency point A of principle are the men who will surrender point B when the time comes, and who, until that time does come, will be honestly shocked at the prophecy of their future conduct. Adherence to come is discounted to meet the consequence of present departure. The principle of the College has been partially surrendered to expediency; no man can say how much more will be given up, nor when. This I said when the Peene legacy was accepted, and I was laughed at. The acceptance of the conditions of that legacy did not drive me from the College, because, after much deliberation, and not a little help from what I now see to be sophism, my love for the College and the life I led in it barred the way with *De minimis non curat lex.*

[1] During the last years of his stay from various causes the proceeds of his chair had fallen off. They had never been great, only one year amounting to nearly 500*l*. His continuing to hold it at a later period, when the returns seldom exceeded 300*l*., and were becoming less, was entirely due to his belief that the institution would fall still lower by his withdrawal.—S. E. De M.

But I ought to have seen that *minimum* is the first step from *nihil* to *totum*; and when St. Denys, with his head under his arm, had made that first step, I ought to have foreseen the second.[1] My self-complacency is comforted by observing that there are even now men of experience and thought who not only cannot foresee the third step, but who affirm it will never be made. 1866. Letter of resignation.

Before proceeding to the most delicate part of the subject I make two remarks.

First, in all that I say I am stating the decision of my own court, by which my own course is determined. It is for me alone to weigh evidence, and for me alone to decide. This distinction is often forgotten; such a letter as the present is treated as appeal to those to whom it is addressed, instead of recorded argument in a decided case. Be it remembered that the first sentence of this letter contains the needful; all the rest is partly respect to the body I am addressing, partly evidence of what is thought by a person who has stood by the College for thirty years, and who is likely to represent the opinions of many.

Secondly, I earnestly protest against being supposed to impute to any one, in or out of the Council, the least wilful or conscious impropriety of reasoning or conduct. I mean to give the offence which, in our thin-skinned day, is always taken at plain and uncompromising attack upon alleged wrong proceedings; but I am free of all intention to be personally disrespectful to any of the promoters. I can never forget the cordial co-operation of thirty years.

In the matter of Mr. Martineau, I am aware of the existence of two cross currents. Since the first vote of the Council I have weighed all that I heard, and have for months been satisfied that there has been an objection to

[1] St. Denys carried his head to Montmartre after his execution. I take the allusion to mean that just as the miracle was complete as soon as the Saint made the first step, so the alienation of the College from its principle was effected at the very earliest departure therefrom.—S. E. De M.

1866. Letter of resignation.

his *psychology* as well as to his *religion*: the first is too far removed from atheism to please the philosopher, the second too far removed from orthodoxy to please the priest. No longer neutral between the disputes of Christians, the College is to apply the abandoned principle in another field. The frontier is to be rectified by putting Theism in the place of Unitarianism, and making God an open question, not to be the basis of any teaching on the human mind. And so it is contrived that one and the same victim, offered on the altar of the Janus Bifrons of expediency, shall appease both the priest and the philosopher, while each votary selects the particular head of the deity to which his offering is made.

I proceed to show that (supposing me willing to remain) I am as worthy to be extruded as Mr. Martineau to be excluded.

I have for thirty years, and in my class-room, acted on the principle that positive theism may be made the basis of psychological explanation without violation of any law of the College. When in elucidating mathematical principles it is necessary to speak of our mental organisation as effect of a cause, I have always referred it to an intelligent and disposing Creator. The *nature of things*, the *eternal laws of thought*, and all the ways by which that Creator is put in the dark corner, have been treated by my silence as philosophical absurdities not worthy to have their silly names intruded upon those who are to be trained to think. Were I to remain under the new system, I should hold it a sacred duty and—ah, poor human nature!—a malicious pleasure to extend and intensify all I have hitherto said on this subject.

Again, for more than thirty years I have been as strong a Unitarian as Mr. Martineau. If I have not raised my voice in this matter, and as strongly as Mr. Martineau[1] has done, it is because I have been deeply

[1] In writing the above (as will be evident to the reader) Mr. De Morgan believed Mr. Martineau to be of the older Unitarian school,

engaged in other things, because I do not care what unreflecting people think they think, and because I have found that the great bulk of reflecting men of all sects keep their Trinitarianism caged in a creed, and are, in every practical application of religion except pelting Unitarians, as truly Unitarian as Mr. Martineau himself. Were I to continue in this College, under even the ghost of a gag, I should soon be heard (without the walls) on a subject to which I have paid long and close attention.[1] I should soon bring the question to issue whether the installed Professor is or is not a subject for such discussion as has arisen about the candidate for admission.

1866.

Letter of resignation.

I hope it will be clear that my absence is as desirable as that of Mr. Martineau. But, for reasons given, I deprecate the supposition of having sacrificed to principle. I have only ceased to sacrifice because the temple has been desecrated. My determination would not be altered by a return to the old principle on the part of the Council. I shall, therefore, not be suspected of any personal motive when I urge the Council to reconsider their suicidal vote, and to re-nail the old flag to the mast.

One point has perhaps been almost overlooked. A teacher of psychology, if he do his duty, expounds all systems of sufficient note, and puts forward the grounds of each. Every one must have his own system, and if one may therefore be suspected of bias, so must another. Mr. Martineau has special reputation as an eclectic teacher. He is noted for ability to prepare students for examinations in which the examiners have no bias towards his views. I have heard it remarked, before this discussion, that he crams his pupils with different systems. Such a man does not cram. It means that those of his students

which receives the New Testament records as literally true. I should not be justified in referring to this mistaken impression if Mr. Martineau's late writings, and his preference of the name of *Theist*, were not well known.

[1] These words, which may seem obscure, must be understood to refer to his Unitarian belief.—S. E. De M.

1866. Letter of resignation.

who desire no better can cram different systems from his lectures. There is more proof of his competency in this respect than in the case of any of the untried candidates.

Return to the old principle. If the College fall, it will fall with honour. No concession of narrow minds, philosophical or theological, will save it. The enemy will give one sneer more, the friend nine cheers less. Thing'embigot, who says that his son shall not enter the College if Mr. Martineau teach there, never meant to send his son in any case. The late vicar of St. Pancras, then a lessee in Gower Street, found the noise of the playground disagreeable, and sent word that if the nuisance were not abated he should withdraw his patronage; he had been an inveterate opponent. He was left to subtract his negative quantity if he pleased. Let Thing'embigot learn the same rule of algebra.

On the other hand, the enemy of religious disqualification, if the present course be persisted in, must decide whether his son shall be educated under selection carried up to its logical extent in the professed fear of God, or exclusion nibbled at up to compulsion of circumstances in the concealed fear of man as to religion, and another *fear of God* as to philosophy. I should myself be puzzled to make a choice, for if there be a tincture of atheism in the second fear of God, there is a tincture of blasphemy in the first. Of the two different ways of putting man in the place of God, I think the world at large would prefer the first.

My best wishes remain with the College which I leave, but I wish to make myself clearly understood on the question which has been opened. I trust that by return to and future maintenance of the sound principle on which it was founded, in which there is more religion than in all exclusive systems put together, the College will rise into prosperity under the protection, not of the Infinite, not of the Absolute, not of the Unconditioned, not of the Nature of things, not of the chapter of accidents, but of God, the

Creator and Father of all mankind.—I am, sir, with much respect, 1866.

Your obedient, humble servant,

A. De Morgan.

Reply of the Council.

The reading of this letter at the Council was (I was told) followed by silence for a minute or two. The minority who could understand its meaning and its motive had already been outvoted. The majority could give no answer, because they were determined not to give the only one it called for, a return to principle. The secretary was directed to inform the writer 'that your letter of November 10, addressed to the Chairman of the Council, was read at a session of the Council on Saturday last, and that your resignation of the Professorship of Mathematics from the close of the current session was accepted.'

The decision and its results gave great dissatisfaction to the friends of religious liberty outside the College. The newspapers, which represented different phases of thought, expressed the variety of opinions held on the subject. By those of the earnest and Liberal school the movement was strongly condemned; among other things it was said that all real Liberals must ask whether it is wise to support a College which, unsectarian in name, can yet be guilty of such religious and philosophical bigotry. Here I may remark upon the expression 'real Liberals.' Every one who has watched the progress of thought, especially during the last half-century, must have seen that its tendency, both in philosophy and in religion, is to the denial, or what amounts to a denial, of God. I am not now attempting to condemn this tendency, but its prevalence has had the effect of confusing formerly well-defined distinctions. The 'Liberal' has frequently gone from liberality to unbelief; and in the case of University College many professed Liberals took the part of intolerance because they preferred atheism to theism. The self-styled *religious* party said that it was now clear that the profession of

1866. University College. desire to preserve the unsectarian character of the College was so much dust thrown in the eyes of the public, 'and that the College had declared itself truly a 'godless College,' as it had long been called by the orthodox. Scoffers laughed, and opined that 'the College' could do without its principles, but the principles could not do without their College.

Meeting of Proprietors. The next meeting of Proprietors was appointed for February 2, 1867. But before the end of the current year, and with reference to the requisition already referred to, the Senate met again and supported the decision of the Council, and in anticipation of the meeting addressed a statement to the Proprietors. Though the real question, which lies in a very few words, has been distinctly stated in Mr. De Morgan's letter of resignation, I should be thought to give an *ex-parte* account of the whole affair if I were to omit the arguments on the other side. The strongest of these may fairly be presumed to be embodied in this statement of fifteen Professors, of whom more than one had belonged to the institution from its foundation. I feel it only right to give the document at length.

Statement addressed to the Proprietors of University College.

A certain number of Fellows and Proprietors of University College, London, have required the Council of the College to convene a Special General Meeting of the Proprietors, 'to consider a recent resolution of the Council declining to appoint the Rev. James Martineau to the Professorship of Mental Philosophy and Logic, after a Report of the Senate that he was the best qualified candidate for the chair;' and a Special General Meeting will be held, in consequence of this requisition, on Saturday, February 2.

The subject which the meeting is convened to consider has attracted much attention, and has been the occasion of many articles in various daily and weekly papers; and as is usually, and indeed inevitably, the case when writers press forward to instruct the public under the influence of a strong preconceived opinion, and with an imperfect knowledge of the constitution and history of the body which they criticise, and an imperfect knowledge of the facts which they discuss, many circumstances have been misrepresented. It is likely that these misrepresentations have influenced some even of the requisitionists; and we may assume without doubt that they have produced some effect, more or less, on other proprietors who will attend the meeting. We wish, therefore, to correct the more important of them, and those especially which relate to the action of the Senate, or body of Professors. 1866.

The gentlemen who have signed the requisition, and who are indignant that Mr. Martineau was not appointed to the vacant chair, have not thought it necessary to make themselves acquainted with the evidence respecting the qualifications of the other candidates. Most of them, in all probability, had no means of knowing it. But as it would not have been seemly to impugn the judgment of such a body as the Council of the College, which had a full knowledge of the case in all its bearings, and considered it very carefully, and which in fact made its final decision after three months' deliberation, merely upon the plea, that knowing only the reputation of one candidate, and not having examined the claims of any other, they had a strong opinion another way, they shelter themselves under the authority of the Senate, and justify their very unusual attempt to pass a censure upon the Council on the ground that the Senate reported that Mr. Martineau was the best qualified candidate for the chair.

The greater number of the requisitionists are Fellows of the College, and were formerly distinguished students; and probably these gentlemen understand the constitutional relation of the Senate to the Council. But in the articles on the subject which have appeared in the public papers there have been expressions as if the privileges of the Senate were invaded by the action of the Council, or at least as if the two bodies were necessarily placed in a position of antagonism by such a difference of opinion. This is not the case. By the charter the power of appointing Professors is given absolutely to the Council. Neither the Senate nor a General Meeting can limit their discretion. But it is wisely provided in our by-laws that, when a Professorship is to be filled up, all the applications of candidates, and all testimonials and other documents which they may present as evidence of their qualifications, shall be submitted to the Senate, and that the Senate shall report thereupon to the Council. But

1866. a report so made is only an expression of opinion. The Council retains perfect liberty of action. The choice of the Council is usually in accordance with the opinion of the Senate. But cases have occurred before in which the Council has exercised an independent judgment, and departed from the recommendation of the Senate, and the Senate has never felt itself aggrieved by the exercise of such an indisputable right.

In this particular case there is especial reason why the Senate should receive with respect the decision of the Council. On subjects of professional learning, and in several departments of science and literature, it is likely that there will be men amongst the Professors better able to form a sound judgment than any members of the Council. But on the question of Mental Philosophy we believe honestly that the members of the Council generally are as fully competent to form an opinion as the members of the Senate.

Moreover, the relative strength of the recommendation of the Senate has been misrepresented. It has been asserted that the Senate 'reported to the Council that Mr. Martineau was *incomparably* the ablest candidate.' The first report of the Senate, in which alone the qualifications of Mr. Martineau were discussed, was not so unjust. The report examined the claims of Mr. Martineau, as attested by his published writings, by the evidence of his pupils, and by his Examination Papers, which he frankly laid before the Council; and the conclusion was in these words:—'All these considerations evidently lead to the conclusion that Mr. Martineau is the most eligible candidate. He appears to be at least equal to the other candidates in ability and learning, while he is superior to them both in reputation, and in experience and success as a teacher.'

A still more flagrant injustice has been done to the other candidates by the language of another writer, who chooses to describe the Council as rejecting a first-rate man in order to look about for 'a safe man with indifferent qualifications as a teacher,' one 'obscure enough to be inoffensive.' It is not necessary to insist upon the qualifications of the candidates whose names are not public, except to protest against the injustice of a writer who knows nothing about the matter using such disparaging language: but in the case of Mr. Robertson, whom the Council have appointed to the Professorship, it is right to state the conclusions of the two reports of the Senate. The first report upon his qualifications ended thus:—'Mr. Robertson is only twenty-four years of age. To judge from his testimonials, his philosophical acquirements are already very uncommon; nor ought we to forget the great public service our College may render by giving a man whose natural vocation, we are assured, is philosophy, an opportunity of following his vocation.' The report then spoke of the objection on the score of youth, and after alluding to drawbacks in the case of certain other of the candidates, concluded, 'But we are inclined to think that, of the disadvantages under which they labour, Mr. Robertson's youth is the least serious, as it is certainly the most remediable.' But the Senate was called upon to make a second report in consequence of new candidates having come into the field, and some of the old candidates having sent in additional testimonials, after the Council had extended the time for filling up the vacant chair. When they drew up the second report, they had before them additional evidence with respect to Mr. Robertson, which gave assurance of the extent of his learning, and of the breadth and impartiality of his views, and of his scrupulous fairness in exhibiting fully contending theories. They expressed their conclusion in these words:—'Upon the strength of this singularly strong testimony we have no hesitation in concluding that Mr. Robertson is exceedingly well qualified to fill the vacant chair; and that, of the candidates whose claims we have examined up to this point, he is the ablest, and, as far as we can judge, the most learned, and the most likely to rise to eminence, and to raise the reputation of the College.' The report then continued, 'But there yet remains upon the list of candidates the name of the Rev. James Martineau. As the Senate has already recommended the appointment of Mr. Martineau, and the Council has declined to appoint him, the Senate does not think it necessary to present a second report concerning him.'

1866.

It would be almost ludicrous, if it were not rather lamentable that parties in a controversy should be unable to conceive that those who differ from them may differ honestly, to see how little able the partisans of Mr. Martineau have been to take in the notion that he may have been rejected upon his merits. Various unworthy motives have been attributed to the Council; but it does not seem to have occurred to their assailants that possibly they did not think Mr. Martineau the best Professor of Philosophy that they could appoint. And yet a zealous advocate of Mr. Martineau, in expatiating upon the soundness of his philo-

1866. sophical views which ought to have recommended his appointment, points out triumphantly that they are opposed on various points to the opinions of Hobbes, and Locke, and Hume, and Paley, and Bentham, and Mill, and Austin, to say nothing of Aristotle. Now it might have occurred to this writer upon his own showing, that some thinkers might, whether rightly or wrongly, at least honestly, think Mr. Martineau's philosophy unsound; and when it is considered from what class of thinkers many of the original founders and many of the late supporters of University College have proceeded, it becomes exceedingly probable that members of the present Council did sincerely and honestly believe that his teaching would be at variance with philosophical truth.

When this explanation of the strange phenomenon was suggested, another argument was set up. We were told that 'the principle' of considering the truth or falsehood of the philosophical opinions of a teacher of philosophy 'is monstrous,' and is a kind of 'philosophical intolerance' almost as bad as religious intolerance. 'The truth of a philosophical doctrine cannot be settled with the same certainty as the truth of a proposition in Euclid. . . . . Mental philosophy is at least as valuable for the intellectual exercise it affords as for the conclusions to which it leads; and the duty of the Council is to choose a particular teacher of the subject, not because he belongs to one or other of the two great metaphysical schools, but because he is the ablest candidate that can be got.' If it were acknowledged on all hands that Mental Philosophy is a subject on which no truth or certainty has yet been arrived at, it would be the duty of the Council to appoint no teacher at all. It would be a subject most worthy of the exertions of the trained student, who still hoped against hope to arrive at truth; but it would be no study for a place of education. It continues to be a branch of education because it is believed that truth and genuine knowledge are involved in it. The directors of a place of education may be mistaken in their estimate of philosophical truth; but surely they are not to be blamed for acting conscientiously on their convictions.

But although it cannot reasonably be doubted that the decision of the Council was determined, in the case at least of some of its members, perhaps of many, by their estimate of Mr. Martineau's philosophical merits, there is ground also for believing that it was affected by a consideration of his position as a leading minister of a religious sect: not, be it carefully observed, by any consideration whether he belonged to this sect or to that, but by the fact that he was a minister of religion, and a minister who was placed in a prominent position. When it became known that he was not appointed to the vacant Professorship, the first outcry raised was that he was rejected because he was a Unitarian, with the additional imputation that he was rejected, not because the Council of University College objected to Unitarians as such, but because they sacrificed their own professed principles in timid and interested subservience to the prejudices and bigotry of the outer world. This calumny was too gross to be long maintained in face of the known characters of the gentlemen who constitute the Council. But it did its work. It was the origin of the agitation which was raised; and probably it is still believed by many persons. And not only was it the origin of the agitation, but the agitators continue to agitate as if their first assertions were true, although they know that the question to be considered is a very different one. 1866.

On August 4 a motion was submitted to the Council in the following terms:—

'That the Council consider it inconsistent with the complete religious neutrality proclaimed and adopted by University College to appoint to the Chair of Philosophy of the Mind and Logic a candidate as minister and preacher of any one among the various sects dividing the religious world.'

This motion was *negatived*; so that no one has a right to assign the principle here laid down as the ground of the subsequent action of the Council *as a body*; although, no doubt, as there was a minority who voted for the motion, it may be assumed that those individual members acted upon this principle in their subsequent votes; and it is possible that other members, who were not present in August, but were present in November, did the same.

It is seldom expedient in any deliberative body to propound a resolution in general terms, when the object is practical action in a particular case; and the Council probably judged wisely in not affirming the general proposition in their minutes: but the argument suggested is one which might be entertained and applied in the particular case, without any departure from the true principles of the College. It is important that the friends of the College should know that such, at least, is the deliberate opinion of the Senate. In the conclusion of their second report,

1866. after stating that they did not think it necessary to report a second time concerning Mr. Martineau, they added these words:—

'We wish, however, to express our opinion upon a question of principle which is supposed to be involved in the matter. If it be thought by the Council that the characteristic principle of the College, impartiality between religious sects,[1] would be violated or endangered by placing in the Chair of Mental Philosophy a prominent theologian and a leader of one school of theological thought, even though the upright and honourable character of the individual gave an assurance that he would not consciously allow his theological opinions to affect his teaching of philosophy, the Senate fully recognises the right of the Council to entertain the objection; and it is not disposed to impugn the discretion of the Council, in whatever way the question may be ultimately determined. In fact, the difficulty has been felt in the Senate as well as in the Council.'

The report including this paragraph was carried, after an adjourned debate, by a majority of 14 to 2; so that it may be fairly taken as expressing the deliberate opinion of the main body of the Professors. It will be observed that the Senate pronounced no opinion upon the case of Mr. Martineau. They desired only to recognise and uphold the perfect right of the Council to consider his ecclesiastical position before they appointed him to the Chair of Mental Philosophy.

The reproach to which the Council is now subjected is of a novel nature. It is something strange that gentlemen professing liberal opinions should make it a reproach to other

[1] This may be accepted as the semi-official declaration of the principle of the College in 1866. It is evident that if it be the right one, my husband's was wrong (see pp. 369–373).

It is also evident that if it were the *full* statement of that principle, divines of all denominations alike might have been, by the Charter, precluded from holding chairs in the College. Had this been announced in the first instance, Mr. De Morgan might possibly have still held a Professorship there, as in any other institution of moderate liberality; but he certainly would have made no sacrifices to retain it, in the idea that he was supporting the sound principle of religious equality defined by himself.

As a matter of fact, the reiteration of the statement that *no* religious qualification or disqualification could be tolerated in the College had become almost tedious.—S. E. De M.

liberals, that they are jealous of clerical influence in education.[1] Such a jealousy is the usual concomitant of religious equality. When the University of Sydney was founded, the authors of it entrusted the selection of a Principal and Professors to a committee of gentlemen in England, informing them that, in consequence of the equality of religious sects in the colony, they would prefer that none of the Professors should be a minister of any religious denomination. Their committee found it impossible to discharge their trust to their satisfaction within this limitation; and as the instruction to them was not imperative, they selected as Classical Professor and Principal the late Rev. Dr. Woolley, who was a clergyman of the Church of England. The appointment of Dr. Woolley was received at first with some suspicion and dissatisfaction; and it was not till the people of Sydney became convinced by personal knowledge of the unsectarian character of his mind that they were thoroughly reconciled to it. We by no means wish, nor would any judicious friend of our College wish, to exclude all ministers of religion from all Professorships; but we mention this instance to show that the advocacy of religious equality does not compel its advocates in all cases simply to shut their eyes and ask no questions as to the religious position of persons with whom they have to do. In the last number of the *Athenæum* there is an article upon the approaching Special Meeting of the Proprietors of University College, in which the principles of the College are thus described :—'There was a universal belief created by every kind of declaration on the part of the promoters, and fostered by an unflinching adherence, that no disqualification on religious grounds was to be tolerated, whether as to teacher or pupil. The best Professors were to be chosen, independently of their faith, and of their notoriety as followers of their faith.' *This is quite true as a general description*, and we trust that it always will be true. But the writer does not perceive that the two clauses of his last sentence may be inconsistent; and if he insists upon the literal application of this rule in all conceivable cases, he is a slave to the letter, and blind to the spirit of the principle which he advocates. The College will appoint a Professor of Anatomy, or a Professor of Latin, or a Professor of Natural 1866.

[1] According to my recollection the reproach against the Council was that they were sensitive only on the score of an *unpopular* clerical influence. It was not implied that they would *as a body* have shrunk from the appointment of a liberal Churchman.—S. E. De M.

1866. Philosophy, without any reference to their theological opinions or their ecclesiastical denomination, because their theology or their denomination will in no way affect their teaching of anatomy, or Latin, or Natural Philosophy. But the College may reasonably and consistently think twice before appointing a professional theologian Professor of Mental Philosophy; because it is not only possible, but not unlikely, that his professional theology will make him a worse teacher of mental philosophy. We will borrow the language of one of the assailants of the Council. 'Such is the nature of speculative inquiry, that of necessity it brings into view the truths of Revelation, and forces up the question whether they agree with the principles which the thinker has reached. If his mind have the logical grasp requisite for the profoundest of all studies, he will not be content with laying down certain doctrines regarding cognition, but will follow their issues through the windings of thought, till they come into contact with theology itself.' True: and if his theology be really his own, good; but if his theology be a foregone conclusion—something which he has accepted independently of philosophical investigation,—if he is trammelled by connexion with a party or a denomination, if he has an ecclesiastical position to maintain, or a theological reputation to lose, there is no small chance that his philosophy will be, or has been, modified by contact with his theology. We put the case generally; but there is a strong *à priori* probability that a layman will be a more unprejudiced, and therefore a better teacher of mental philosophy than a minister of religion; and, if so, it is no part of the duty of the Council to ignore the distinction.

If the general principle be sound, there is no force in the objection that our late Professor, Dr. Hoppus, was a minister of religion. If the general rule be a safe one, the fact that no harm followed from one departure from it[1] is no argument for lightly departing from it a second time; and still less is it an

[1] The suggestion that the election of Dr. Hoppus was at the time regarded as a departure from a general rule, shows that the writer of this passage had not been connected with the College from its foundation. The elections of two clergymen of the Church of England, among the first appointed, were looked on rather as affirmations of a distinct principle than as departures from an unexpressed 'general rule.'

When Dr. Hoppus's name first appeared in the list of Professors, a remark was made in my hearing on his being a minister of religion. The reply, given by an influential member of the institution, was, 'We do not consider these things.'—S. E. De M.

argument to debar the Council from their right to consider all circumstances likely to affect a Professor's teaching. 1866.

We are assured on very good authority, that when the first appointment to the Professorship of Mental Philosophy was to be made in the College, there were two instances of eminent ministers who thought of offering themselves as candidates for the office, but were withheld by the conviction at which they themselves arrived, after further consideration, that it would not be consistent with the duty of the Council to appoint them.[1]

The preceding argument is perfectly general; but in the particular case of Mr. Martineau there is one point of another kind which deserves consideration. No one will question that the authorities of our College are bound by the strongest obligations to avoid carefully any act by which they would induce our students, or even put facilities in their way, to submit themselves to one form of religious teaching rather than another. Mr. Martineau is described as 'Professor of Mental, Moral, and Religious Philosophy' in Manchester New College; and we have every reason to believe that he discharges faithfully, zealously, and ably the duties of his office. In his letter of application for the Professorship he stated that, if he were appointed, he should transfer his lectures on logic and mental philosophy entirely to University College. Moral philosophy is not included in the duties of our Professorship. We may fairly conclude, therefore, that, if Mr. Martineau had been appointed, he would have lectured on logic and mental philosophy in University College, and would have continued to lecture on moral and *religious* philosophy in the institution called Manchester New College, all the business of which is done in the building of University Hall, which is immediately contiguous to the College. A very natural consequence would have been, that students who were attracted by his lectures in the College would have been led to attend the further part of his course in the Hall. At present, if any students seek the instruction of Mr. Martineau, they are at perfect liberty to do so, but the College is in no way responsible.

The Council has been attacked for having proceeded to elect Mr. Croom Robertson to the vacant chair at the same meeting at which they received the requisition for a Special General Meeting, without waiting for the answer to their consultation of

[1] This statement would have had more force if substantiated by particulars. According to my recollection these gentlemen had other reasons for not coming forward.—S. E. De M.

1866. the high law officers whether a meeting could legally be convened for the purpose named in the requisition, and, if it could, without waiting for the result of it. In the first place, the Council had a perfect right to do this. By the very charter of the College the power of appointing Professors is committed to the Council, and not to a General Meeting; manifestly because the Council is a body capable of judicial action, and a General Meeting is a popular body, and not a fit instrument for such a purpose. The Council is elected by the proprietors to discharge this and other duties; but they are not mere delegates of the proprietors, and it has never been considered that they were bound to receive instructions from them. No doubt the General Meeting may express an opinion upon the acts of the Council; but the duty of the Council was to do what they thought best for the College, and then to await the judgment of their constituents. If the Council had delayed to act, for the purpose of deferring to a vote of the meeting, they would have abdicated their proper function.

Our sole purpose is to uphold the legitimate authority of the Council, and to vindicate their right (and, indeed, we might insist upon it as their duty) to consider all circumstances which make a candidate more or less fit to discharge the duties of the office which he seeks. When the first report of the Senate was drawn up, the Senate named Mr. Martineau as the best candidate; and although the objection presented by his ministerial position was discussed, it was not insisted upon, and therefore it was not mentioned in the report. When the second report was framed, which was rendered necessary by the reception of additional evidence, the relative position of Mr. Martineau and Mr. Robertson was certainly not left what it was before; and, moreover, the attacks upon the College made it incumbent upon us distinctly to recognise in the Council, and members of the Council, the right to consider all points in the position of Mr. Martineau, or any other candidate, which seemed to them likely to affect his fitness for the Professorship. We learn now, on what we must consider good authority (the *Spectator* of January 26), that the issue to be submitted to the meeting is substantially the same as the proposition of the writer in the *Athenæum*, which we have discussed above. The resolution announced runs thus:—'That in the opinion of this meeting any candidate, who is otherwise the most eligible for any chair or other office in this College or the School, ought not to be regarded as in any manner disqualified for such office because he is also eminent as a minister or preacher of any religious church or sect.' The writer in the

Athenæum, and the framer of the resolution, under cover of a specious general proposition, thrust out of sight the certain fact that there are some departments of human knowledge, and mental philosophy is eminently one of them, in which the prepossessions of a theologian and the habits of a theological teacher may make him a wòrse qualified candidate than another man; and in such a case those who have to appoint the teacher are bound to take cognisance of the fact, and it is very unwise to fetter their discretion. 1866.

We pass no judgment on the particular case of Mr. Martineau. We are quite ready to assent to the general proposition, that no candidate ought to be regarded as *ipso facto* disqualified because he is a minister or preacher. But we desire to maintain the right of the Council to examine all the circumstances of every case that comes before them; and we earnestly entreat every true friend of the College not to concur in any vote which would seem to inflict a censure upon the Council for the legitimate exercise of their discretion; and, above all, not to concur in any vote which would impose a restraint upon their freedom of judgment in future.

This was signed by fifteen Professors, of both Faculties.

On reading this document, my husband said the principal part of the question was left out altogether; for had he ever understood that the profession of religious impartiality made by the founders of the College was only to be understood 'as a general description,' his name would never have been connected with it. He drew a distinction between the part taken by the older Professors, who, from their long connection with University College, could not fail to know that its very life consisted in the entire rejection of all religious distinctions, and that of those more recently appointed, who, he thought, might and probably did believe that the Council was not bound by any condition except that of making the appointment which might seem to them best for the worldly prosperity of the institution. From this latter point of view, it is not difficult to understand why a candidate believed to be prominent in an unpopular sect should have met with disfavour in the eyes of the Council.

The special meeting of Proprietors was held early in

1867. February 1867, and a few weeks later the annual meeting took place. On both these occasions those who contended for the old principle were beaten, and the College proceeded to work in its new character. Whether it is held in higher estimation since then I have no means of knowing. I only know that it is changed since I heard the conversations of Lord Brougham, Thomas Campbell, and my father at the time of its foundation.

My husband told me that during the session in which he worked after his resignation was sent in he met his colleagues as before in the Professors' room. Not one of them ever spoke on the subject of his retirement, and he left the place without one word of acknowledgment for all he had done for it. Only once, after the end of the session, he paid a visit to his old lecture-room. He went to bring away the note-books and manuscripts which he had used in his lectures. The visit was a painful one, but was so cheerfully borne that I should hardly have known all he felt if I had not said something to the effect that I hoped he would not suffer for the trial. He said, 'Oh, I shall do very well. I felt all the time to-day that the College had left me, not I it. It was no longer the old place. But then,' he added sorrowfully, 'all my thirty[1] years' work has been thrown away.' The answer to this was of course easy. I said that no such efforts as his could ever be without result; that his teaching had trained many strong and honest minds; and that if the College had done nothing more, its establishment might have helped in the opening of the two Universities. He acquiesced, but the blow was struck.

In the spring of 1867, after the efforts of many of the best of his old pupils and friends to retrace the false step had failed, some of these gentlemen, desirous that he should not leave the scene of his work without taking

[1] It will be observed that sometimes more than thirty years are mentioned. He was Professor from 1828 to 1833, and again from 1836 to 1866—in all more than thirty-five years.

with him some memorial of their respect and friendship, asked me whether I thought he would refuse a testimonial in money.[1] I answered without consulting him. He had strongly objected to the system of testimonials, which of late years had grown to such a height, and I was quite certain that his answer would be in the negative. He soon after received the following, enclosed in a letter from our friend Mr. Jacob Waley:— 1867.

'May 7, 1867.

'DEAR SIR,—Many of your old pupils, at whose request we address you, desire, upon your resignation of a chair which for upwards of thirty years you have filled with so much distinction, to give some appropriate expression to the high estimation in which they hold you. Proposed memorial bust.

'Our admiration for your philosophical views of education, your skill in the art of instruction, and your scientific attainments, as well as our cordial regard and esteem for you as our old teacher and friend, render us desirous of recording these feelings in some substantial shape.

'We understand, however, that you feel you cannot consistently accept any testimonial of intrinsic value. But we hope that you may be persuaded to gratify your pupils by sitting for a picture or bust to be placed in the library of our old College. We remain, sir,

'Yours faithfully,

'JACOB WALEY. H. M. BOMPAS.
W. A. CASE. R. B. CLIFTON.
J. G. GREENWOOD. J. M. SOLOMON.
G. JESSEL. H. COZENS HARDY.
RICHARD HOLT HUTTON. THEODORE WATERHOUSE.'
WALTER BAGEHOT.

[1] The year before his death several old pupils and friends kindly obtained for him a pension of 100*l*. from Government. On hearing of this his first impulse was to decline it with thanks. I entreated him to receive the kindness as it was meant.

1867. It gave a pain to my husband to refuse a request so kindly and cordially made. His reply was as follows:—

'MY DEAR WALEY,—I acknowledge your kind letter of the 7th with the cordial and gratifying inclosure, signed by eleven old pupils, whose dates represent the time which has elapsed since I rejoined the College in 1836.

'The inclosure is in itself a testimonial. It has all the meaning and all the value. And to those who hold that the mind of the teacher counts for something in the making of the pupil, the string of names appended to it will be no mean presumption that I have in some degree a claim to the terms in which I am described.

'I am asked to sit for a bust or picture, to be placed in what is described as "our old College." This location is impossible; our old College no longer exists. It was annihilated in November last.

'The old College to which I was so many years attached by office, by principle, and by liking, had its being, lived, and moved in the refusal of *all religious disqualifications.* Life and soul are now extinct.

'I will avoid detail. I may be writing to some who think that the recent transaction is a reparable dilapidation, or even to some who approve of it. To me the College is like a Rupert's drop with a little bit pinched off the small end; that is, a heap of dust.

'I can never forget that I have been usefully employed, though I now wish my life had been passed in any other institution. I have worked under the conviction that I was advancing a noble cause, until every letter in the sentence "Augustus De Morgan, Professor of Mathematics in University College, London," stands for 234 hours of actual lecturing, independent of all study and preparation; and all this under a banner which is now shown to have been either shamfully raised or shamefully deserted.

'So much is necessary that my old pupils may understand my mind, and the repugnance I feel towards any proceeding which must record my connection with University College. I am happy to say that the circumstances have not created any personal bitterness of feeling; individuals are to me what they were before. But if force of will can succeed, the institution is to pass away from before my mind, and to become as if it had never existed. 1867.

'You will see that I am altogether averse to lending aid or countenance to any scheme which will tend to remind others that I was a teacher in the College which did homage to the evil it was created to oppose.

'But I am even more sensible to my old pupils' remembrance than I should have been if I could have accepted the result of their most acceptable good opinion. Such remembrance would have been, in any case, a treasure. It has now the additional value of a treasure saved out of the fire.

'You will, of course, communicate my answer, and with my warmest thanks and most heartfelt regards,

'I am, my dear Waley,

'Yours sincerely,

'A. De Morgan.'[1]

He often spoke with satisfaction of the uninterrupted friendly relations which had for thirty years subsisted between himself and his colleagues. From his declining health and other circumstances he saw but little of them latterly, but this was in no case due (on his part at least) to personal feeling created by the question which had caused his withdrawal.

Crabb Robinson.

One of his social pleasures during the last few years had been in the acquaintance and friendship of Mr. H. Crabb Robinson, one of the first active promoters of the establishment of University College. Through a life of nearly ninety-one years Mr. Robinson had been the steady friend of

[1] This letter was printed after Mr. De Morgan's death for circulation among friends who had been asked to join in an injudicious attempt to found a scholarship under his name in University College.

1867. Mr. Crabb Robinson.

freedom and progress, but his influence, which was considerable, had been felt chiefly in his conversation and social intercourse with other minds, for his writings were few and comparatively unimportant. In December 1866 he had voted in the minority in the Council on the question of Mr. Martineau's appointment, and on the next meeting, when the cause was lost by a majority of one, the chairman giving the casting vote, Mr. Robinson was absent from illness. This, and the adaptation of principle which afterwards ensued, was a cause of great concern to him. During the winter of 1865–66 Mr. De Morgan helped him in the task of arranging and sorting his books, a miscellaneous but very valuable collection. My husband, who was interested in the work, said that it was a very slow process, because every book or pamphlet looked at gave occasion for some literary or historical anecdote, and this sort of gossip was pleasant to his hearer, who knew much of books and of men; for Mr. Robinson had been the contemporary of all—the friend of many—of the eminent, political, and literary characters whose life and work made the history of the end of the eighteenth and much of the nineteenth centuries. He had been the friend of Goethe, Wordsworth, and Coleridge. He remembered and knew the political trials of Horne Tooke and his friends, and told me incidents connected with my father's trial at Cambridge of which I had never heard.[1] His Sunday morning breakfasts were, I suppose, occasions of much pleasant intercourse among many intellectual men of various opinions. At these my husband used to meet the Rev. F. D. Maurice, the Rev. J. J. Tayler, the Rev. J. Martineau, and many others; and it was at these parties of friends that his acquaintance with Mr. Martineau was chiefly formed.

Early in 1867, shortly after the trouble at the College, the kind-hearted, consistent old man left this world. Mr.

[1] I first saw Mr. C. Robinson at Mrs. Barbauld's. I was then twelve years old.

De Morgan visited him daily, and saw the day before he died that his end would be without pain. In the *Diary and Memoirs*, published by Dr. Sadleir, a little sketch by Mr. De Morgan gives a portrait of the subject, which shows how actively his mind was still at work, and his interests alive to the last. 1867.

The last work of any importance undertaken by my husband was a large calculation, I think, for the Alliance Assurance Company. But his health had begun to fail. Every one who saw him observed the change which had passed over him, and before the great sorrow which came to us at the end of 1867 he was no longer the strong, vigorous man, full of hope and activity, which he had been before his alienation from the institution to which so much of the work of his life had been devoted.

George's death.

In October our dear son George was taken from us. He had worked hard during the winter, and even late into the spring, both in giving lessons and in examining the papers for the degrees of the University of London. He was at that time Vice-Principal of University Hall, Gordon Square, but was almost every day with us in Adelaide Road. His father, who only saw his cheerfulness and the seeming improvement in his health when, after a short time at Herne Bay, he parted from him to join us at Bognor, did not realise his state, and hoped against hope to the last. George went on with one of his sisters and myself to Ventnor. He was still warmly interested in the success of the Mathematical Society. As his name belonged to it as one of the secretaries, his father was anxious that, if possible, the first diploma given should have his signature. For this purpose parchment was placed before him, and he evidently recognised its import, passing his finger over the words Mathematical Society. But he was too weak to hold the pen, and died two days after. His father, already enfeebled in health, had been at home with two of our daughters, and could not come in time to see him while he would have been recognised.

1868. He bore the blow as bravely and firmly as he had borne other trials, expressing his full confidence in the wisdom which had removed from among us one who seemed intended to tread in his own steps on earth. But another trial awaited us in the illness of our third daughter. I had to leave home with her for several weeks, and she recovered so far as to remove much anxiety. In the year 1868 my husband's own health, which had continued steadily to decline, broke down entirely. A sharp attack of congestion of the brain, the result of so much intense mental suffering, left him so prostrated that it was evident he never again would be equal to sustained effort.

We moved in the summer to 6 Merton Road, near Primrose Hill, a house which he said was the most comfortable he had been in since our marriage. We dreaded this moving on account of his weak state, but all was ready to receive him, and he did not suffer.

In his enfeebled condition the task of placing his books was a heavy one. The room destined for them was much smaller than the one he had had in Adelaide Road, which he said was a palace. It was a work of time for him to measure the walls, and to direct the placing of the new shelves, but it was done, with intervals of rest. A large number of the books had been sold, but about 3,000 remained, and I feared he could not get them all in, and of course begged him to have help. He said, with his old spirit, 'They *shall* all go in, and I will put them all in myself;' and so he did. The work was done gradually, and I do not think it hurt him. He always liked looking through his treasures, and showing to any friend any special rarity.

During the last few years of his life my husband occupied himself a good deal in reading the Greek Testament, and comparing the different versions and translations. I regret much that many comments which he made on this subject were not preserved, as he did not write them. He also compiled a sort of history of his family and biography of himself—not in a connected form—to be left as materials for his Life, and from this book I have taken much of the earlier part of this Memoir. He also rearranged and added to his *Budget of Paradoxes*, which, however, was not published till after his death. 1869.

Free Christian Union.

One of the last subjects which afforded him interest was the proposed formation of a society to be called the Free Christian Union. The idea, a beautiful and attractive one, was the formation of a union for the promotion of good in various directions of men of all religious beliefs and opinions, on the common ground of the Fatherhood of God and the brotherhood of man. But there was some inaccuracy in the designation, for under the simple, universal principle, professed Jews, Hindoos, and other Easterns were eligible as members; indeed, my husband said Christ himself and the Apostles might belong to the society, which they could not perhaps have done to many associations taking the name of Christian. Either the designation or the conditions of membership must be abandoned; and on the former proposal, several persons of well-defined orthodox opinions left it. Mr. De Morgan hesitated before giving his name, feeling that in the present uncertain and unsettled state of opinion among the best meaning persons a union based upon anything but absolute and simple theism was impossible. This would exclude the use of the word Christian, leaving a common basis of belief so broad as not to satisfy men of deep religious thought, while it would not admit Comtists and others whose philanthropic views and desires to benefit mankind were as wide and earnest as those of the founders themselves.

He also desired to learn to what the designation 'Christian' applied—what were the opinions of the founders with reference to the work and mission of Christ. The writings of some of these, friends whom he valued and respected, had led him to suspect that in their view what is called the supernatural element in the Gospels, the account of the miracles and resurrection of Christ, were

1869. due either to the exaggerations of Eastern fancy and expression, or to the interpolations of superstitious times. My husband, who believed fully in the account of the resurrection of Christ as given in the Gospels, wished to ascertain the views of those who held what are called 'advanced' opinions on this head. He wrote and inquired, but told me he could not make out what their ideas were. I once said to him that I thought one element in the question had been generally overlooked, the 'opening of the (spiritual) eyes' of the witnesses, as mentioned in the Gospels on other occasions. This would give some apparent subjectivity to the fact, but it is nowhere stated that *all* present saw the rising of Christ. He said, 'Very possibly, but there *was* a rising; the history is clearly given and well attested, and the rejection of it would be to cut away the root from the tree. And the accounts given of this and the other miracles cannot be taken from the history without throwing a discredit on the narrators' character that would make all their statements worthless. They say,' he said of the Rationalistic school of interpreters, 'that it is the character of Christ that commands reverence, and proves his mission from God. You cannot separate the two. He himself claimed extra-natural powers, given by the Father. If this was false He was false, and His character would not have been what it was; and the men who could invent fictions about His works could not have described the character as they did.' It was with reference to this society that we spoke of public prayer. In his letters on Christian union he speaks of a basis on which people might meet and pray together. He had always said to me that Jesus Christ had not enjoined public prayer; and though He had not forbidden it, the tenor of His teaching was strongly in favour of privacy and seclusion in this most internal and sacred communion. 'Enter into thy closet, and when thou hast shut thy door, pray to thy Father which is in secret,' &c. But though he felt this strongly himself, he knew that all did not feel with him.

He himself felt the happiness of prayer, but he said, 'I regard it rather as a luxury than a duty.' 1870.

In reference to the vision of the apostles I may mention that he had always been interested in cases of the kind, especially those in which departing persons, while fully conscious, assert the presence of those who have gone before. Such instances, he said, were so common that one could not believe them to be all illusion; but whatever they were, they should be recorded carefully.

Death of our daughter Christiana.

In August 1870, seven months before his own release, our daughter Christiana was taken. She had stayed at Bournemouth on her return from Madeira, and died there. I came home the day after her death to find her father so weak that he had that day fallen on the floor, and was unable to rise without help.

From this time the decline in his health was very apparent, but he did not seem to suffer, except from weakness and sleeplessness. The physical state was a complicated one, chiefly owing to nervous prostration, and traceable in the first instance to the shock of the College disappointment, and afterwards to anxiety and sorrow on our children's account.

In March 1871 he became still weaker, and talked very little. The only word I remember relating to his own state was, after saying that any way all would be right, 'But I shall be glad when I have got it over.' When I expressed a hope that he would not be taken yet, he told me to 'leave it all in God's hands,' and he then waited quietly for the end. 1871.

During the last two days of his life there were indications of his passing through the experience which he had himself considered worthy of investigation and of record. He seemed to recognise all those of his family whom he had lost—his three children, his mother and sister, whom he greeted, naming them in the reverse order to that in which they left this world. No one seeing him at that moment could doubt that what he seemed to perceive,

1871. was, to him at least, visible and real. After this he said very little, only on the last morning of his life asking me, as he had been used to do, 'if it was time to get up.' On being told that it would soon be, he seemed to be carefully dressing himself. Then he lay quite still till, just after midnight, he breathed his last. The state of mind in which he had lived, and in which he died, is shown by a sentence in his will:—

I commend my future with hope and confidence to Almighty God; to God the Father of our Lord Jesus Christ, whom I believe in my heart to be the Son of God, but whom I have not confessed with my lips, because in my time such confession has always been the way up in the world.

SECTION XI.

CORRESPONDENCE, 1867–70.

To Sir John Herschel.

91 Adelaide Road, March 25, 1867.

MY DEAR SIR JOHN,—'How do you bear this trying weather?' asked a friend during the late *froid d'enfer*, as a Frenchman might say. 'Trying weather!' said I; 'convicting weather! sentencing weather! penal servitude weather!' 1867.

The question between me and the College is simple. I entered that College on what all the world knows was its loudly vaunted principle, that the creed of neither teacher nor student was to be an element of his competence to teach or to learn. After forty years of existence the College, for worldly reasons, has decided that a teacher must not be too well known to be heterodox: he must not be conspicuous as a Unitarian. Breach of faith, surrender of principle, and D. I. O.

But this is not all. Between ourselves, the candidate who has been refused the chair of Mental Philosophy because he is so very wicked a Christian in religion, is also excluded because he is *too much of a theist* in philosophy. He cannot help founding his psychology on a moral Governor of the universe.

Now, I would not have objected to leaving the existence of God and His action on the minds of men an open question for the best qualified candidate to treat in his own way; but the interference of the College as a college, and a settlement of that question *officially*, is a step in which it concerns me, with my way of thinking, to take a part. The public knows nothing about this view of the question, but the Council have been roundly charged with it by one of themselves in debate, and by me in my D. I. O. I have told them *totidem verbis* that they had acted from *fear of God* in philosophy and *fear of man* in religion. I am only here till the end of the current session. . . .

Yours sincerely,

A. DE MORGAN.

To J. S. Mill, Esq., M.P.

July 3, 1867.

1867. My dear Sir,—A person described by you as a remarkable working man, and your correspondent, is one whose case is more than usually worth looking into. He had better write to me direct, and state in some detail what he knows, and, as well as he can, what he wants. I dare say I shall be able to shorten his route. He must specify arithmetic, his knowledge and habits, geometry, algebra, physics, if any. You need not tell *him* that the glimpse I shall get of his mind is one of my data. I hope you are lifelike in spite of Reform debates. 'Confound this rope!' said the Irishman who was hauling in the slack, 'sure somebody has cut off the other end of it!' Do you not begin to suspect that somebody has stolen the third reading?

Yours very truly,

A. De Morgan.

To J. S. Mill, Esq., M.P.

91 Adelaide Road, August 2, 1867.

My dear Sir,—As touching your proposal to me to join the committee of the National Society for Women's Suffrage, I cannot accede. I never join political agitations, or associations for procuring changes in the political machine. I remember signing a petition which, as I understood it, was for franchise to be granted to single women having the property qualification. Your Society, as its title is worded, contemplates a full female suffrage—*e.g.* a vote for a man and another for his wife. Supposing me willing to join a political agitation, I should hardly be ready for such a one as this. I should think better of two votes given to the couple jointly—*i.e.* the two to agree upon the two. I almost thought this was the meaning of the phrase 'compound householder,' when I first heard people mention it.[1] I got as far as joining the Decimal Coinage, but this was for the

[1] I cannot help thinking my husband wrote this for the sake of playing on the expression 'compound householder,' as he can scarcely have missed seeing that the result of one vote to each of two people has the exact effect of two votes to both if they agree, except only in the case of one of the two not voting at all.—S. E. De M.

sake of education, and the furtherance of arithmetic among the labouring classes. 1867.

Yours very truly,
A. DE MORGAN.

To Sir John Herschel.

91 Adelaide Road, August 8, 1867.

MY DEAR SIR JOHN,—Many thanks for the Latin original of Schiller's poem.[1] I always take the older language as being the original. I see Spaziergang—in my dictionary it is *Spatziergang* in the German-English and *Spaziergang* in the English-German—means a walk. Now, I have not taken a *walk* for many a year. Had I done so, I might have started off something at this kind of pace:—

Good morning, Mr. Mountain, with the light upon your top,
Just the rubbing of Apollo's eyes before he opens shop;
And you, you 'daisy-spangled meads,' and you 'resounding grove,'
Where the feathered songsters make a row; I'd feather 'em, by Jove, &c.

But *seniores priores*: Schiller has the start. I hope your hexes and pents show that you are in a good condition. I see you don't care for the dissyllabic ending. No more did I when I was at school; and I was reprimanded, which I should not have cared for; but I was then remanded to set it right, and this was a bore.

Of all the verses I made at school, I only remember one couplet. The subject was poetic inspiration, and was very classically intended. It pleased me to take it that *dinner* was the thing, and I have always been inclined to support my thesis. The pair of lines I remember is—

Gustat Virgilius—procul o procul este Camoenæ
Conclamat vates—hoc mihi numen erit—

If you look into the *Athenæum* of Saturday week, you may chance to see a little account of the last mare's-nest at Paris—the discovery that Pascal preceded Newton in the theory of gravitation. The letters, if genuine, prove nothing. $\frac{m}{p\,2}$ was guessed to be the law before Newton or Pascal by Bouillaud. But the funny point is that Pascal is made to talk of a *tasse de café*, years

[1] A Latin translation from Schiller by Sir John.

1867. and years before coffee was known in Paris, so far as has been stated. The first coffee-shop was really started nine years after Blaise's death by a man named *Pascal*, and the first started in England was in 1652 by *Pasqua*, a Greek. In England in 1657, 'Coffa—see *Caupha*,' is in Phillips's dictionary, and of 'Cauphe' it is said, 'it is much used in these parts.'

I have nothing to say about myself or my people. The world wags as usual. I should like to hear of you. I have just found out that Dr. Pearson began life as a junior partner in Sketchley & Pearson, who kept a school at Fulham for boys from four to ten. Here he had been for some years in 1800. I picked up a sensibly written prospectus—they said *plan* then—of this establishment. He founded his great school at Sheen in 1811. I make out that he was not a graduate of Oxford or Cambridge. He was the undoubted projector of the Astronomical Society, and his *dinner* there again set it going. Our kind regards to Lady H , &c.

Yours sincerely,

A. De Morgan.

To Sir John Herschel.

91 Adelaide Road, August 15, 1867.

My dear Sir John,—A country clergyman (I think: how is one to address where the party is a suspected clergyman? May one put down the probabilities as

Rev. [$\frac{5}{6}$] John Smith, Esq. [$\frac{1}{6}$]?)

has written to me to know why, in the Runic almanacs, or very many of them, the days of the week begin from Monday. I could only suspect that their almanac makers had the notion I had when a child. They told me the week had seven days, and that the seventh day was to be kept holy; and they kept Sunday holy, and called it Sabbath. So I thought Monday was the first day, and I well remember taking it as of course that the day the women went to the sepulchre was Monday.

But this puzzled me more. They always made the 'Scriptures,' when mentioned in the New Testament, mean Old and New both. So when I saw (Acts) that the Bereans searched the Scriptures, I thought they would find Bereans who searched,

and who found Bereans who searched, &c., and *ad infinitum*. 1867.
Query: Is this a convergent or a divergent series?

Hard rain—great relief.

Yours sincerely,
A. DE MORGAN.

To Rev. Wm. Heald.

August 20, 1867.

MY DEAR HEALD,—My bit of news is my retirement from University College, after two terms of service, 1828–31 and 1836–67. The world knows, or takes note of, only one side of the cause. I was meditating retirement in a session or two on account of the general decadence of the College, which made the emoluments wholly out of reasonable proportion to the time the duties took. In November last came a course of conduct which made me glad to escape at the end of the current session.

University College, as you know, was founded on the principle of giving secular education without reference to religion, which was left to the parents. The best men who could be found independently of creed, being of good fame and conduct, were to teach all who were willing to be taught equally without reference to creed. From this principle there was never a departure. At the very outset, indeed, there was a circumstance of this kind. Dr. Southwood Smith was proposed for a chair of mental philosophy, with some mixture of moral philosophy. Zachary Macaulay read extracts from a work of his, I think with the name 'On the Divine Government,' so heterodox that he, Z. M., declared he would take no further part if S. S. were chaired. The Council gave way. But, for other reasons, I fancy, no one ever complained of infraction of principle, and the case made no noise.

The principle was put to a very severe test when Francis Newman, then actually Professor of Latin, published his *Phases of Faith* —an attack, it was said, on Christianity. No one proposed that he should be called to account for a work the title of which did not state his connection with the College. So it seemed pretty certain that the College would always hold to its declared principle of perfect indifference to the creed of a teacher.

The Professor of Logic and Mental Philosophy, Dr. Hoppus, resigned in the spring of 1866. The best candidate beyond a doubt to succeed him was Mr. James Martineau, a leading man among the Unitarians, but not thoroughly in accordance with

1867. the bulk of them. And thereby hangs a tale. The Unitarians in general are highly intellectual, but have a practical dislike of the *spiritual*. Mr. Martineau is a strong spiritualist, not merely in religion, but in psychology. He neither can nor will teach psychology, the structure and action of the mind, without a distinct recognition not merely of *a* God, but of God and His action upon the minds of men. This he teaches in his lectures at University Hall—an unconnected appendage to University College, in which Manchester New College is located, the students of which attend lectures on secular subjects at University College, and learn Theology at the Hall.

The Senate of University College (*i.e.* the Professors) reported that Mr. Martineau was the best candidate. The Council (which is composed of a small body of *philosophical* men, whose creed no one knows; a larger body, perhaps one-third of the whole, of Unitarians, and a full half of the miscellaneous Churchmen, &c., whom one finds making up the mass of all public bodies) rejected him on the ground that a very distinguished Theologian, no matter of what sort, would injure the College, as giving an appearance of breach of its neutrality. This influenced many, but all the world knew that it was his sect being *Unitarian* that was objected to, and fear of the unpopularity of the Unitarian doctrine was of considerable effect. But it was very well known in the College that the *philosophical* party was only making a tool of the anti-Unitarian party. Their objection was to Mr. M.'s theism in psychology. There is a school of philosophers who cultivate what they call *sensational* philosophy. They are driving at the doctrine that thought is a secretion of matter, and they want to get rid of all but matter and its consequences. . . . The fact then was, as I told the Council in my letter of resignation, in these words, that Mr. M. was rejected because he was too far from orthodoxy to please the priests, and too far from atheism to please the philosophers; that he was offered up to the Janus Bifrons of expediency, each member of the majority of the Council choosing the head of the idol to which his offering was to be made.

To myself, who never will have anything to do either with religious exclusion or with atheism, the proceeding was a call to resign, which I immediately obeyed. I knew it to be an abandonment of the principle of the College done in the worst way; a pretence of fearing heterodoxy, with the fear on the minds of the leaders of nothing but theism. Not a soul of the Council or of the Professors has ventured to deny to me the truth of what I told them. I did not publish my letter of resignation because I did not wish to hurt my old friends in the College. If they can get a mess of pottage for their birthright I should be very glad they had it. I resigned in November, but remained until the end of the session, of course. All that time *not one* of the old hands among my colleagues made the slightest allusion to the fact of my resignation, and I am sure they felt that they had better let the subject alone. Two of the younger ones, indeed, undertook to instruct me that I was wrong about the principle of the College, which I had studied before they were born. 1867.

I believe, from observation, that both in colleges which profess exclusion, and in those which profess perfect neutrality, there is a concealed under-current of, let us say, *philosophy*, veiled under formalism in one case and toleration in another. Get your smelling-salts ready, for I am going to tell you that a certain section of your order are in earnest about nothing but the endowments. These men see danger in all but formal adherence to religion, and would rather have a world of concealed philosophers than one of earnest believers in actual Providence and guidance. This, I say, I glean from observation; there are easy means of verifying what I say. Of course, the neutral places have their share of this. But a place like University College—and not alone—has its share of the philosophers who are really earnest about their system—*religious* atheists—this phrase comes nearer than anything else; men who believe that the moral ends of the universe, so far as there are any, are better answered by their concoction of reasoned right and wrong than by any reliance on higher government. To one or two perhaps of these men University College is indebted for its rise or fall, whichever it shall turn out to be. I hope it will be a rise, for they may as well have the profits of duplicity as others.

Perhaps I shall never write as much about University College in all my life to come.

I am no way surprised at the money part of the business—I mean the fear of diminution of pupils from a distinguished Unitarian. Twelve years ago or more a Mr. Peene left about 1,500*l.* to be a fund for buying books, and to be selected by Professors of Latin and Mathematics, *being members of the Establishment.* To my surprise they caught at the money. But they did not venture to acknowledge openly what they were doing. They did not, as they ought to have done, write to me to know

1867. if I could take the office. The secretary said (without application to me) that he knew I would not take it. The other Professors declined on principle, though they could have held it, and some old students were appointed. But in time the other Professors came into it quietly, without my being told. I smiled and shut my eyes.

I have got over all the disgust of the matter, and am fast losing the remembrance of the place. To my surprise I went to my publisher's to-day, opposite the College, and was there half an hour. When I had come away I remembered that I had quite forgotten the College, and had not any recollection of having seen it.

I suppose you have your share of the ritualism in Leeds. I hear of it from my girls, who sometimes go to one of the show-places. I read a book of essays about it a month or two ago. There is the Roman system complete; doctrine quite full on all points; a strong aspiration for the time when men will be *prevented* from undermining the orthodoxy of their neighbours, which can mean nothing but penalties for expression of heterodox opinion in private life. We are getting on in every point, the matter is coming to a crisis, and the hierarchy has no courage; one would suppose they back the winner.

Neate is with his family at Baden, or some such place. I have hardly heard of Mason since he changed his living. I believe he finds the change has done him much good. I have lately renewed acquaintance with a man who was an infant when I last saw him—Samuel S. Greathead, of Trinity College. He is the nephew of S. Maitland, and is rector of Corringham, near Romford. He tells me that a large quantity of Maitland's papers were destroyed by an ignorant executor.

Yours sincerely,

A. De Morgan.

To Sir John Herschel.

91 Adelaide Road, N.W., October 18, 1867.

My dear Sir John,—Be no way prepared to start when you see no black border; it is a thing which I never will use. Black is not my colour of death. I followed to Kensal Green yesterday the remains of my second son, George, who died of phthisis in the throat on Monday at Ventnor, after three years of alarm as to his lungs. A voyage to the West Indies two years (+) ago appeared to restore the lungs, but the laryngeal affection came on slowly, and ended this world for him without any great suffering, he being perfectly satisfied until within forty-eight hours—in fact, until he began to wander—that he was as strong and well as need be—the phrase he used when he could not rise in his bed. 1867.

I bear it well, and so does my wife. Many condoling friends have found out that the great and special force of the blow is that he was the son who was to follow in my footsteps, and had made some beginning. To which I assent; but, truth to speak, I did not remember this until I was told, nor did it produce any effect. I am peculiar, I suppose. I remember with satisfaction that he and a young fellow-student were the projectors of the Mathematical Society, which seems to have taken firm root; but this is only the general love of memorial which belongs to our nature. Any other instance would do as well. A strong and practical conviction of a better and higher existence does much better for every purpose, and reduces the whole thing to emigration to a country from which there is no way back, and no mail packets, with a certainty of following at a time to be arranged in a better way than I could do it.

Our kind regards to Lady Herschel and the family. You have been through the same Valley of the Shadow, and know all about it.

Yours sincerely,

A. De Morgan.

To Miss Sheepshanks.

91 Adelaide Road, July 23, 1868.

My dear Miss Sheepshanks,—All are at Esher, and I have sent your letter down there. 1868.

. . . . The Bishops of Oxford and Cape Town are a pair of opposites: C. T. foolish, and believed to be sincere; O. sharp, and suspected of a sort of slyness. Colenso won my good opinion before he became a heretic—when he would have got it of course—by showing that he understood one part of the New Testament which it is a rule to hide under the cushion. What did the Gentile and Jewish converts do who had several wives? Did they break their contract with all but one? There would have been a rule laid down if they had, and a controversy. They kept their wives, St. Paul ordaining that only the husband of *one* wife should be a

1868. bishop or a deacon. This is the meaning of the regulation. Now Colenso saw this, and did *not* require his converts to reduce themselves to one wife. Not much noise was made about it, for the divines at home did not like to raise the question.

Colenso is founding a church in Natal with success. He has both a clerical and a lay body of followers. There will be curious consequences, which will find their way home. One Pope at Rome and an opposition Pope at Avignon did something towards the sowing of Reformation seed. But which is Rome and which is Avignon in this case seems not quite clear. When Dr. Philpotts got up his diocesan synods against Gorham, he declared very strongly that every diocese is a separate church, with its own right to pronounce a doctrine. He was quite right; according to ante-papal Christianity nothing but a general Council could override any one bishop and his synod. But he did not see he played into the hands of the Independents. So the synod was held, and it will not be the last.

The wind is getting up, and the day is cloudy and comparatively cool; there is a synod of clouds.

Yours sincerely,

A. DE MORGAN.

To J. S. Mill, Esq., M.P.

91 Adelaide Road, September 3, 1868.

MY DEAR SIR,—You are, I suppose, in this interval, as likely as you will be to make a note about your Logic; so I send a couple of corrections (edit. 6, pp. 9, 71, vol. i.).

(P. 9.) You will be taken to mean—perhaps you meant—that the phrase *ars artium* is due to Bacon. It was an old technical definition of logic. Ludovicus Vives (in the only word of praise he gives to a Schoolman, or nearly) commends Petrus Hispanus (ob. 1277) for making it his definition, and corrects those who think it only an hyperbole of praise, explaining it as the art which treats of arts.

(P. 71.) A wrong quotation may be defensible when it enhances a joke; at any rate, Sterne's recording angel would erase the record with a grin, as he did Uncle Toby's oath with a tear. But a defect of quotation which converts humour into dry gravity is one over which the angel, if he shed a shower of tears, would take care none of them should fall. In p. 71 you say that a pedantic physician in Molière accounts for the fact that 'l'opium endormit' by the maxim 'parcequ'il a un vertu soporifique.' From whom do you get your quotation marks? Not from Molière. You know the original at the end of the *Malade Imaginaire*:— 1868.

Mihi a docto doctore.
Domandatur causam et rationem quare
Opium facit dormire.
A quoi respondeo
Quia est in eo
Virtus dormitiva,
Cujus est natura
Sensus assoupire.

I never read this exquisite satire without wishing for a Molière to expose the school of thinkers of our day who invert the process; and having settled that opium has not and cannot have a *virtus dormitiva*, will deny the sleep, or else declare that it is only a coincidence. Eighteen years' experience has told me that infinitesimal doses, so called, meet my symptoms as well as the finite doses of the eighteen years preceding; but the *docti doctores* assure me that it cannot be, because there cannot be a *virtus curativa* in doses so small. I think the Schoolmen were the more rational of the two.

I cannot understand how you liken the *virtus dormitiva* to a case of the 'scholastic doctrine of occult causes.' In fact, I have never been able to arrive at such causes in the Schoolmen. I know that these offenders are charged in our day, and since the time of Bacon, with upholding certain things called *occult causes*, but I cannot find any. *Virtutes occultæ* and *occult qualities* I find enough of. The following is my account of the matter.

The class of inquirers who cultivated magic, a large part of which was mysterious physics, upheld the existence of many qualities which do not show on the surface of things, and cannot be inferred from the sensible qualities. Many of these were fictions and many were truths. The sources of these things were hidden in a sense in which they presumed more common qualities were not. Thus Cornelius Agrippa (*De Occulta Philosophia*) says that though heat in the stomach digests food, yet the external heat from fire, for instance, will not do it. Accordingly the stomach has a *virtus quædam occulta, quam ignoramus.* As the dead stomach will not do, we say it is an effect of vital force, and laugh at the Schoolmen for their hidden cause.

1868. The school of magicians has a great number of such occult qualities, as plants which will produce certain dreams, or will repel poison when worn next the skin, &c. They had also the remarkable stone which attracts iron, and we have it too. Put all these things together, make the school of magical writers include *all* the Schoolmen, and you may attribute to their philosophy the treatment of occult causes.

Their budget of facts was hampered with an immense number of wonders handed down from the ancients. They were not enough of our spirit to deny all they could not understand, so they declared that the *virtutes* were of an *occult* character. Thus the story, not rejected, of the Ark being still in existence on Mount Ararat, made them pronounce that the glue which fastened the timbers had an occult virtue; and if you and I believed the fact, we should say the same. The great error of the Schoolmen was too easy belief in antiquity; the great crime they are charged with is declaring that what they did not know was hidden from them.

When Leibnitz attacked gravitation he called it an occult *quality*, not *cause*. If you can put me on the scent of any doctrine of occult *causes*, I will follow it up.

When Agrippa wrote *De Vanitate Artium* against all that he had explained in the *De Occulta Philosophia*, and against everything else, his chapters against logic and sophism have not a word about the matter. Ludovicus Vives, who also satirised everything, is equally free. I do not recollect any satire on the subject in any old writers, however fierce they may be against the scholastic writers.

You say elsewhere that the following proposition is not intelligible: 'Abracadabra is a second intention.' Literally, 'animal *is* a second intention' may be held *false*, not unintelligible. For a second intention *is* a subjective use of a name. Probably you mean that the proposition 'Abracadabra is a (name of) second intention' is unintelligible. But why more than 'animal'? If you mean that Abracadabra is a mere sound, you do less than due honour to the name of a medical instrument of 1,200 years' life. I suspect you are not aware that no less an authority than Serenus Samonicus, in the *Carmen de Medicina* (perhaps you don't care for his authority), lays it down that the word thus treated—

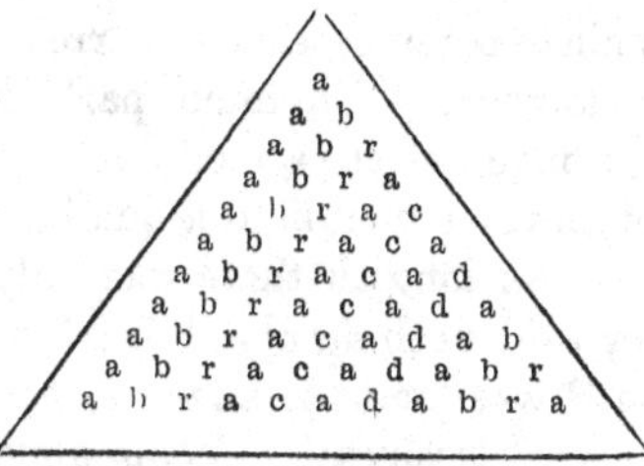

will cause fever gradually to abate if it be hung round the neck of the patient. This was the Abracadabra, and it was a class of objects, and could be a name of second intention.

This is a *sextant* (60°). Agrippa prefers the *octant*,—

a
a b
a b r
a b r a, &c.

I wish every voter in the country would hang one round his neck until this election is over. Perhaps what would abate would prevent.

I hope you are not thoroughly knocked up with heat and politics.

Yours very truly,

A. De Morgan.

(The following is from Mr. J. S. Mill's papers, with the above letter.)

Virtus Dormitiva.

In the article 'Physique' of the *Dict. de Phil. Schol.* of the Abbé Migne's collection, after noting the *virtutes* and *essentiæ* as scholastic faults, which is only true of their abuses, the author proceeds thus:—

'Arnauld lui-même, Arnauld le Cartésien, pratiquait les vieux erremens de la scholastique, lorsqu'il disait à Malebranche, "Il est insensé de se demander pourquoi l'âme humaine pense à l'infini et au nécessaire. Elle y pense parce que c'est dans son essence d'y penser."

'Aujourd'hui encore l'école écossaise et l'école éclectique expliquent exclusivement les phénomènes psychologiques par des facultés qu'on multiplie et qu'on distingue parfois avec une ridicule subtilité; et on s'imagine qu'en plaçant ainsi sous les faits intimes des facultés que la conscience n'a jamais perçues on a fait de la science.

1868. 'L'école rationaliste commet la même erreur dans la question de l'origine du langage. L'homme parle parce qu'il a la faculté, donc il a pu inventer la parole.'

The Schoolmen never generalised a quality until they had at least *two* instances. As long as there was only A which had a certain *virtus*, they said nothing about it; it was *occult*, i.e. *unknown*. But when B was found to have the same they had such knowledge as comes of classification, being almost all they had.

The moderns invented a name upon *one* instance, and made it a cause. They said that *magnetism* was the explanation of the *magnet*. The Schoolmen would have waited until the amber showed its quality, and then the distinction of *magnetism* and *electricity* would have been specific knowledge, the genus being *virtus attractiva*. It is something to know two phenomena with a generic agreement and a specific difference.

If the medical candidate had known the mind of those who classed, he would have said, I do not know *why* except in that I can refer the phenomenon to a class. We note agreements and differences and arrange them. Arnauld, &c., might have a similar answer made for them, but not for those who inferred power of invention of languages from possession.

To J. S. Mill, Esq., M.P.

September 20, 1868.

My dear Sir,—Seeing you at Avignon again reminds me of a question I intended to ask long ago. If a mathematician were asked what Avignon reminds him of, I do not know what he would answer, except *the Avignon edition of Gardiner's Logarithms.* Gardiner published a very celebrated folio of logarithms in 1742, with a very solid subscription list of 120 persons, of whom two-thirds are now known in the history of science—that is, to a close inquirer. I greatly doubt that an old list could be found except this of which the same proportion could be recovered. Gardiner corrected all the errors with his own hand in all the copies. My great-grandfather, James Dodson, who also in 1742 published his *Anti-logarithmic Canon*, with 1,100,000 *computed* figures, did the same thing. I suspect there was some concert as to this excellent plan between the two.

To proceed, in 1770 appeared at Avignon the reprint of Gardiner, in folio, 'Chez T. Aubert, Imprimeur, Libraire Rue de l'Epicerie.' There is a printed *Avis*, signed T. Aubert, which has a list of errata, and an announcement that the corrections are all made in their proper places. But in my copy the correction is thus made. The correct figures are cut out and pasted over the incorrect ones, with a *written* announcement *in English*, signed G. Townsend, stating what has been done; he says it was done in the edition, but perhaps he only means the copy. The editors have modestly concealed their names. They were three industrious priests, Pezenas, Dumas, and Blanchard, who would probably have remained unknown if Lalande, who had been a pupil of Dumas, and who was in communication with them, had not preserved their names in his gossip. 1868.

If there should be any local science at Avignon I dare say you will know some one who will be able to tell whether any tradition of the three editors remains. Of course there will be a copy in some public library, and it will be seen whether G. Townsend's performance appears there. Pray do not trouble yourself further than to make any local antiquary acquainted with what is wanted. I suppose Avignon must now be what people would call an out-of-the-way place; but such places very often have people who, like Captain Clutterbuck, spend their whole time in illustrating their locality.

A nice job you will have made for the courts. Some ladies have actually passed the revising barrister because there was no opposition. The R. B. was right; he is not bound to know that Jane Smith is a woman, nor could he raise the question. I have a cousin whose wife is *David*. When the poll clerk sees a female claimant, I suppose he will be bound to say, 'Madam, you cannot be the Jane Smith on the list, for the law says that voters are all men. I must wait until some man comes forward and declares he is the person described.' Then the poll clerk may perhaps be subjected to an action. But if he should admit the claim, there may be a scrutiny demanded, and perhaps a petition against the return. The question will raise some logic. The world of concepts being divided into man and non-man, if man mean male person, and only man can vote, non-man equally excludes Jane and her pussy and her pianoforte. They all come under the contra-positive—All voters are men. All voters are men, *i.e.* all non-men are non-voters. There is but one answer to Jane, the cat, and the pianoforte, *i.e. non-man.* I hope you will push the point and get rid of the bother; it infests the house. But, in justice, let no woman be placed on the register

1868. except on her *demand*. To be a voter is sometimes dangerous. A man ought to face the danger, but you have no right to enforce it on women; in principle you might as well enforce the militia on them. Many women think exemption from politics is one of their rights.

Yours very truly,
A. DE MORGAN.

To the Rev. J. J. Tayler.

91 Adelaide Road, N.W., April 9, 1869.

1869. MY DEAR SIR,—I have at last found head to complete the reading of your two tracts, for which I am much obliged. I think I also have to thank you for a copy of Mr. Martineau's tract, of which we were speaking, and which I received within thirty-six hours of our conversation. Your propositions for a free Christian Union are brought to a point at which no opinion can be given until more comes out. It seems clear that the freedom extends to a rejection of all direct interference of Deity in old time, that is by those who choose, with liberty to others to retain it, and to dealing in the same manner with the actual existence of Jesus. All that seems to be required is Mr. Martineau's *triad* (*absit Trinitas*) of (p. 19) belief in God, piety, and charity. It seems to be required that the morality of Jesus shall be acknowledged. But whether because it finds a response in the human heart, or because it is in some unexpressed way sanctioned by God, does not appear. In fact, there are as yet many points which are only seen through a glass darkly, but the one on which expectation must now wait, which is in the field but without illumination, is the question of *worship*. Is a joint worship contemplated? I cannot make out. The moment the plan is sketched out a hundred points will arise. There are two classes of persons with whom I should hold that neither you nor Mr. Martineau should refuse communion:—

1. The old-fashioned Christian—the man who starts with Peter.

2. Those who know no more whether there be a personal and moral directing God than whether He have an Anointed. If these men stipulate for another o, and adopt the creed of love of go(o)d and love to man, they must either be admitted under general agreement as to what *is* good—or a reason of inclusion

must be found in the assumed necessity of appeal to *God* to know what is *good*. And if it come to that I wait for further information as to how the appeal is made. If man be to settle morals his own way, I would not bind myself to his law; for though I do not go quite the length and all the strength of St. Paul, I am clearly of opinion that the heart of man is as deceitful as most things, and is very much, not a little, wicked. 1869.

But further, supposing the intermediates could fraternise with the extremes, could the extremes fraternise with one another? This is the old difficulty of compromise, under which many an attempt at political and religious concentration has failed. We know that if A and B coincide with C they coincide with one another; but if A and B should happen to be within x feet of C, all we can positively affirm is that they are within $2x$ feet of one another. And x may do where $2x$ will not. These are the first things which strike an old thinker on the subject, who does not feel equal to more than two sheets. There may be comfort in the Scotch proverb that those who pluck at a gown of gold may get a sleeve of it. In the meanwhile the name is too bold. 'Free' 'Christian' 'Union'! Until you tell us how free and how Christian you mean to be, no one can tell how united you will become.

You have your sand, and you aspire to make rope. Michael Scott's devils failed, but they did not know that a very easy process would make their sand ropes into *glass*, which makes very good thread. What have you got to mix with your sand? That is what I am curious to see. But the attempt is praiseworthy, and must be most useful in any case of result.

Yours sincerely,

A. De Morgan.

To Sir Frederick Pollock.

[Not in his own handwriting.]

91 Adelaide Road, June 13, 1869.

My dear Sir Frederick,—I should have said, till now, that it would have been your business to receive me when I got to the gate of the other world, but now I hold it not so sure. I had an attack of congestion of the brain on Wednesday, which kept me several hours in a condition of which I had afterwards no recollection. I think I may put this against your greater age, and consider myself as a candidate of equal preten-

1869. sions with yourself. It was not what they call the real apoplexy, because it had no lasting effect on any organ; but Mr. Weller, senior, would have admitted it for a genuine *applepleaxy*. I am going on very well now, and if anything happens to the contrary, my wife will write.

Give our kind regards to Lady Pollock and all the family.

Yours sincerely,

A. DE MORGAN.

To Sir F. Pollock.

[Written in pencil.]

91 Adelaide Road, June 19, 1869.

MY DEAR SIR FREDERICK,—I enterprise a letter of my own writing. I am glad to hear you have got hold of H. C. R.,[1] whom I allow to be *Herr Conversations Rath*, though I cannot answer for the German correctness of the title.

I suppose you will find a host of reminiscences—even *I* find a large lot. The only thing I have done for many weeks is to write—from the sheets—a little notice of his works for a theological Review; so I have picked it all through, and made extracts. They say autobiography is always readable. Three thick volumes, intermixed with letters, is a severe test, but I think many will get through it. Certainly, it will be a repertory of facts tending to literary history, in which the smallest bit of personal biography is sometimes clinching.

As to myself, I progress. The medical men are agreed that nothing is *apoplexy* except what leaves injury to some power or organ. The *congestion*, which they make out to be common, is voted no disorder at all. All which is very true, as to the great superiority of an attack which leaves no consequences over one which does leave them. But congestion is congestion, after all.

Our kind regards to Lady Pollock and all the family.

Yours sincerely,

A. DE MORGAN.

To Sir F. Pollock, Bart.

6 Merton Road, N.W., July 24, 1869.

MY DEAR SIR FREDERICK,—As we neither of us are strong on the legs, and yet can use our fingers, I employ mine to beg you

[1] Henry Crabb Robinson's *Diary*, &c.

will employ yours, at any leisure you like, in answering a question. 1869.

I want a tolerably distinct account of the reading of a senior wrangler of your day. How much had he read, and in what books? A very general view will do. Anything as to distinction between algebra and geometry will be valuable.

The people are all gone who could give any information. With me it will be perpetuated in some shape.

I gain strength pretty fast, but am not without warning that *head-time* is not come back yet. There are all kinds of legends current about old reading. A trustworthy account would be of historical value. With kind regards all round,

Yours sincerely,

A. De Morgan.

On looking at the list of *seniors*, I see that you are not only the last left, but that you come *at the turn.* In 1810 comes Maule, who was in communication with Babbage about functional equations, and all kinds of novelties. Maule would have been conspicuous, among the moderns, with Herschel, Peacock, Babbage, if he had held on. Had I not known you, I should, on these circumstances, have applied to you, as the only chance left for information essential to historical knowledge of Cambridge.

From Sir F. Pollock, Bart.

Hatton, Hounslow, July 29, 1869.

My dear De Morgan,—I am glad to hear you 'gain strength pretty fast.' I lose it *slowly*; but I *lose* it. I shall write in answer to your inquiry, *all* about my books, my studies, and my degree, and leave you to settle all about the proprieties which my letter may give rise to, as to egotism, modesty, &c. The only books I read the first year were Wood's Algebra (as far as quadratic equations), Bonnycastle's ditto, and Euclid (Simpson's). In the second year I read Wood (beyond quadratic equations), and Wood and Vince, for what they called the *branches.* In the third year I read the *Jesuit's* Newton and Vince's Fluxions; these were all the *books*, but there were certain MSS. floating about which I copied—which belonged to Dealtry, second wrangler in Kempthorne's year. I have no doubt that I had read less and seen fewer books than any senior wrangler of about my time, or any period since; but what I knew I knew thoroughly,

1869. and it was completely at my fingers' ends. I consider that I was the last *geometrical* and *fluxional* senior wrangler; I was not up to the *differential* calculus, and never acquired it. I went up to college with a knowledge of Euclid and algebra to quadratic equations, nothing more; and I never read any second year's lore during my first year, nor any third year's lore during my second; my *forte* was, that what I *did* know I *could produce at any moment* with PERFECT *accuracy*. I could repeat the first book of Euclid word by word and letter by letter. During my first year I was not a '*reading*' man (so called); I had no expectation of honours or a fellowship, and I attended all the lectures on all subjects—Harwood's anatomical, Woollaston's chemical, and Farish's mechanical lectures—but the examination at the end of the first year revealed to me my powers. I was not only in the first class, but it was generally understood I was *first* in the first class; neither I nor any one for me expected I should get in at all. Now, as I had taken no pains to prepare (taking, however, marvellous pains while the examination was going on), I knew better than any one else the value of my *examination qualities* (great rapidity and perfect accuracy); and I said to myself, 'If you're not an ass, you'll be senior wrangler;' and *I took to 'reading' accordingly*. A curious circumstance occurred when the Brackets came out in the Senate-house declaring the result of the examination: I saw at the top the name of Walter *bracketed alone* (as he was); in the bracket below were *Fiott, Hustler, Jephson.* I looked down and could not find my own name till I got to Bolland, when my pride took fire, and I said, 'I must have beaten *that man*, so I will look up again;' and on looking up carefully I found the nail had been passed through my name, and I was at the top bracketed *alone*, even above Walter. You may judge what my feelings were at this discovery; it is the only instance of two such brackets, and it made my fortune —that is, made me independent, and gave me an immense college reputation. It was said I was more than half of the examination before any one else. The two moderators were Hornbuckle, of St. John's, and Brown (Saint Brown), of Trinity. The Johnian congratulated me. I said perhaps I might be challenged; he said, 'Well, if you are you're quite safe—you may sit down and do nothing, and no one would get up to you in a whole day.'

This is confirmed by what the Bishop of Gloucester told me Brown said at Lord Lonsdale's table at Lowther. The examination in the Senate-house became the subject of conversation.

Brown said *no one answered any question that I did not also answer*, and *better*. 1869.

I have no doubt Walter and Jephson had read more books than I had, and knew (*after a sort*) *more*. Maule was a man of great ability and *immense* acquirements; he reminded me of Porson more than anybody else did.

My experience has led me to doubt the value of competitive examination. I believe the most valuable qualities for practical life cannot be got at by any examination—such as steadiness and perseverance. It may be well to make an examination part of the mode of judging of a man's fitness; but to put him into an office with public duties to perform merely on his passing a good examination is, I think, a bad mode of preventing mere patronage. My brother is one of the best generals that ever commanded an army, but the qualities that make him so are quite beyond the reach of any examination. Latterly the Cambridge examinations seem to turn upon very different matters from what prevailed in my time. I think a Cambridge education has for its object to make good members of society—not to extend science and make profound mathematicians. The tripos questions in the Senate-house ought not to go beyond certain limits, and geometry ought to be cultivated and encouraged much more than it is.

Euclid and conic sections studied geometrically improve, enlarge, and strengthen the mind; studied analytically, I think not. But I must have exhausted your patience—a virtue which may be *tried* but not *examined*. My best regards to Mrs. De Morgan and your family.

Ever sincerely yours,

FRED. POLLOCK.

P.S.—Looking over what I have written, I fear you will find little that you want; but I am still ready to answer any specific questions.

To Sir F. Pollock.

6 Merton Road, Adelaide Road, N.W.,
August 1, 1869.

MY DEAR SIR FREDERICK,—Your letter has better Cambridge history than any 100 pages of the Esq. Bedell—I forget his name—who quoted Wm. Frend as saying that the market women complained of being *scotched* a quarter of their wages, and quoted the word three times in italics to call attention to it. Mr. Frend's

1869. word was *sconced*—the old word for fined. If you can recall at this rate, you will rival H. C. R.

Your letter suggests much, because it gives possibility of answer. The *branches* of algebra of course mainly refer to the second part of Wood, now called the theory of equations.

Waring was his guide. Turner—whom you must remember as head of Pembroke, senior wrangler of 1767—told a young man in the hearing of my informant to be sure and attend to quadratic equations. 'It was a quadratic,' said he, 'made me senior wrangler.' It seems to me that the Cambridge *revivers* were Waring, Paley, Vince, Milner.

You had Dealtry's MSS. He afterwards published a very good book on fluxions. He merged his mathematical fame in that of a Claphamite Christian. It is something to know that the tutor's MS. was in vogue in 1800–1806.

Now—how did you get your conic sections?

How much of Newton did you read?

From Newton direct, or from tutor's manuscript?

Surely Fiott was our old friend Dr. Lee.[1]

I missed being a pupil of Hustler by a few weeks. He retired just before I went up in February 1823.

The echo of Hornbuckle's answer to you about the challenge has lighted on Whewell, who, it is said, wanted to challenge Jacob, and was answered that he could not beat if he were to write the whole day and the other wrote nothing.

I do not believe that Whewell would have listened to any such dissuasion.

I doubt your being the last fluxional senior wrangler. So far as I know, Gipps, Langdale, Alderson, Dicey, Neale, may contest this point with you.

I go on fairly. With kind regards all round,

Yours sincerely,

A. De Morgan.

From Sir F. Pollock.

Hatton, Hounslow, August 7, 1869.

My dear De Morgan,—You seem not to know the story of Gunning's book (the Bedell you allude to). He really kept a sort

[1] Much of this is not perfectly clear to me; but I insert the letter as it stands, as it may have interest for old Cambridge men. For the same reason I have departed from my general rule, in inserting Sir F. Pollock's letter in reply.—S. E. De M.

of diary, in which he put all the scandal of *every* sort he met with 1869.
in Cambridge society—much about Porson; Mansel's epigrams and verses; his contest with Sir Busick Harwood; much of the vulgar drollery of Jemmy Gordon, who did not survive till your day: in short, a collection of very low and ribald stuff, mixed with what was worth preserving (but Gunning could not distinguish between the two). Some one persuaded him to burn it, and the book he published was what he remembered when his memory was gone and the real book burnt. His son Frederick was my pupil, and did well as a barrister, from whom I had this. You have put together as *revivers* five very different men. Woodhouse was better than Waring, who could not prove Wilson's (Judge of C. P.) guess about the property of prime numbers; but Woodhouse (I think) did prove it, and a beautiful proof it is. Vince was a bungler, and I think utterly insensible of mathematical beauty. Milner was incomparable. The Claphamite Christians are a class to be found in every form of religion; and when they are not *too* intolerant (which generally they are) they have much of my sympathy, though I don't agree with them.

Now for your questions. I did not get my conic sections from Vince. I copied a MS. of Dealtry's. I fell in love with the cone and its sections, and everything about it. I have never forsaken my favourite pursuit; I delighted in such problems as two spheres touching each other and also the inside of a hollow cone, &c. As to Newton, I read a good deal (men *now* read nothing), but I read much of the notes. I detected a blunder which nobody seemed to be aware of. Tavel, tutor of Trinity, was not; and he augured very favourably of me in consequence. The application of the Principia I got from MSS.

The blunder was this: in calculating the resistance of a globe at the end of a cylinder oscillating in a resisting medium they had forgotten to notice that there is a difference between the resistance to a globe and a circle of the same diameter.

The story of Whewell and Jacob cannot be true. Whewell was a very, *very* considerable man, I think not a *great* man. I have no doubt Jacob beat him in accuracy, but the supposed answer *cannot* be true; it is a mere echo of what actually passed between me and Hornbuckle on the day the Tripos came out—for the truth of which I vouch. I think the examiners are taking too *practical* a turn; it is a waste of time to calculate *actually* a longitude by the help of logarithmic tables and lunar observations. It would be a fault not to know *how*, but a

1869. greater to be handy at it. Oh dear! I longed to get among the Fellows; but when I did, I was utterly disgusted at the rubbishing conversation that prevailed, and I then longed to get away. You see I linger over Cambridge recollections; but no particular time has been the happiest of my life, certainly not *school*. Best regards to Mrs. De Morgan and your family.

Sincerely yours,
FREDERICK POLLOCK.

To the Rev. W. Mason.[1]

Adelaide Road, August 13, 1869.

DEAR MASON,—As touching myself I get stronger gradually. I am slowly getting my books into order, which is a long job. I have no more information of any very decided character than is to be found in my wife's book, *From Matter to Spirit*. I retain my suspense as to what the phenomena mean, but I am as fully persuaded as ever of their reality.

The presence of the dead is a thing widely felt, but by certain temperaments. Bishop Jebb is an instance of no very forcible kind, because the two worlds had been in constant connection in his mind. I will give you a more curious one.

An actuary, a man of science and a keen searcher after old printing, married a second cousin of mine. He was a cheerful and kind-hearted man, but to all appearance as thoroughly unspiritual as a man could be. I never heard a word drop from him which made it appear that another life was his familiar thought. He was, though moderate in drinking, rather fond of eating, and skilled in it. The ladies of his acquaintance who had dinners to give would consult him on all details. His wife, to whom he was devoted, died, and he himself fell into a weakly state. I used to sit with him by the hour. A few weeks before his death I found him debilitated by a long conference he had had with a lady about a dinner she had to give: this merely to show that his mind was not turned to the subject of death by anything

[1] This was in answer to a letter in which Mr. Mason asks him, if able, to give him 'some information on the interesting subject to which you alluded in your last.' 'I have long thought,' Mr. Mason says, 'that departed spirits are often with those they left at death. When Bishop Jebb had been for some time under a paralytic seizure, he said, on his recovery, that in the prospect of death he had felt that he should be as truly with his friends after death as he was then when speaking to them.'—S. E. DE M.

external. He suddenly turned to me and said, 'De Morgan, my wife is often with me.' I was astonished, not at the phenomenon, but at *his* being the recipient. 'Often?' said I. 'Every evening,' said he, 'and oftener.' 'Do you see her?' said I. 'No,' said he, 'but I *feel her presence.*' By these three words hangs a long tale. 1869.

With kind regards to your family,

Yours sincerely,

A. De Morgan.

To Rev. W. Heald.

August 21, 1869.

My dear Heald,—I think I shall be able to keep up the institution of a summer letter, though I may not be so long as usual. It is the forty-fourth observance.

You think, one letter of yours says, that I am feeling the effects of hard work; in fact, that I have been working too hard. Rid your mind of the idea. I have never been *hard* working, but I have been very *continuously* at work. I have never *sought* relaxation. And why? Because it would have killed me. Amusement is real hard work to me. To relax is to forage about the books with no particular object, and not bound to go on with anything.

You remember that my amusement used to be Berkeley and the like. Quite true. I did with Trinity College library what I afterwards did with my own—I foraged for relaxation. I used to shock you with my reading of Voltaire, who existed in that library in about eighty quarto volumes. So you called me an atheist vagabond, fancying that Voltaire was an atheist: he was, in fact, theistic to bigotry, and anti-revolutionist to the same extent.

I read an enormous deal of fiction—all I could get hold of—so my amusement was not all philosophical. I have never worked hard—never got so far as a headache. If I felt tired I left off.

My illness is well enough explained by the following chain of events.

1866. Discovery that University College was going to betray its principles, and abandonment of the place in 1867.

1867. Long illness and death of my second son, with all the anxiety occurring during the turmoil of the College affair. In the meanwhile my third son had taken refuge from illness on board ship, and was away for eighteen months in very fluctuating

1869. health. One of my daughters was also labouring under the same symptoms, and was four months with her mother at Hastings after the death of my son. All this I could have borne. It attacked the spirits, but I could have held on as I have always done in sure reliance on the higher and wiser management which 'shapes our ends.' But the heats of 1868 broke down the physical force, and gave a cough and weakness which was followed up by the consequences you know of. Does it, upon the preceding showing, require the hypothesis of thirty years of overwork to explain an attack of diabetes which yielded to the first remedies, or rather to diet alone, *in a week*, and a stroke of congestion of the brain which left no result but weakness? I have heard of overwork on all sides, and have seen people stare at their own omission of all the misfortunes—so called—which have come upon me in the last two years.

I am now weak enough, but I gradually improve. I shall soon get all the way upstairs foot *over foot*, that is, *sans* both feet on one step at once, and without the banisters. At present it is, after half is done, either a tug at the banister, or bring up the second foot before you remove the first. Three weeks ago it was this alternative the whole way. The stairs are a beautiful dynamometer.

I am very anxious about Arthur Neate. He has an ugly cancerous tumour on the lower side of the left cheek, which opinion decides variously cancer or no cancer; but those who think it cancer think it a very serious case. He is not aware of the dangerous opinions. If they be correct the matter will be beyond doubt in a few months. As yet there is no serious internal symptoms, and such things have sometimes passed off. Neate is about sixty-four. I turned the grand climacteric (sixty-three) in June last (27th).

August 30.—So much for delays and feelings of inability. I get on fairly, head and arms, but the legs do not thrive in proportion. I sawed a plank of wainscot (hard wood); and a man who can do that ought to walk three miles, but I do not do more than one.

September 1.—My wife, who was at the sea with my second daughter and a bronchial cough to be got rid of, came home on Saturday, and my daughter's cough nearly gone.

A man cannot have the sort of attack I had without some amount of evidence of it. I had two well-marked consequences for several days, an inability and a delusion. 1. I could not for a week master the word *congestion* as the name of my own attack.

I had it at once for every other purpose. If I got hold of *con*, it was *confusion*, *conglomeration*—anything but *congestion*. If I got hold of *gestion*, it was digestion, suggestion, &c. Several times, and days after I had recovered my senses, I used to amuse myself by trying, and was at last obliged to ask what had been the matter with me. 1869.

2. When I woke to recollection of the universe, and for days afterwards, I was possessed with the idea that before the seizure I had received a letter from Ireland, written on the supposition that I was a clergyman, and offering me a great lot of Irish preferment. If there be one political subject on which I had never thought or cared, it is the Irish Church and its management.

My idea was that some poor patron, in a hurry to induct some one into the benefices, by way of securing some vested interest before the final disendowment, had taken it into his head to select me as the holder of the profits for the rest of their term. I was very anxious to set him right, not knowing how much consequence a day might be of. But as I got nearer to the letter-writing state, the vision became fainter, and when I at last looked, more to see what could have suggested it than with any idea of finding, I could not get a trace of any such letters. Besides these, I had not any consequences whatever of the loss of consciousness.

I think this must go as it is. I hope your family are well, and yourself. Do you know, or can you find, anything about H. Parr Hamilton, the Dean of Salisbury? Kind regards to Mrs. Heald.

Yours sincerely,
A. De Morgan.

To Sir J. Herschel.

6 Merton Road, October 20, 1869.

My dear Sir John,—Surely I sent you my card, which you will find within the envelope.[1] This envelope arrived on Monday with the pie of πs which you see. But the Leverrier has not come yet: no doubt it is hunting me all over N.W., with a

[1] A card on which he had printed a small map of Merton Road and the immediate neighbourhood as a guide to friends. Unless the 'pie of πs' means the number of circles stamped by the Post Office on the envelope, which appears to have travelled half over London before reaching him, I cannot interpret it.—S. E. De M.

1869. change of air by a jaunt to Merton Road, S.W. In a few days it will be time to look up the dead-letter office. I am afraid I am not strong enough for this yet. Have you anybody you could ask who goes near the P.O. often?

I thrive—and the cold weather is bracing me up like a bundle of asparagus, having been no better than a rope of onions. A week's cold weather last winter would have kept me from striking my flag. Two or three days of half-cold told me so, and then took leave.

I shall be glad to see the Leverrier account. If it should come, I will write at once. With kind regards all round your circle,

Yours sincerely,
A. DE MORGAN.

To Sir J. Herschel.

6 Merton Road, November 8, 1869.

MY DEAR SIR JOHN,—The Queen used up so much of the fine weather on Saturday, that the chief clerk of that office says he can issue no more until he gets a further supply. So I am rain-bound for to-day, and can clear off obligations. I, therefore, return your Leverrier with many thanks. What a miserable mess has been made by Chasles, Lucas, and Co.! I am obliged to give up Chasles until he clears himself, which I have small hope of his doing. The different accounts he has given at different times are such as must be reconciled, or otherwise explained. If there be no explanation except sub-human credulity, then arises the question which is so important in lunacy inquiries, *When* did this defect begin? For Chasles has a lifetime of memoirs full of references to MSS., many of them unseen as yet except by himself. It will be unsafe to quote him—at least to a *better-not* extent.

I have lately lost my friend Libri, and of course, he being removed, the accusations which he put down begin to revive. I wrote a short article in the *Athenæum* of the mortuary character, and the Parisians, quite forgetting the beating they got, are pleased to be excessively astonished at the revival of a defence which silenced them fifteen years ago. There is a little knot of subscribers in England who try to act privately on editors and contributors. *Ex. gr.*: A person who described . . . .

(*Nov.* 11.—Sunshine came out, and drew me out also, and I have not been able to resume until now. I walked $1\frac{3}{4}$ miles yesterday !^{1} !$^{\frac{1}{2}}$!$^{\frac{3}{4}}$ I catch up the unfinished sentence—),

. . . . himself as a known book-collector (N.B.—No less a person than wax-chandler to the Queen, &c., very rich, and collects no end of elegantly bound large paper; all this I learnt afterwards) came to me in a neat carriage and a heavy shower, and as he was doing a wabbling preamble about nothing, I cut him short sternly with, 'Pray, sir, what is the upshot of all this?' He answered that, seeing my article in the *Athenæum*—which it was very impertinent to assume was mine—he could prove in two minutes that Libri was guilty of all that was imputed to him. 1869.

'What do you know of the matter?'

'I have read all the pamphlets.'

'So have I,' said I, 'and some of them before they were pamphlets.'

'Oh! I thought perhaps you had not investigated.'

He then produced 'Vapereau,' a French biographical dictionary of first-rate size and tenth-rate accuracy, and, opening at *Libri*, said, 'Have you read that article?' 'I have,' said I, 'in former days, before I found out what a worthless affair Vapereau was.' 'I assure you,' said he, 'the people at Paris are much astonished at your article.' 'No doubt,' said I: 'they are the parties whom Libri's defence incriminates.' 'I thought perhaps you were not aware of the facts, and that by coming to you we might avoid a *polemic*.' 'Now,' said I, 'you must go to the editor of the *Athenæum*, and polemic with *him*. Do you really suppose you will prove to me that one of my dearest friends was a robber by an extract from "Vapereau" and Parisian opinion?' So he went away, and there has been no polemic yet.

A matter of this kind brings out the hidden fun of the world. So with kind regards all round,

Yours sincerely,

A. De Morgan.

To Rev. J. Martineau.

6 Merton Road, Adelaide Road, N.W.,
December 19, 1869.

Dear Sir,—Many thanks—to you I suppose I am indebted—for the reprints of the journal memoir of J. J. Tayler. He is well recalled, which in his case is not very easy to do in writing. It appeared to me that his treatment of controversy in conversation allowed the wave to pass over the reef without breakers. A congregation of such men could have realised his plan of a

1869. scheme of joint worship in which the party were agreed upon everything except accurate definition of what they were agreed about.

I have your two tracts—one of J. J. T., and the 'anniversary' of 1869. If there be any more I should like to know of them. For I am interested in the attempt, which, hopeless as it seems to me to the extent proposed, may yet originate a sect in which people may pray together without each man being fettered to his neighbours.

But there must be some little definiteness of statement. I tried hard to get from J. J. T. whether his Christianity had a supernatural element. His final information was that he thought it most likely the apostles had a supernatural element which we do not understand.

I intend to keep watch on the attempt to couple supernaturalists and anti-supernaturalists, for that is what is aimed at. When I get something definite about its indefiniteness, I intend to write about it.

Dubius sed non improbus. This is what Sheffield said of his own religion, and the scholars (Dean Stanley included) make him 'sceptical, but not wicked.'

Improbus means one who declares against the proof—one to whom there cannot be proof. As when Pliny says that Hipparchus counted the stars—*rem Deo improbam*—a thing unproved by the gods; and Virgil, *Labor omnia vincit, improbus*—toil yet unproved, or untried, conquers all things.

Yours very truly,
A. De Morgan.

To Sedley Taylor, Esq.

6 Merton Road, December 26, 1869.

My dear Sedley,—Many thanks for your pamphlet,[1] which I shall join on to some of Martineau's, &c., in one notice. I think you will produce some effect on people who begin to have a cranny through which the light comes. I saw Jas. Martineau a day or two ago, and he tells me that his organisation is contemplating the circulation of your pamphlet in aid of their view. I wish their view were a little less of a dissolving view when

[1] *On Clerical Subscription.* (Macmillan, 1869.) Mr. De Morgan believed that this pamphlet hastened the disruption of the Free Christian Union.

you come to look closely at it. I cannot make out whether they 1869.
have a super—— religion or not. I do not know how to fill up the word. Now I tolerate everything except passing off one thing under the name of another. There are people who can detect in the foundation of Christianity a third alternative, 'Super—— or Imposture.' I cannot. I am content they should, but I want them to be explicit. I am very much afraid they want a common worship with the above question left open. No! there is no objection to leaving it open if people will, but they do not want it openly open, but secretly open, under a cloak of some indefinite pretension of divine origin. I hope you will *follow up*.

Yours sincerely,

A. De Morgan.

(The only letters to friends after this time that have come into my possession are two to Sir John Herschel, in my husband's own handwriting, the first bearing date
June 25, 1870. In this he says, 'I am creeping along, and 1870.
shall get right about as soon as the blessed St. Alcuin's snail got to dinner. It is one of the pleasant problems in the works of that holy man that the sparrow asked the snail to dinner at a league distance. Now the snail moved half an inch a day. How long, the Saint asked, will it take him to get to dinner?' The second is in an extremely feeble hand, merely describing his own state. Sir John himself died within the year. In his letter to me on receipt of mine telling him of my husband's death, he wrote, 'I have been expecting as much. The last letter I received showed me too clearly that the lamp was flickering in the socket, and it is consoling to know that the end was so peaceful and so painless, and so full of hope. Looking back on our long friendship, I do not find a single point on which we failed to sympathise; and I recall many occasions on which his sound judgment and excellent feeling have sustained and encouraged me. Many and very distinct indications tell me that I shall not be long after him; and I can only hope that my own end may be such

1870. as you describe.' His last surviving correspondent was the Rev. W. Heald, who died three years after him. All his other old University friends had gone before.

I have, with respect to domestic matters and details, done what I know my husband himself would have wished, for he never liked making known what nearly concerned his family. Moreover, those to whom he wrote at length and on questions of general interest were friends with whom he did not get frequent opportunities of conversation. Consequently, as we were almost always together, his correspondence with myself and our sons and daughters was fragmentary, and not suited for publication. I trust, however, that the foregoing selections will not be thought insufficient to show the character of one to whom letter-writing was a pleasure and a relaxation, and among whose leisure occupations it always held so prominent a place.—S. E. De M.)

LIST OF WRITINGS.

1828. Bourdon's Elements of Algebra, translation.
Elements of Arithmetic, 1st edition, 1831.
,, ,, 2nd edition, 1832.
,, ,, 3rd edition, 1833.
,, ,, 16th 1,000, 1857.

ARTICLES IN QUARTERLY JOURNAL OF EDUCATION, FROM 1831 TO 1835.

1831. (1) Account of the Polytechnic School of Paris, vol. i.
(2) Notice of Tables for facilitating Calculation, vol. i.
(3) Rev. of Pinnock's Catechism of Geometry, vol. i.
(4) On Mathematical Instruction, vol. i.
(5) Walker's Theory of Mechanics (rev.), vol. i.
(6) Darley's System of Popular Geometry (rev.), vol. ii.
(7) Bayley's Elements of Algebra (rev.), vol. ii.
1832. (8) A Plan for Conducting the Royal Naval School, vol. iii.
(9) Study of Natural Philosophy, vol. iii.
(10) A Preparation for Euclid, vol. iii.
(11) Barlow's Mathematical Tables, vol. iii.
(12) On some Methods Employed for the Instruction of the Deaf and Dumb, vol. iii.
(13) James Wood's (D.D., of Ely) Algebra (rev.), vol. iii.
(14) Quetelet on Probabilities (rev.), vol. iv.
(15) Young's Elements of Mechanics (rev.), vol. iv.
(16) State of Mathematical and Physical Sciences in the University of Oxford, vol. iv.
(17) Von Turk's Phenomena of Nature (rev.), vol. iv.
1833. (18) On Teaching Arithmetic, vol. v.
(19) Cunningham's Arithmetic (rev.), vol. v.

(20) On the Method of teaching Fractional Arithmetic, vol. v.
(21) On the Method of teaching the Elements of Geometry, vol. vi.
(22) The School and Family Manual (rev.), vol. vi.
(23) Method of teaching Geometry, No. 2, vol. vi.
(24) Busby's Catechism of Music (rev.), vol. vi.
(25) Geometry without Axioms (rev.), vol. vii.
(26) Ritchie's Principles of Geometry (rev.), vol. vii.
(27) Elementary Works by M. Quetelet (rev.), vol. vii.
(28) Cambridge Differential Notation, vol. viii.
(29) Gravitation, Airy's article, *Penny Cyclop.* (rev.), vol. viii.
(30) Peacock's Treatise on Algebra (rev.), vol. ix.
(31) Peacock's Treatise on Algebra, No. 2, vol. ix.
(32) Progress of Physical Science, vol. x.
(33) Ecole Polytechnique, vol. x.

MEMOIRS IN THE CAMBRIDGE PHILOSOPHICAL TRANSACTIONS.

(1) On the General Equation of Curves of the Second Degree, *read Nov.* 15, 1830.
(2) On the General Equation of Surfaces of the Second Degree, *read Nov.* 12, 1832.
(3) Sketch of a Method of introducing Discontinuous Constants into the Arithmetical Expressions for Infinite Series, where they admit of several values, *read May* 16, 1836.
(4) On a Question in the Theory of Probabilities, *read Feb.* 1837.
(5) On the Foundations of Algebra, *read Dec.* 9, 1839.
(6) On the Foundations of Algebra, No. 2, *read Nov.* 29, 1841.
(7) On the Foundations of Algebra, No. 3, *read Nov.* 27, 1843.
(8) On Divergent Series, and Points connected with them, *read March* 4, 1844.
(9) On the Foundations of Algebra, No. 4, On Triple Algebra, *read Oct.* 24, 1844.
(10) On Divergent Series, and on Various Points of Analysis connected with them, *read* 1844.
(11) On a point connected with the Dispute between Keill and Leibnitz, *read Jan.* 1846.
(12) On the Structure of the Syllogism, and on the application of the Theory of Probabilities to questions of Argument and Authority, *read Nov.* 9, 1846.

(13) Method of Integrating Partial Differential Equations, *read June* 1848.

(14) On the Symbols of Logic, the Theory of the Syllogism, and in particular of the Copula, and the application of the Theory of Probability to some questions of Evidence, *read Feb.* 1850.

(15) On some Points of the Integral Calculus, *read Feb.* 1851.

(16) On some Points in the Theory of Differential Equations, *read* 1854.

(17) On the Singular Points of Curves, and on Newton's Theory of Co-ordinated Exponents, *read* 1855.

(18) On the Question: What is the Solution of a Differential Equation, Supplement to No. 3, *read April* 1856.

(19) On the Beats of Imperfect Consonances, *read Nov.* 1857.

(20) A Proof of the Existence of a Root in every Algebraic Equation, *read Dec.* 1857.

(21) On the General Principles of which the Composition or Aggregation of Forces is a Consequence, *read* 1859.

(22) On the Syllogism, No. 4, and on the Logic of Relations, *read* 1860.

(23) On the Theory of Errors of Observation, *read Nov.* 1861.

(24) On the Syllogism, No. 5, and on various points of the Onymatic System, *read* 1863.

(25) On the Early History of the Signs + and −, *read* 1864.

(26) A Theorem relating to Neutral Series, *read* 1864.

(27) On Infinity and the Sign of Equality, *read May* 1864.

(28) On the Root of any Function, and on Neutral Series, No. 2, *read May* 1866.

(29) On a Theorem relating to Neutral Series, *read Oct.* 1868.

In the Philosophical Magazine.

(1) On Taylor's Theorem, 1835.

(2) On the Relative Signs of Co-ordinates, 1836.

(3) On the Solid Polyhedron, 1838.

(4) On the Rule for finding the value of an Annuity for three lives, 1839.

(5) Suggestion on Barrett's Method, 1841.

(6) On Fernel's Measure of a Degree, Nos. 1 and 2, 1841.

(7) On the reduction of a Continued Fraction to a Series, 1841.

(8) On Fernel's Measure of a Degree, Nos. 3 and 4, 1842.

(9) On Leonardo da Vinci's Use of + and −, 1842.

(10) On Torporley's Anticipation of part of Napier's Rule, 1843.

(11) On the almost total disappearance of the earliest Trigonometrical Canon, 1845.
(12) Derivation of the word Theodolite, 1846.
(13) Derivation of Tangent and Secant, 1846.
(14) Account of the Speculations of Thos. Wright, of Durham, 1848.
(15) On the Additions made to the Second Edition of the Commercium Epistolicum, 1848.
(16) On a Property of the Hyperbola, Jan. 1848.
(17) On Anharmonic Ratio, 1849.
(18) On the Early History of Infinitesimals in England, 1, 1852.
(19) On Indirect Demonstration, 1852.
(20) On the Authorship of the Account of the Commercium Epistolicum, 1852.

CAMBRIDGE MATHEMATICAL JOURNAL.

(1) On the Perspective of the Co-ordinate Plane, pp. 92, 93, 1841.
(2) On a simple property of the Conic Section, pp. 201–3, 1841.
(3) Remarks on the Binomial Theorem, pp. 61, 62, 1843.
(4) On the Equation $(D+a)^n 3 = x$, pp. 60–62, 1845.
(5) On a Law existing in the successive approximations of a Continuous Fraction, pp. 97–99, 1845.

CAMBRIDGE AND DUBLIN MATHEMATICAL JOURNAL.

(1) On Arbogast's Formulæ of Expansion, pp. 238–255, 1846.
(2) Suggestion on the Integration of Rational Fractions, Nov. 1848.
(3) On a Point in the Solution of Linear Differential Equations, 1849.
(4) Extension of the word *Area*, May 1850.
(5) Application of Combinations to the explanation of Arbogast's Method, Feb. 1851.
(6) On the Mode of using the signs + and – in Plane Trigonometry, May 1851.
(7) On the Connection of Involute and Evolute in Space, Nov. 1851.
(8) On Partial Differential Equations of the First Order, Feb. 1852.
(9) On the Signs + and – in Geometry, and on the interpretation of the Equation of a Curve, Nov. 1852.
(10) Mathematical Notes, pp. 93, 94, 1853.

Quarterly Journal of Mathematics.

(1) On the Dimensions of the roots of Equations, 1857.
(2) On the Fractions of Vanishing or Infinite Terms, 1857.
(3) Historical Notes on the Theorem respecting the Dimensions of Roots, 1857.
(4) Notes on Euclid i. 47, 1857.
(5) On the Integrating factors of $Pd + 2dy + Rdz$, 1858.
(6) On the Classification of Polygons of a given number of sides, 1858.

Central Society of Education.

On the Mathematics; their value in Education.
On Professional Mathematics, 1837–38–39.

The Mathematician.

(1) Remarks on General Equations of the Second Degree, pp. 242–246, 1850.
(2) Organised Method of making the resolution required in the integration of Rational Fractions, pp. 242–246, 1850.
(3) Remarks on Horner's Method of solving Equations, pp. 289–291, 1850.

British Almanac and Companion.

1831. On Life Assurance.
1832. On Eclipses.
1833. On Comets.
1834. On the Moon's Orbit.
1835. Halley's Comet.
1836. Old Arguments against the Motion of the Earth.
1837. Notices of English Mathematical and Astronomical Writers between the Norman Conquest and the year 1600.
1838. On Cavendish's Experiment.
1839. Progress of the Problem of Evolution.
1840. On the Calculation of Single Life Contingencies.
1841. On the use of small tables of Logarithms in Commercial Calculations, and on the practicability of a Decimal Coinage.
1842. On Life Contingencies, No. 2.
1843. References for the History of the Mathematical Sciences.
1844. On Arithmetical Computation.

1845. On the Ecclesiastical Calendar—Easter.
1846. On the earliest printed Almanacs.
1847. Recurrence of Eclipses and Full Moon.
1848. On Decimal Coinage.
1849. Short Supplementary on the First Six Books of Euclid's Elements.
1850. On Ancient and Modern Usage in Reckoning.
1851. On some points in the History of Arithmetic.
1852. A Short Account of some recent Discoveries in England and Germany, relative to the Controversy on the Invention of Fluxions.

(See Life of Newton in Knight's 'British Worthies,' Arts. *Commercium Epistolicum* and *Fluxions*, P. Cyc., *Dispute between Keill and Leibnitz*, Cambridge Memoirs.)

1853. On the difficulty of correct descriptions of Books.
1854. On a Decimal Coinage.
1855. The Progress of the Doctrine of the Earth's Motion between the times of Copernicus and Galileo, being notes on the Ante-Galilean Copernicans.
1856. Notes on the History of the English Coinage.
1857. Notes on the State of the Decimal Coinage Question.

In Smith's Classical Dictionary.

Diophantus.
Eucleides.
Heron.
Hipparchus.
Sosigenes.
Theon.
Ptolomæus.

Tracts Published by the Society for the Diffusion of Useful Knowledge.

(1) Study and Difficulties of Mathematics, vol. i., 1836.
(2) Arithmetic and Algebra, 1831, vol. i., 1836.
(3) Examples of the Processes of Arithmetic and Algebra, vol. i., 1836.

Dublin Review.

Laplace on Probabilities, No. 1, April 1837.
Laplace on Probabilities, No. 2, July 1837.
On Legislation for Life Assurance, Aug. 1840.
Review of Jones on the value of Annuities (published by Diffusion Society), August 1841.

Peyrard's Elements of Euclid, Nov. 1841.
Weights, Measures, and Coinage, May 1841.
Science and Rank, Nov. 1842.
Baily's Repetition of the Cavendish Experiment, March 1845.
Speculators and Speculations, Sept. 1845.
Book-keeping, Dec. 1845.
Mathematical Bibliography, Sept. 1846.
Helps to Calculation, March 1847.

Encyclopædia Metropolitana.

Calculus of Functions, 1835.
Theory of Probabilities, 1837.

List of Articles in 'Penny Cyclopædia.'

(This list of articles in the 'Penny Cyclopædia' is taken partly from a copy in the British Museum, in which all the articles have the names of their authors appended by the donor; and partly from the marked copy in Mr. De Morgan's possession. The constellations and planets not included in this list are also by him, if the Museum copy is correct.)

Abacus, 2
Abatis
Abauzit, Firmin
Abbreviation, mathl.
Abel, Niels Henri
Aberration
Aberration (in optics)
Abscissa
Absurdum, reductio ad
Accelerated motion
Accent (mathl.)
Achromatic
Acoustics
Acronychal
Act (University)
Actuary
Addition
Addition of ratios
Advowsons, value of
Aeolipyle
Aero-dynamics
Aerostatics
Agnese, Maria
Air
Air-gun
Air-pump
Aliquot part
D'Alembert
Algebra
Algebraic
Algebraic geometry
Algorithm
Almacanter
Almagest
Almanac
Alonsine tables
Alternate
Altitude
Amphiscii
Amplitude
Analysis
Anaxagoras
Anaximander
Anaximenes
Anemoscope
Angle
Anker
Annuity
Annulus
Anomalistic year
Anomaly (astronl.)
Antecedent
Antecedentia
Antilogarithm
Antinomians
Antipodes

Antiscii
Antoeci
Aphelion
Appian, Peter
Apogee
Apollonius Pergæus
Apparent (astron.)
Apparent magnitude
Apparent motion
Approximation
À posteriori, à priori
Apsides
Aratus, astronomer
Arbogast, Louis Fred.
Arc
Arch
Archimedes
Architrave
Archivolt
Arctic circle
Are
Area
Argument
Aristarchus
Aristoxenus
Arithmetic
Arithmetical complement
Arithmetical mean
Arithmetical progression
Arithmetical proportion
Armillary sphere
Arroba
Ascension, ascensional difference
Aspect (astronl.)
Asterism
Astrolabe
Astrology
Astronomy
Asymptote
Atmosphere
Atmospheric air
Attraction in physics
Attwood, George
Aurora borealis
Autolycus
Automaton
Auzout, Adrien
Average
Avoirdupois
Axiom
Axis
Azimuth

Bacon, Roger
Bailly, Jean Sylvain
Bainbridge, John (astron.)
Balance
Ballistic pendulum
Balloon
Barlaam
Barlowe, William
Barometer
Baroscope
Barrel
Barrow, Isaac
Barter (arithmet.)
Bartholinus
Bassantin, James
Bayer, John
Bearing (nautical)
Beccaria, Giovanni Battista
Belidor
Bellows
Berkeley, Geo. (the latter half)
Bernard, Edward
Bernoulli
Berosus (the latter half)
Binomial
Binomial theorem
Biquadratic
Bombelli, Raphael
Bonnycastle, John
Borda, Jean Charles
Borel and Borelli, part
Boscovich, R. J.
Bossut, Charles
Bouguer, Pierre
Bouillaud, Ismael
Boyle, Robert
Bradley, James
Brahé, Tycho
Briggs, Henry
Brouncker
Burning-glass
Bushel

Cagnoli
Calculating machine
Calculus
Calibre
Calippus
Callet
Camera lucida
Campani, M. and J.
Campanus, J.
Camus
Cancer (sign)
Canon (mathl.)
Canton, John
Capacity
Capillary attraction
Capricornus (sign)
Carat

Cardan
Carnot
Cassini
Castelli
Catalogue (astron.)
Catenary
Cause
Caustic (optics)
Cavalieri
Cavendish, Henry
Celsius
Centigrade
Centre
Centrifugal forces
Centripetal force
Chain (surveying)
Chaldron
Chance
Chappe d'Auteroche
Characteristic (of a logarithm)
Chastellet
Chiliad
Chord
Church rates
Circinus
Circle
—— of declination, &c.
Circular parts
Circulating decimals
Cissoid
Clairaut
Clavius
Cleomedes
Clepsydra
Cloud
Cocker, Edward
Coefficient
Cohesion
Collins, John
Collision
Colson, John
Combinations and Permutations
Comet
Comet of Biela
Commandine
Commensurable
Commercium Epistolicum
Common measure
Compass, azimuth
Compass, mariners'
Compasses
Complement
Composition
Compound quantities
Concave and convex
Concentric
Conchoid
Condition (mathcs.)
Condorcet
Cone
Conic sections
Conical projection
Conjugate
Conjunction and opposition
Conoid
Conon of Alexandria
Constant
Constellation
Construction
Contact
Content
Contrary and contradictory
Convergent
Converse
Co-ordinates
Copernicus
Cord
Corollary
Corvus
Costard, George
Cotes, Roger
Coulomb
Craig, John
Cramer, Gabriel
Crux
Ctesibius
Cube
Culmination
Cunitz, Maria
Curtate
Curvature
Curve
Cusp
Cycle
Cycloid
Cylinder

Data, Datum
Day
Decagon
Declination
Deferent
Definition
Deflection
Degree of an equation
Deism
Demoivre's hypothesis
Demonstration
Deneb
Denominator
Density
Departure
Depression
Derivation
Determinate
Development
Diagonal
Diagonal scale

Diagram
Diameter
Differential calculus
Differential co-efficient
Digit
Dimension
Direct and retrograde
Direction
Directrix
Disc
Discontinuity
Discount
Dividend (arithl.)
Division
Dodecagon
Dome
Drachm, or Dram
Duodecimals
Duplicate ratio
Duplication of the cube
Dynamics
Dynamometer

Earth (astronl.)
Earth, density of
East
Easter, method of finding
Eclipse
Elimination (algebc.)
Ellipse
Elliptic compasses
Ellipticity
Elongation (astronl.)
Encke's Comet
Epoch (astronl.)
Equal
Equation
Equation (astronl.)
Equation of payments
Equations, algebraical
Equations, differential
Equations, functional
Equator and ecliptic
Equiangular, Equilateral
Equimultiples
Equinoctial
Equinoxes
Euclid
Excentricity

Factor, in algebra
Fall of bodies
False position
Fellowship (in arithmetic)
Ferrei and Ferrari
Figure (in geometry)
Firkin
Flamsteed
Flexure, contrary
Fluxions
Force
Forces, impressed and effective
Forces, parallelogram of
Fourier
Fractions, common and decimal
Fractions, continued
Fractions, vanishing
Functions, calculus of
Functions, theory of

Gage
Gallon
Generating functions
Geometer
Geometrical
Geometry
Globe
Globular projection
Globular sailing
Gnomon
Gnomonic projection
Golden number
Graduation
Gravity, centre of
Guldinus
Gunter
Gyration, centre of

Hachette
Halley
Halley's Comet
Harmonic proportion
Harmonics
Height, measurement of
Heliacal
Herschel, William
Heteroscii
Hevelius, partly
Hogshead
Homogeneous and Heterogeneous
Homologous
Horary
Horizon
Horologium
Horrocks, Jeremiah
Hour, Hour-circle, Hour-line
Huyghens
Hyperbola
Hyperbole
Hypothenuse
Hypothesis

Impact
Impenetrability
Impulse
Inclined plane
Incommensurables, theory of
Increment and decrement
Indefinite
Indeterminate
Induction (mathl.)
Inequality
Inertia
Infinite, Infinity, Infinitesimal calculus
Integer
Integration, Integral calculus
Interest
Interpolation
Interpretation
Invariable
Inverse, Inversion
Involute and Evolute
Involution and Evolution
Irrational quantity
Irreducible case
Isochronous

Julian period

Kalendar

La Caille
Laplace
League
Leap Year
Least squares
Legendre
Lemma
Lemniscata
Lens
Lever
Libration
Life, mean duration
Light, barometrical
Light-equation
Lights, Northern ?
Line
Limits, theory of
Litre
Logarithms
Logarithms, Gauss's hyperbolic
Logarithmic curve and spiral, use of
Longitude and latitude
Lubienietske

Maclaurin
Magic square
Magnitude
Map
Mariotte
Maseres, Francis
Maskelyne, Nevil
Mass, in physics
Mathematics
Maurolico, or Marullo
Maxima and Minima
Menelaus
Mensuration
Mercator, Nicholas
Mercator's projection
Meridian
Meton, Metonic cycle
Metre
Middle latitude
Mile
Milky Way
Momentum, or Moment
——— or Moment of inertia
Money
Monge
Moon
Mortality, law of
Motion
——— direction of
——— of the earth
Moving force
Multiple, Submultiple
——— Points
Musa, Mohamed Ben

Napier's bones, or rods
Negative and impossible quantities
Node
Nothing
——— differences of
Nucleus of a comet
Number
Numbers, appellations of
Numbers, theory of
Numbers of Bernoulli
Numeral characters
Numeration
Numerator
Numerical

Oblate
Oblique
Obliquity
Oblong
Observation and experiment
Occultation
Octans
Offsets
Operation
Orb, Orbit
Order
Ordinate
Oscillation and centre of oscillation

Pace
Parabola
Paradox
Parallelepiped
Parallelogram
Parallels
Parameter
Penumbra
Percussion, centre of
Perimeter
Periodic functions
Periods of revolution
Piccolomini, A.
Piling of shot
Pipe
Planet
Planisphere
Point of contrary flexure
Pole, Polar
Polygon and polyhedron
Pond, John
Power (mechanics)
Practice
Precession and nutation
Pressure
Prime
Primum mobile
Principia
Probability, theory of
Problem
Process ?
Progression
Projection
Projection of mathematical diagrams
Proportion
Proportional logarithms
Proportional parts
Ptolemaic system
Pyramid

Quadrant
Quadrature
Quadrature of the circle
Quantity
Quarter
Quartile
Quintal
Quintile

Radix
Ratio
Ratios, composition of
Ratios, prime and ultimate
Reciprocal
Rectangle
Rectification
Recurring series
Reduction
Regular figures, polygon, solids, polyhedrons
Relation (mathl.)
Residual phenomena
Resistance ?
Reversion
Reversion of series
Revolution
Rhomb, Rhombus, Rhomboid
Rhumb, or Rumb
Right angle
Root
Rotation
Round
Rule, Ruler

Sagittarius
Salient
Saros, Neros, Sosos
Satellite
Scale, musical
—— mathematics
Scholium
Scruple
Seasons, change of
Segment
Series
Sexagesimal
Sextans (constell.)
Side
Sidereal
Sign, astronomy
Sign, mathematics
Similar, similar figures
Sine and Cosine
Sine and Cosine, curves of

Vernal and Equinox
Vibration
Vieta
Viga Ganita
Virtual velocities
Vitello
Volume
Volute
Vortex
Wager
Wallis, John
Weight
—— of earth
—— of observations
Weights and measures
Whiston, W.
Wingate, Edward
Wrangler
Wright, Edward
Year
Young, Thomas

Separate Works.

Elements of Arithmetic, 8vo, 1835.

Algebra, Prelim. to the Differential Calculus, 8vo, 1835.

Connection of Number and Magnitude: an attempt to explain the Fifth Book of Euclid, 8vo, 1836.

Essay on Probabilities, and on their Application to Life Contingencies and Insurance Offices, Cabinet Cyclopædia, small 8vo, 1838.

The Differential and Integral Calculus, one vol. 8vo, pp. 770, 1842.

Arithmetical Books, from the invention of printing to the present time, being brief notices of a large number of works, drawn from actual inspection, 8vo, 1847.

First Notions of Logic, preparatory to study of geometry, 8vo, 1839.

Formal Logic, or the Calculus of Inference necessary and probable, 8vo, 1847, pp. 336.

The Globes, Celestial and Terrestrial, for Malby's Globes, 8vo, 1845.

Syllabus of a Proposed System of Logic, 8vo, 1860.

Trigonometry and Double Algebra, 1849.

The Book of Almanacs, with an index of reference, by which the almanac may be found for any year up to A.D. 1000, with means of finding the day of any new or full moon from B.C. 2000 to A.D. 2000, compiled by A. De Morgan, 8vo, 1850.

Miscellaneous.

Correspondence of Scientific Men of the Seventeenth Century, by S. P. Rigaud. Contents and Index by A. De Morgan, 1841.

Treatise on the Problems of Maxima and Minima, by Ramchundra, Calcutta, edited, with Introduction, 1850.

Decimal Association Proceedings, with Introduction, 1854

Debate on the Decimal Coinage Question, with remarks on the speech of the Member for Kidderminster, 1855.

Reply to Facetiæ of the Member for Kidderminster, 1855.

Journal of a Tour in the Unsettled Parts of North America, by Francis Baily, edited with Preface, by A. De Morgan, 1856.

Preface to From Matter to Spirit, by Sophia De Morgan, 1863.

Notes on Colonel W. H. Oakes's Table of the Reciprocals of Numbers, 1865.

Seven Figure Logarithms of Numbers 1 to 108,000, corrected, with a description of the Tables added by Prof. De Morgan, 8vo. (Trans. of Ludwig Schroen Siebenstellige. Gemeine Logarithmen der Zanten.) 1865.

The Eleventh Chapter of the History of the Royal Society, with additions, by Prof. De Morgan, 1849.

General Information on Subjects of Chronology, Geography, Statistics, &c., References for History of Mathematical Science, 1842.

Statement in answer to an assertion made by Sir W. Hamilton in reference to a discovery of a new principle in the Theory of Syllogism, 1847.

English Science, Report of the British Association, vols. i. and ii., written for a Review, edited by Mr. Beaumont (title forgotten), 1835.

Tables, C. Knight's English Cyclopædia, a new edition of Useful Knowledge, London, 1861.

In addition to the above there were numerous contributions to the 'Memoirs' and 'Obituary Notices' of the Astronomical Society, to the Journal of the Institute of Actuaries and Insurance Record. I do not find any contribution to the North British Review except the Review of 'Brewster's Life of Newton.' Among the smaller writings which I know to be omitted from the above list are some to Charles Knight's 'Library of Anecdote,' biographies in the same publisher's 'Portrait Gallery,' &c., &c.

The voluminous contributions to the 'Athenæum,' 'Notes and Queries,' &c., I have been obliged to omit on account of their number.

INDEX OF NAMES, ETC.

Printed in the United States
By Bookmasters